Microbes in Agriculture and Environmental Development

Microbes in Agriculture and Environmental Development

Edited by
Chhatarpal Singh, Shashank Tiwari,
Jay Shankar Singh and Ajar Nath Yadav

CRC Press
Taylor & Francis Group
Boca Raton London New York

CRC Press is an imprint of the
Taylor & Francis Group, an **informa** business

ISBN: 9780367524135 (hbk)
ISBN: 9781003057819 (ebk)

Typeset in Times
by Deanta Global Publishing Services, Chennai, India

Contents

Foreword

The rapid increase in human population is creating excessive pressure on the existing cultivable land area for food, fuel, and raw materials. The indiscriminate anthropogenic interventions to various li vi ng and non- li vi ng components of the eco-systems are affecting the environment and causing distress to nature. Our primary food source for survival on this planet is agriculture, thus agricultural cultivation practices should be given due attention. The excess use of synthetic chemicals and climatic disturbances are continuously spoiling the soil health and water quality. Likewise, improper management of industrial waste is also causing serious problems related to the environment and human health. Therefore, viable potential solutions for management of agriculture and environmental sustainability are required at present. The microbes and their bioactive compounds (metabolites) could be a beneficial and viable option for the well-being of the agriculture and environment. This book explores the possibility of agriculture and environmental quality development via application of microbes and their beneficial secondary metabolites. The use of microbes is widely known in solving several issues but only few are identified to be effective for the management of various emerging problems in agriculture and environment. The indigenous microbial-based inoculants and technology may enhance the crop health in nutrient-poor and disturbed soils. Furthermore, microbial resources may be exploited for the removal and remediation of pesticides, heavy metals, xenobiotics, and other industrial pollutants from contaminated water and soils. Careful selection of microbes and efficient designing of testing tools are the key steps in developing new technologies for effective utilization of microorganisms for sustainable agriculture and environmental protection. The waste-water treatment and recycling of agricultural and industrial wastes are other important issues that need to be addressed via new viable microbial tools and technology. This book covers these issues and provides a detailed outlook about recent trends in microbial application in plant growth promotion. Soil fertility and sustainable environment are achieved through bioremediation, biodegradation, and bio-sorption processes. The editors invited leading subject experts from across the globe to contribute relevant articles in the area of microbial mediated agro-environmental development. *Microbes in Agriculture and Environmental Development* will enrich the knowledge of various stakeholders about the use of microbes in agricultural production and environmental sustainability.

Sanjay Singh
Vice-Chancellor
Babasaheb Bhimrao Ambedkar University
Lucknow, India
April 06, 2020

Preface

According to current reliable information, to satisfy the increasing Indian population, agricultural production is required to enhance the current situation significantly. At the same time, it should be kept in mind that sustainable agricultural practices without compromising the resources for future generations are essential to meet the present agricultural demands. However, this agricultural demand satisfaction would also be associated with the huge exploitation of non-renewable natural resources with the emission of greenhouse gases causing global climate disturbances. The present research challenges are to meet sustainable agriculture, environmental, and economic issues without compromising the crop yields. Therefore, contemporary and eco-friendly agriculture farming is being adopted on a global scale to meet the environmental and sustainability issues with minimal exploiting the natural resources. In this context, there is a need to discover promising effective approaches to full fill the current requirements.

Currently, the role of soil microbial communities and their beneficial services has been recognized as a potential option in sustainable and healthy crop production, while taking care of the well-being of the environment. In fact, agriculturally beneficial soil microorganisms and their services play crucial roles in agriculture productivity, particularly improving the plant health and soil nutrient status. As a result, several strategies to exploit the beneficial microbial services are being advised as a cost-effective tools in sustainable agriculture production and environmental safety. Our final scientific goal should be to optimize the activity and contribution of existing soil microbiome under various environmental constraints to sustain the soil fertility and crop productivity. Since the role of soil microbial communities to crop health are influenced by diverse ecological and agronomic drivers, the scientific investigations and information describing the impact of environmental factors on the microbial services must be considered. In particular, the current scenario of climatic/ environmental disturbances, as they may influence the management of beneficial soil microbial activity in soil functioning, should be addressed properly.

This book aims to gather up-to-date effective information along with beneficial microbial services that may address the current challenges in sustainable agriculture and environmental stability. After rigorous analysis of various challenges and limitations, this book focuses on future perspectives and opportunities related to enhancing our knowledge of the microbial services to agro-environmental well-being and improving the potential of soil microbiome to alleviate various environmental negatives to crops. The experts in the subject area have contributed informative articles for this very book. The book *Microbes in Agriculture and Environmental Development* strives to be instrumental in providing the latest updates to both graduate and postgraduate students, scientists/researchers of agriculture, environmental microbiology, and environmental science in combating the related problems and challenges.

Chhatarpal Singh
Shashank Tiwari
Jay Shankar Singh
Ajar Nath Yadav

Editors

Chhatarpal Singh is a President of Agro Environmental Development Society (AEDS), Majhra Ghat, Rampur, Uttar Pradesh, India. Dr. Singh is currently working in the field of Agricultural and Environmental Microbiology; his research interest is agro-environmental development through various innovative and scientific approaches. He has been published in various scientific research and review papers, book chapters, and magazine articles in the field of agricultural and environmental microbiology. He earned his Ph.D. degree in Environmental Microbiology with the specialization of Agricultural Microbiology from Babasaheb Bhimrao Ambedkar University Lucknow, U.P., India. He has been honored by various organizations for his outstanding contribution in the field of Agricultural Microbiology. Dr. Singh has organized various international conferences and training programs.

Shashank Tiwari earned his M.Sc. and Ph.D. degree in Environmental Microbiology from Babasaheb Bhimrao Ambedkar University, Lucknow, India. Currently, Dr. Tiwari is working in the field of methanotrophs ecology (methane-oxidizing bacteria), which is the sole entity responsible for the oxidation of potent greenhouse gas CH_4. He has also been involved in the assessment of soil microbial biomass and methanotroph diversity across different land use changes and their impact on methane oxidation at Vindhyan plateau, India. His research and review papers have been published, in addition to book chapters and magazine articles in the field of agricultural and environmental microbiology, along with journals and magazines of national and international repute.

Jay Shankar Singh is presently working as a faculty member in the Department of Environmental Microbiology at Babasaheb Bhimrao Ambedkar University in Lucknow, India. Dr. Singh has contributed significantly to the subject of restoration ecology and natural resource management. He has published his research output in international journals with high impact factors on Scopus and other scientific databases. He is also actively serving as a member of various scientific committees, holding editorial responsibilities for journals, such as *Microbiology Research*, *PLoS ONE*, etc. He has published several books from Springer and Elsevier, among others.

Ajar Nath Yadav is an Assistant Professor in the Department of Biotechnology, Akal College of Agriculture, Eternal University, Baru Sahib, Himachal Pradesh, India. He has four years of teaching and ten years of research experience in the field of microbial biotechnology, microbial diversity, and plant–microbe interactions. He obtained his doctorate degree in Science (Microbial Biotechnology) in 2016, jointly from Indian Agricultural Research Institute, New Delhi, and Birla Institute of Technology, Mesra, Ranchi, India. He has made pioneering contributions in the areas of microbial biotechnology; microbial biodiversity; microbial ecology; plant-microbe interaction; agricultural microbiology; and environmental microbiology. Dr. Yadav has 138 publications, which include 67 research and review articles, 17 books, 1 laboratory manual, 53 book chapters, and 1 patent.

Contributors

Ram Naresh Bharagava
Laboratory for Bioremediation and
  Metagenomics Research (LBMR)
Department of Microbiology (DM)
Babasaheb Bhimrao Ambedkar
  (Central University)
Lucknow, India

J. Chapla
S. N. Vanita College of Pharmacy
Hyderabad, India

Indra Jeet Chaudhary
School of Environment and Sustainable
  Development
Central University of Gujarat
Gandhinagar, India

Rishabh Chitranshi
Department of Environmental
  Microbiology
School for Environmental Sciences
Babasaheb Bhimrao Ambedkar
  (Central University)
Lucknow, India

Rajeswari Das
Department of Soil Science
Dr. Rajendra Prasad Central
  Agriculture University
Samastipur, India,

D. K. Dash
Department of Fruit Science and
  Horticultural Technology
OUAT
Odisha, India

Namo Dubey
Division of Biotechnology
CSIR-Institute of Himalayan
  Bioresource Technology
Palampur, India

Luiz Fernando R. Ferreira
Institute of Technology and
  Research
Tiradentes University
Campus Farolandia
Aracaju, Brazil

R. Y. Hiranmai
School of Environment and Sustainable
  Development
Central University of Gujarat
Gandhinagar, India

Raj Kapoor
Department of Testing and
  Certification
National Collateral Management
  Limited
Dada Nagar, Kanpur, India

K. M. Karetha
Department of Horticulture
College of Agriculture
JAU
Junagadh, India

Robinka Khajuria
Department of Biotechnology
Harlal Institute of Management and
  Technology
Greater Noida, India

Mukesh Kumar
Department of Soil Science
Dr. Rajendra Prasad Central
    Agriculture University
Samastipur, India

V. K. Mishra
College of Agriculture
Rani Lakshmi Bai Central
    Agricultural University
Jhansi, India

Subhrajyoti Mishra
Department of Fruit Science
College of Horticulture
JAU
Junagadh, India

Ajay Neeraj
School of Environment and
    Sustainable Development
Central University of Gujarat
Gandhinagar, India

Umesh Pankaj
College of Horticulture and
    Forestry
Rani Lakshmi Bai Central
    Agricultural University
Jhansi, India

Abhay Raj
Environmental Microbiology
    Laboratory, Environmental
    Toxicology Group
CSIR-Indian Institute of
    Toxicology Research
    (CSIR-IITR)
Lucknow, India

P. Rajarao
Department of Botany
    and Environmental
    Science
Osmania University
Hyderabad, India

P. M. Sameera
University College of Technology
    Osmania University
Hyderabad, India

Ganesh Dattatraya Saratale
Department of Food Science and
    Biotechnology
Dongguk University
Seoul, Republic of Korea

Mohd Arshad Siddiqqui
School of Environment and Sustainable
    Development
Central University of Gujarat
Gandhinagar, India

Shalini Singh
Department of Microbiology
School of Bioengineering and
    Biosciences
Lovely Professional University
Punjab, India

Snigdha Singh
School of Environment and Sustainable
    Development
Central University of Gujarat
Gandhinagar, India

Sunil Soni
School of Environment and Sustainable
    Development
Central University of Gujarat
Gandhinagar, India

M. Soniya
College of Agriculture
Rani Lakshmi Bai Central Agricultural
    University
Jhansi, India

A. Suresh
University College of Technology
    Osmania University
Hyderabad, India

Usha
College of Agriculture
Rani Lakshmi Bai Central Agricultural
    University
Jhansi, India

B. Vijayakumari
Department of Botany
AvinashilingamInstitute for Home
    Science and Higher Education for
    Women
Coimbatore, India

Ashutosh Yadav
Laboratory for Bioremediation
    and Metagenomics Research
    (LBMR)
Department of Microbiology (DM)
Babasaheb Bhimrao Ambedkar
    (Central University)
Lucknow, India

and

Environmental Microbiology
    Laboratory
Environmental Toxicology Group
CSIR-Indian Institute of Toxicology
    Research (CSIR-IITR)
Lucknow, India

R. Hiranmai Yadav
School of Environment and Sustainable
    Development
Central University of Gujarat
Gandhinagar, India

Pooja Yadav
Environmental Microbiology
    Laboratory
Environmental Toxicology Group
CSIR-Indian Institute of Toxicology
    Research (CSIR-IITR)
Lucknow, India

1 Plant–Microbe Association Leading to a Sustainable Agroecosystem

Rajeswari Das, Subhrajyoti Mishra, and Mukesh Kumar

CONTENTS

1.1   INTRODUCTION

There is a need to understand precisely the way in which microbes are associated with plants (metaorganisms) and their interactions. Plants secret root exudates which contain enzymes, water, H^+ ions, mucilage, and carbon-containing primary and secondary compounds, and help develop a microbial colony. It has been observed that microbial density is several times greater in the rhizosphere as compared to soil (Mendes et al. 2013; Baetz and Martinoia 2014). Plants are specific in the selection of a rhizosphere microbial colony, which may be the reason behind their coexistence (Ciccazzo et al. 2014; Lareen et al. 2016).

Since the dawn of the Industrial Revolution, and various associated human activity, there has been an aggravation of the climate, which directly or indirectly affects crop production worldwide. Biotic and abiotic stress factors also contribute to crop loss in terms of millions of dollars (Hua 2013; Suzuki et al. 2014). Abiotic stress involves extreme temperature, drought, waterlogging, light, and salinity as major parameters (Majeed et al. 2015). Infection and damage by various insect pests, pathogens, and other organisms are covered under biotic stress. Many plant-associated microbes play a crucial role in reducing crop losses by mitigating various stresses. They enhance plant growth and physiology by various means, such amelioration is regarded as promoting rhizobacteria (PGPR). PGPR are often found to produce 1-aminocyclopropane-1-carboxylate (ACC) deaminases, indole acetic acid, siderophore, and solubilizing phosphorus and thereby enhancing nutrient uptake and the growth of plants (Egamberdieva et al. 2015). As the role of PGPR is very prominent in the plant lifecycle in enhancing its physiology and growth aspects by providing tolerance of various forms of environmental stress, there is a need to focus attention on the study of plant microbiomes (Prashar et al. 2013; Ho et al. 2015). This chapter therefore explores the plant–microbe relationship. Several reports have discussed the effect of plant microbiomes on growth, productivity, and host survival in various symbiotic associations. This knowledge will guide efforts to move towards a sustainable agroecosystem by understanding plant–microbe intercommunication.

1.2   IMPACT OF MICROBIAL ECOLOGY OF
RHIZOSPHERE ON TERRESTRIAL SYSTEM

Dynamic changes in soil microbial ecology, as a result of functional alteration of the ecosystem, have an impact on crop ecology. A few stabilizing mechanisms have been postulated for the maintenance of species diversity and coexistence (Cheson 2000). Microbially mediated positive and negative feedback might play a crucial role in the whole plant ecosystem and contribute to plant–microbe intercommunication (Bever et al. 2010). Rhizospheric microbes have the potential to alter the nutrient

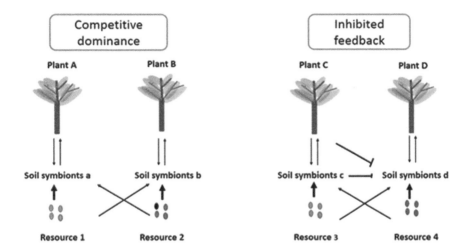

FIGURE 1.1 Schematic figure showing feedback mechanism among plant soil community.

uptake from soil to plant and can affect plant–plant interaction (competition) through resource partitioning which drives plant community dynamics.

The coexistence of plant species can be affected by an indirect feedback mechanism (i.e., competition or inhibition of symbionts) of rhizospheric microbes (Figure 1.1). To explain the mechanism that produces low diversity plant communities, a few hypotheses have been proposed, among which the empty niche hypothesis and degraded mutualist hypothesis are explained here. The former suggests the inhabitation of novel symbionts in the areas invaded by invasive plants, whereas the latter suggests the inhibiting ability of both the invasive plant and their symbionts towards a native symbiotic community to acquire resources, indirectly reducing the performance of native plants (Stinson et al. 2006; Vogelsang and Bever 2009). Ecological linkages of plant-soil feedbacks might enhance the growth and survival of exotic seedlings near efficient native symbionts.

A plant microbiome is a complex network, demonstrating the inhabitation of plant-associated microbes (Berendsen et al. 2012; Bulgarelli et al. 2012). Thorough research and analysis are a prerequisite for dealing with the impact of rhizospheric microbes on the terrestrial ecology.

1.3 FACTORS AFFECTING RHIZOSPHERIC MICROBIAL ASSOCIATION

A compatible phyto-microbial interaction is key to successful endophyte colonization in plants. A host plant recognizes the endophyte invading it through the crosstalk of signal molecules (plant–microbe intercommunication). Various studies have shown the response of endophytes to various chemical secretions from the plant root viz., chemotactic movement of endophytes in response to root exudates. Several other chemoattractants are secreted by plants (flavonoids). This plays an important

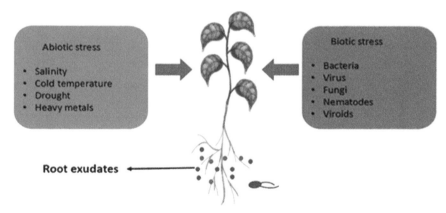

Factors affecting plant-microbe interactions.

FIGURE 1.2 Factors affecting plant–microbe association in the agroecosystem.

role in endophytic interaction with root hair also being used as bioformulations, including the successful infection of legume roots by rhizobia. Flavonoids are also found to be crucial in non-rhizobial endophytes. Other than chemoattractants, there are known to be only a few signal molecules, including lipo-chitooligosaccharides (LCOs) and strigolactones (SLs), which initiate and enhance microbial colonization with the roots of a higher plant. LCOs are also known as nod-factors, which activate a common symbiotic pathway in both rhizobium legume association and arbuscular mycorhizal association (Compant et al. 2010; Brader et al. 2014).

In contrast, such plant–microbe intercommunication is frequently affected by various factors (Figure 1.2). These include various forms of abiotic and biotic stress (Nguema-Ona et al. 2013; Chagas et. al. 2017; Majeed et. al. 2015). Extreme temperature, drought, waterlogging, and salinity are major parameters included in abiotic stress factors, whereas pathogen infection and insect pest attack in plants are categorized as biotic stress factors (Arora and Mishra 2016). Biotic stress is found to be controlled by a few PGPRs, as bioagents and abiotic stress are mitigated by root and shoot growth promotion. This is due to the secretion of plant hormones (IAA), nutrient solubilization and the uptake by production of siderophores, the solubilization of the phosphorus, and the fixation of nitrogen and performance of other plant growth-promoting activity (Prashar et. al. 2013).

1.4   INTERACTION BETWEEN BENEFICIAL MICROORGANISMS AND CROP PLANTS

The study of the interaction between beneficial microorganisms and crop plants is crucial for predicting their role in the development of a sustainable agroecosystem. Microorganisms and plants interact with each other in a number of ways: for nutrient availability, growth promotion, survival, and many more benefits for both plant and microbe. For this reason, there is subsequently less impact on both plant–microbe interaction and intercommunication.

1.4.1   Interactions Leading to Atmospheric Dinitrogen Fixation

Nitrogen, being a key macro element, is utilized in a broad spectrum by all living organisms. Higher plants and bacteria can utilize diverse organic and inorganic nitrogenous compounds, and some prokaryotic organisms can use N_2 via biological N_2 fixation. Biological nitrogen fixation (BNF) is carried out by a specialized group of prokaryotes, which utilize the nitrogenase enzyme as a catalyst for the conversion of atmospheric dinitrogen (N_2) to ammonia (NH_3), thence readily assimilated by plants to synthesize nitrogenous biomolecules. Prokaryotes responsible for nitrogen fixation are further categorized as aquatic organisms (i.e. cyanobacteria), free-living soil bacteria (i.e. Azatobacter), associative soil bacteria (Azospirillum) and most importantly symbiotic bacteria (i.e. Rhizobium and Bradyrhizobium).

1.4.1.1   Symbiotic Nitrogen Fixation

Symbiotic nitrogen fixation, which is one of the crucial biological processes for the development of sustainable agriculture, is mediated by bacteria inside the cells of legume root nodules. The symbiotic relationship with rhizobia (soil bacteria collectively) gives a unique characteristic to legumes. The interdependency of host plant and rhizobia (used for Rhizobium, Bradyrhizobium, Sinorhizobium, and Mesorhizobium) is beneficial for both partners. The former provides carbon and energy sources for growth and the function of the latter, and on its return it fixes dinitrogen and provides the source of reduced nitrogen to the former (Udvardi and Poole 2013; Jensen et al. 2012). It has been illustrated that this symbiotic process offers an ecologically sound and economically feasible means of resource utilization.

The nodule symbiosome is the site to conversion of atmospheric dinitrogen to ammonia, which provides a perfect microanaerobic environment due to its tissue buffering, along with high respiratory bacteroid activity to protect nitrogenase, which is an oxygen-sensitive enzyme complex catalyzing the whole conversion process. Moreover, a monomeric protein like the human hemoglobin molecule (known as "leg hemoglobin") binds the oxygen molecule.

Now, scrutinizing the biochemistry of nitrogenase enzyme: it is composed of two metalloproteins viz., a dinitrogenase and a dinitrogenase reductase. Whereas the former is molybdenum iron protein (Fe-Mo) composed of two distinguishable subunits (α and β), the latter is a dimer of Fe protein formed from indistinguishable subunits (Rubio and Ludden 2008). This enzyme complex is co-involved with ATP and electron donors by a series of reactions for the reduction of dinitrogen. MoFe protein is considered the actual site of N_2 reduction.

As per several estimates, symbiotic nitrogen fixation produces approximately 200 million tons of nitrogen annually (Peoples et al. 2009). If all agriculturally important crops were able to fix nitrogen, both humanity and environment would benefit hugely.

1.4.1.2   Asymbiotic Nitrogen Fixation

Apart from legume production in the cereal production system, asymbiotic nitrogen fixation (ANF) is a potential alternative source of biological nitrogen fixation.

Veritable ANF by diazotrophic bacteria (Azatobacter, Azospirillum) are major contributors to the nitrogen economy of the biosphere (30–50% of total nitrogen in crop fields (Rogers and Oldroyd 2014; Ormeno-Orrillo et al. 2013). Being found among alphaproteobacteria, betaproteobacteria, gammaproteobacteria, firmicutes, and cyanobacteria diazotrophs correlates to being trivial components of the ecosystem.

Though there is a limited contribution of nonlegumes to nitrogen fixation, a few bacterial species have promoted indicative enhancement in N content in crops (Beijerinckia spp. In maize hybrids). The application of genetically modified bacteria is also found to be promising in plant growth enhancement through nitrogen fixation (ammonium-excreting Azospirillum enhanced nitrogen supply to wheat crop). (Geddes et al. 2015; Ambrosio et al. 2017). Nitrogen fixation is highly variable due to frequent changes in the associated diazotroph and plant variety. Therefore, there are possibilities for increasing their efficiency in N fixation by favoring their population by regular inoculation.

1.4.2 Interactions Leading to Nutrient Dynamics and Crop Growth

The rhizosphere is the critical zone of interactions among plants, soils, and microorganisms. Plant roots can greatly modify the rhizosphere environment through their various physiological activities, particularly the exudation of organic compounds such as mucilage, organic acids, phosphatases, and some specific signaling substances, which are key drivers of various rhizosphere processes. The chemical and biological processes in the rhizosphere not only determine the mobilization and acquisition of soil nutrients as well as microbial dynamics, but also control the nutrient-use efficiency of crops, and thus profoundly influence crop productivity (Richardson et al. 2009).

1.4.2.1 Plant-Soil-Microbe Interaction in Phosphorus Dynamics

Phosphorus (P) has rapidly depleted in the rhizosphere due to root uptake, resulting in a gradient of P concentration in a radial direction away from the root surface due to its low solubility and mobility in soil. In spite of total soil P content usually exceeding the plant requirements, its low mobility in soil can restrict its availability to plants. Soluble P in the rhizosphere soil solution should be replaced 20–50 times per day by P delivery from bulk soil to rhizosphere, in order to meet its demand in plants. Therefore, P dynamics in the rhizosphere are mainly controlled by plant-root growth and function, and also highly related to the physical and chemical properties of soil. Because of the unique properties of P in soil, such as low solubility, low mobility, and high fixation by the soil matrix, the availability of P to plants is mostly controlled by two key processes; one is spatial availability and acquisition of P in terms of plant root architecture (as well as mycorrhizal association), and another is bioavailability and acquisition of P based on the rhizosphere chemical and biological processes. To improve P-use efficiency in agricultural systems, a better understanding is essential of the relative importance of plant mechanisms, compared to microbial processes for P mobilization in the rhizosphere.

1.4.2.1.1 Mycorrhizal Associations

Mycorrhiza literally translates as "fungus-root". Mycorrhiza defines a (generally) mutually beneficial relationship between the root of a plant and a fungus that colonizes the plant root. The fungus facilitates water and nutrient uptake in the plant; the plant provides food and nutrients created by photosynthesis to the fungus in a mutually beneficial relationship. This exchange is a significant factor in nutrient cycles and the ecology, evolution, and physiology of plants. Mycorrhizal symbioses by extending the nutrient absorptive surface with the formation of mycorrhizal hyphae, and can increase the spatial availability of P. Arbuscular mycorrhizal fungi (AMF) form symbiotic associations with the roots of about 74% of angiosperms. In the symbioses, nutrients are transferred by AMF via their extensive mycorrhizal mycelium to plants, while in return the fungi receive carbon from the plant. AMF not only influence plant growth through increased uptake of nutrients (e.g., P, zinc, and copper), but may also have non-nutritional effects in terms of stabilization of soil aggregates and alleviation of plant stress caused by biotic and abiotic factors. The beneficial effects of AMF and other microorganisms on plant performance and soil health can be very important for the sustainable management of agricultural ecosystems. A primary benefit of AMF is improved P-uptake conferred on symbiotic plants. In low-P soils, mycorrhizal plants usually grow better than non-mycorrhizal plants as a consequence of enhanced direct P-uptake of plant roots via the AM pathway. However, plant growth can be suppressed even though the AM pathway contributes greatly to plant P-uptake. The growth inhibitions might be caused by the down-regulation of the direct root P-uptake pathway. Some of the recent gene expression study shows the plants induce some common sets of mycorrhiza-induced genes but there is also variability, showing that there exists functional diversity in AM symbioses. Regulation of direct uptake pathways through epidermis and root hairs and AM pathways requires further investigation, because the differential expression of symbiosis-associated genes among different AM associations is related to the fungal species, plant genotypes, and environmental factors (Smith et al. 2017).

1.4.2.1.2 Role of Phosphorus-Solubilizing Microbes

Some rhizosphere microorganisms, except mycorrhizal fungi (for example, plant-growth-promoting rhizobacteria, particularly P-solubilizing bacteria [PSB] and fungi [PSF]) can also enhance plant P acquisition by directly increasing solubilization of P to plants, or by indirect hormone-induced stimulation of plant growth (Richardson et al. 2009). P-solubilizing microorganisms (PSM) account for approximately 40% in P-solubilization potential. The PSB or PSF may mobilize soil P by the acidification of soil, the release of enzymes (such as phosphatases and phytases), or the production of carboxylates such as gluconate, citrate, and oxalate.

It is commonly proposed that to improve P-use efficiency through the use of specific inoculants, microorganisms play an important role. Isolates of Penicillium spp. appear to have high potential for development as inoculants, based on their capacity to solubilize P under various laboratory conditions, to be mass-produced, and to readily and nonspecifically colonize the rhizosphere of a range of potential host plants. Stimulation of root growth or greater root hair elongation by specific microorganisms may enhance plant P nutrition indirectly by allowing greater exploration of soil, rather than by direct increase in the soil P availability.

1.4.2.2 Potassium-Solubilizing Microbes Activity in the Rhizosphere

After nitrogen (N) and phosphorus (P), potassium (K) is the most important plant nutrient that has a key role in the growth, metabolism, and development of plants. In addition to increasing plant resistance to disease, pests, and abiotic stress, K is required to activate over 80 different enzymes responsible for plant and animal processes Examples are: energy metabolism, starch synthesis, nitrate reduction, photosynthesis, and sugar degradation (Almeida et al. 2015; Hussain et al. 2016). Minerals containing K are feldspar (orthoclase and microcline) and mica (muscovite and biotite). The non-exchangeable form of K makes up approximately 10% of soil K, and is trapped between the layers or sheets of certain kinds of clay minerals. Solution K is the form of K that is directly and readily taken up by plants and microbes in soil.

It has been reported that some beneficial soil microorganisms, such as a wide range of saprophytic bacteria, fungal strains, and actinomyces, are capable of solubilizing the insoluble K from soils by various mechanisms. Some of these mechanisms include the production of inorganic and organic acids, acidolysis, polysaccharides, complexolysis, chelation, polysaccharides, and exchange reactions. Among these microorganisms, K-solubilizing bacteria (KSB) have gained the attention of agricultural scientists as soil inoculum to promote plant growth and yield. The KSB are effective in releasing K from inorganic and insoluble pools of total soil K through solubilization (Meena et al. 2014; Zhang et al. 2013). KSB are also known as plant-growth-promoting bacteria (PGPRs). A considerably higher concentration of KSB is commonly found in the rhizosphere in comparison with non-rhizosphere soil (Padma and Sukumar 2015). An important aspect regarding K availability in soils is the solubilization of K by KSB from insoluble and fixed forms. The ability to solubilize the silicate rocks by B. mucilaginous, B. circulanscan, A. ferrooxidans, Arthrobacter sp.; Enterobacter hormaechei, Paenibacillus mucilaginosus, P. frequentans, Cladosporium, Burkholderia, and Paenibacillus glucanolyticus has been reported. Among the soil bacterial communities, B. mucilaginous, B. edaphicus, and B. circulanscan have been described as effective K solubilizers (Meena et al. 2015, 2016).

1.4.2.3 Interaction of PGPRs Resulting Micronutrient Availability via Redox Mechanism in Phytomicrobiome

Micronutrients are important for plants because they are active in organic structures. They also play important roles as components or activators of enzymes. Moreover, they work as electron carriers or osmoregulators. Micronutrients function in the regulation of metabolism, reproduction, and protection against abiotic and biotic stresses. Sometimes plants may show deficiencies of mineral nutrients even when they are present in adequate amounts in soil, due to their non-availability. Plants depend for their micronutrient needs on their availability in the rooting zone. The acquisition of micronutrients from the labile pool is mostly affected by biological activities in the soil, physical factors of the environment (temperature, pH, light intensity, etc.), and cultural practices. All micronutrients form stable complex with organic ligands. When complexed to proteins, they function as biological catalysts (metalloenzymes).

Microorganisms play an important role in enhancing micronutrient availability to plant roots. Solubilization of mineral nutrients such as Fe, Mn, Cu, B and Zn by PGPR makes them more readily available for plant uptake, and this should be considered as a mechanism for enhanced plant growth (Glick 1995). Microorganisms are the main agents in the natural nutrient element cycle. Mineral nutrient solubility may be increased due to PGPR, which releases organic and sugar acids to the rhizosphere and creates acidic conditions by CO_2 (respiration). In general, acid-producing bacteria readily accumulate in the rhizosphere because of favorable habitat for its growth. There are numerous studies on many PGPR strains and species, including their ability to absorb some micronutrients, especially Fe, Zn, Cu, Mn and B. PGPR incorporate these micronutrients via the production of organic acids such as citric, glutamic, succinic, lactic, oxalic, malic, fumaric, and tartaric acid (Pratiwi et al. 2016). PGPR may also increase the mobility and availability of micronutrients by producing siderophores (Ghavami et al. 2016; Sharma et al. 2015). In plants applied with PGPR strains, the nutrient element amount of the plant may provide important information about the effect of bacterial inoculation in the nutrient element intake. Rhizobacteria treatments affect not only the availability of Fe but also other micronutrients. Generally, the enhancements in micronutrient contents such as Fe, Zn, Mn, Cu, and B were more pronounced in organic acid-releasing and phytosiderophore-producing bacterial inoculations.

1.4.2.4 Enhancement in Crop Growth – Consequence of Plant Growth-Promoting Activities of Microbes

There have been studies of two specific mechanisms of plant growth promotion; one process is that of the direct action mechanism, either by regulating their hormone levels or by providing the plant with essential minerals; and another is the indirect action mechanism by acting as biocontrol agents (reducing the deleterious effect of pathogens on plant growth) (Glick 2012). A plant grows in a field condition being associated with a well-structured and coordinated complex microbial community called the phytomicrobiome (Bulgarelli et al. 2015; Smith et al. 2017). The phytomicrobiome – along with the plant – forms a holobiont.

Various plant structures (flowers, fruits, stems, roots, leaves) are associated with elements (microbes) of phytomicrobiome. However, the microbial population varies i.e., the microbial community, in association with plant roots, have been found to be populous as compared to other plant parts. Nitrogen-fixing bacteria colonizing the legume root is the most suitable example. Along with various parts, plant roots controls over the rhizomicrobiome composition as it produces root exudates and signaling compounds which serve as the source of reduced carbon to microbes and regulate genetic and biochemical activities of specific species within the phytomicrobiome (Smith et al. 2017) (Figure 1.3).

1.4.2.5 Nutrient Acquisition by PGPR

As compared to conventionally managed soils, soils with dynamic microbial ecologies and high organic matter typically have lower fertilizer requirements. For example, bulk microbial activity in soils is often taken into consideration when managing the application of organic nutrient sources. Research on plant–microbe association

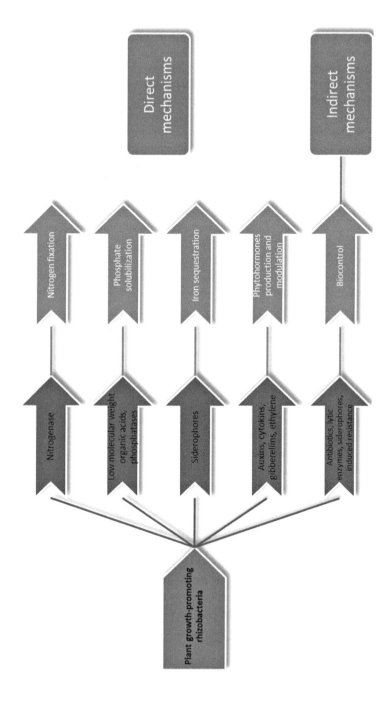

FIGURE 1.3   Direct and indirect mechanisms of plant growth-promoting rhizobacteria.

is beginning to reveal specific plant–microbe interactions that are directly involved in plant nutrition. Microbes that assist in plant nutrient acquisition (biofertilizers) act through a variety of mechanisms including nitrogen fixation, P-solubilization, augmentation of the surface area accessed by plant roots, siderophore production, and HCN production. Therefore, managing microbial activity has great potential for providing crops with nutritional requirements. N-fixing symbiosis between rhizobia and legumes is considered to be the most extensively studied and exploited beneficial plant-bacteria relationship.

1.4.2.6   Plant Hormones Produced by PGPR

Phytohormones are considered "important factors" in regulating plant growth and development as they function as molecular signals in response to environmental factors. Many rhizospheric bacteria have been reported to excrete hormones for root uptake in order to boost growth and stress response in higher plants. Many PGPR can produce auxins to exert particularly strong effects on root growth and architecture. The most widely studied auxin produced by PGPR is indole-3-acetic acid (IAA), as it is involved in plant–microbe interactions. PGPR that produce auxins have been shown to cause transcriptional change in hormone, defense-related, and cell-wall-related genes, induce longer roots, increase root biomass and decrease stomata size and density, and activate auxin response genes that enhance plant growth. Many PGPR produce cytokinins and gibberellin but the role of bacterially synthesized hormones in plants, along with the bacterial mechanism of synthesis, are not yet completely understood. Some strains of PGPR can promote relatively large amounts of gibberellins, leading to enhanced plant shoot growth.

1.4.3   Microbe-Mediated Induced-Stress Tolerance in Crop Plants

The mechanisms that regulate stress tolerance in plants are complex and intricate. Improving stress tolerance in crop plants through genetic engineering is associated with ethical and social acceptance issues, while conventional breeding is a long and capital-intensive process. Therefore, the role of beneficial microorganisms is gaining importance in stress management.

1.4.3.1   Contribution to Phytoremediation

The term "phytoremediation" is derived from the Greek "phyto" (plant) and Latin "remedium" (cure, remedy, or heal) (Cunningham and Berti 1993). Phytoremediation is widely considered a low-cost and environmentally friendly plant-based technique for the reclamation of contaminated sites. Besides being an aesthetically pleasing technology, phytoremediation also contributes to atmospheric CO_2 fixation, increased biodiversity, reduced erosion, and the production of energy through biomass incineration (Thijs et al. 2017)

Two parameters are particularly useful to assess the potential of a given plant species for phytoremediation: the bioconcentration factor (BCF) and the translocation factor (TF). The BCF indicates the capability of a plant to absorb the contamination from the growing substrate and accumulate them in its tissues. The TF denotes the plants' aptitude to transport contaminants from the root to its aerial parts. the BCF and TF can be calculated as follows:

$$BCF = C_{plant} / C_{soil}$$

$$TF = C_{shoot} / C_{root}$$

Where C_{plant}, C_{soil}, C_{root}, and C_{shoot} are the pollutant concentration in the plant, soil (ideally in labile forms), root, and shoot, respectively.

Because of significant developments in the last 25 years, phytoremediation has branched into a number of sequential processes that relate to the type of contaminant and reclamation mechanism (Table 1.1). PGPR have been the focus of an increasing number of studies during the last decade which typically entail the use of different strains of rhizobacteria that are either isolated from mine tailings, or selected foe-known advantageous traits. In both cases, bacteria are usually inoculated alone or in association, and assessed for their resistance to metal toxicity and other plant-growth-promoting traits (Ma et al. 2011).

Generally, PGPR have very limited influence on the metal bioavailability of mine lands. The most pronounced effects of PGPR-aided phytoremediation of mine soils are: clearly enhanced metal tolerance, plant growth, and biomass yield. This increment of biomass leads to greater harvestable amounts of metal that could favor phytoextraction. Still, results thus far have not shown any outstanding increase in the levels of harvestable metal and, as such, Phyto stabilization seems to be the subcategory of phytoremediation that benefits the most from PGPR. An exception can be made to studies concerning serpentine soils, where, in the case of PGPR, may be instrumental in the improvement of the performance of nickel (Ni) hyperaccumulators (Novo et al. 2018).

TABLE 1.1

List of Sequential Mechanisms of Phytoremediation

Process	Description
Phytostabilisation	Fixation of contaminants in rhizosphere by action of roots and microbes
Phytoextraction	Accumulation of contaminants in the above ground plant tissues
Phytodegradation	Degradation of pollutants by plant enzymes
Rhizodegradation	Degradation of pollutants in the rhizosphere through the action of rhizospheric microbes
Phytovolatalisation	Transformation of contaminants to volatile form, followed by their release to the atmosphere
Phytofiltration	Sequestration of contaminants in water by the roots of plants
Phytodesalinisation	Reduction of surplus salt from saline soils through halophytes

1.4.3.2 Contribution of Rhizospheric Microbe Towards Rhizomediation

Plant enzymes implant the degradation of pollutants during phytoremediation, whereas the microbial population performs degradation during natural attenuation or bioaugmentation. In many of these studies, an important contribution to the degradation of pollutants is attributed to microbes present in the rhizosphere of plants used during the phytoremediation of plants, which are emerging as natural vegetation on a contaminated site. This contribution of the rhizomicrobial population is referred to as rhizoremediation (Figure 1.4). Although the importance of the rhizosphere community for degradation of pollutants has been recognized, very little is known about the exact composition of the degrading population.

Studies of the most suitable plant species for rhizoremediation showed that various grass varieties and leguminous plants such as alfalfa are suitable, due to their ability to harbor large numbers of bacteria on their highly branched root systems (Kuiper et al. 2001). Success also depends on factors such as primary and secondary metabolism, and the establishment, survival, and ecological interaction with other organisms. The use of plants in combination with microbes has the advantage of causing an increase in microbial population numbers and metabolic activity in the rhizosphere. It can also create improvements with regards to the physical and

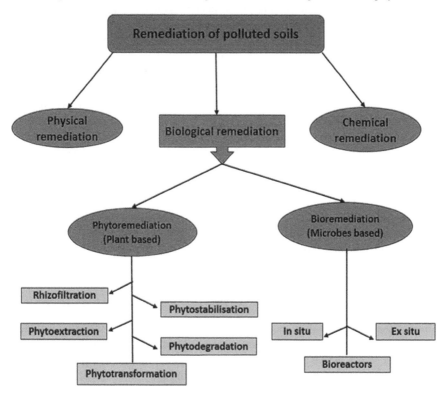

FIGURE 1.4 Flow chart showing various ways for remediation of polluted soils.

chemical properties of contaminated soil, along with an increase in contact between the microbes associated with roots and the contaminants in soil (Kuiper et al. 2001).

1.4.3.3   Microbial Association Improving Tolerance of Crops to Abiotic and Biotic Stress

1.4.3.3.1   Abiotic Stress

All living organisms are affected by environment factors such as abiotic stress. Some plants can overcome stress, while others have internal mechanisms to cope with stress. Abiotic stress factors include: water deficits, excessive water, extreme temperatures, and salinity. The association of PGPR with certain plants can help plants combat certain abiotic stresses and prevent the plants from dying. Grover et al. (2011) have reported that bacteria belonging to different genera, including Rhizobium, Bacillus, Pseudomonas, Paenibacilllus, Burkholderia, Azospirillum, Microbacterium, and Enterobacter, endow host plants under different abiotic stress environments. Some studies stated how maize plants exposed to low temperatures show reduced shoot and root growth that has been attributed to severe oxidative damage induced by cold stress, whereas a cold-tolerant PGPB Methylobacterium phyllosphaerae strain IARI – HHS2 – 67, isolated using a leaf-imprinting method from the phyllosphere of wheat (Triticum aestivum L.,), showed improved survival, growth, and nutrient uptake compared to a non-inoculated control at 60 days under low-temperature conditions (Baek and Skinner 2012; Saeindnejad et al. 2012; Verma et al. 2015; Singh et al. 2018, 2017a, b, c, 2019; Tiwari et al. 2018, 2019a, b; Kour et al. 2019a).

The effect of global warming in recent years can be felt with the increase in the global temperature. Thermo-tolerant plant growth-promoting has been proven to be beneficial for the growth of crops (Pseudomonas putida strain AKMP 7 in wheat) under heat stress (Ali et al. 2011) by significantly increasing the root and shoot length and dry biomass of wheat as compared to uninoculated plants. When comparing salinity stress, approximately 20–50% of crop yields are lost to high soil salinity (Shahbaz and Ashraf 2013). Crop plants are very sensitive to soil salinity, as it is the harshest environmental factors that limit a crop's productivity. Tank and Saraf (2010) reported salinity-resistant bacterial species (pseudomonas fluorescens, p. aeruginosa and p. stuterzi) ameliorated sodium chloride stress in tomato plants. Salt-stressed Arabidopsis plants treated volatile organic compounds (VOCs) from B. Amyloliquefaciens GBO3 showed higher biomass production and less sodium (Na^+) accumulation compared to salt-stressed plants without VOC treatment (Mathew et al. 2015).

1.4.3.3.2   Biotic Stress

Biotic stress in plants mainly includes damage caused by other living organisms such as insects, bacteria, fungi, nematodes, viruses, viroids, and protists. The PGPR can resist biotic stress in two different ways: by the direct promotion of plant growth by the production of phytohormones, or by facilitating the uptake of certain nutrients. The indirect promotion of plant growth occurs when PGPR reduces the deleterious effects of phytopathogen. For example, *P. fluorescens* produces 2,4 – diacetyl phloroglucinol, which inhibits the growth of phytopathogenic fungi and extracellular chitinase and laminarinase produced by *P. stutzeri* causing the lysis of mycelia of root rot causing *F. solani* (Keel et al 1992; Lim et al. 1991) and *P. fluorescens* CHAO as a bio agent protected tomato plant from *F. oxysporum* f.sp. *lycopersici* by induced resistance.

As more and more pests are becoming immune to pesticides, hence pest management has become an issue over time. The development of entomopathogenic bacteria for pest management is a new approach to handle resilient pests. Species belonging to the genera *Aschersonia*, Agerata, Sphaerostibe, Verticillium, Laterosporus action have been reported to be effective against insects such as Coleoptera, Lepidoptera, nematodes, and phytopathogenic fungi (De Oliveria et al. 2004; Saikia et al. 2011). Hence, microbial interaction could enhance plant growth even under conditions of stress, due to plant growth promotion activities, as shown in Figure 1.5.

1.4.4 CONCEPT OF BIOREMEDIATION MICROBIAL ROLE IN SOIL QUALITY ENHANCEMENT

Bioremediation is defined as the use of biological processes to degrade, transform, and essentially remove contaminants or soil and water quality impairment. Bioremediation is a natural process relying on microbes and plants to alter contaminants as these organisms carry out their normal life functions. Bioremediation technology exploits various naturally occurring mitigation processes, such as natural attenuation, biostimulation, and bioaugmentation. Biostimulation utilizes indigenous microbial populations to remediate contaminated soils. Biostimulation consists of adding nutrients and other substances to soil to catalyze natural attenuation processes, whereas bioaugmentation involves the introduction of exogenic microorganisms capable of detoxifying a particular contaminant, sometimes employing genetically altered microorganisms.

Three primary ingredients for bioremediation are: (1) presence of a contaminant, (2) an electron acceptor, and (3) presence of microorganisms that are capable of degrading the specific contaminant. Generally, if the contaminant is a naturally occurring compound in the environment, or chemically similar to a naturally occurring compound, it is more easily and quickly degraded because microorganisms capable of its biodegradation are more likely to have evolved. Petroleum hydrocarbons are naturally occurring chemicals; therefore, microorganisms which are capable of reducing the effect of hydrocarbons are found in the environment. The most suitable example is the development of biodegradation technologies of synthetic chemicals; such DDT is dependent on outcomes of research that seek natural or genetically improved strains of microorganisms to degrade such contaminants into less toxic forms. Therefore, bioremediation has the potential to provide a low-cost, reclusive as well as natural method to seize toxic substances in the soil, thus rendering them less harmful, or even harmless, over time.

1.5 SUSTAINABLE AGROECOSYSTEM: AN OUTCOME FROM HEALTHY PLANT–MICROBE ASSOCIATION

Plant growing under field conditions is a complex interaction between plant and rhizosperic microbes (Lundberg et al. 2012). The well-defined microbial community is

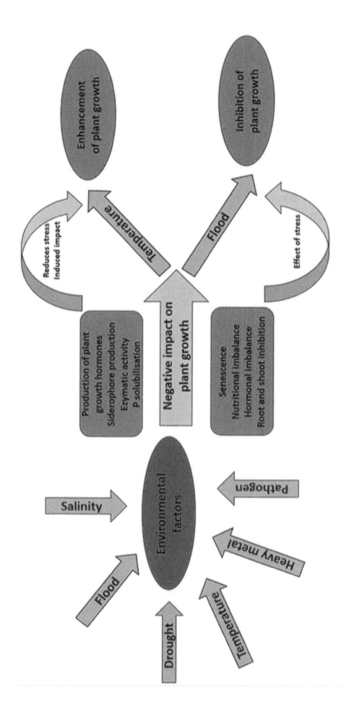

FIGURE 1.5   Various stress factors affecting plant growth and PGPRs combating environmental stress.

TABLE 1.2
Seed Yield (kg ha^{-1}) of Red Gram Influenced by Native Rhizobium Inoculation (Sethi 2019)

Treatments	CHRS-7*	RAN-1	RAB-1	MEAN-N
N_0#	902	682	759	781
N_{50}	1529	1001	968	1187
N_{100}	1793	1331	1463	1529
N_{150}	1749	1309	1441	1500
MEAN-S	1493	1081	1158	
LSD(P=0.05)	S=52	N=60	SXN=103	

* CHRS-7(native strain), RAN-1 and RAB-1 are isolated Rhizobium strains
# Level of N application (N_0 – No nitrogen, N_{50} – 50%, N_{100} – 100%, and N_{150} – 150% of N requirement)

always associated with the crop and these communities are phytomicrobiomes; the phytomicrobiomes with a plant is the holobiont (Bulgarelli et al. 2015; Smith et al. 2017). This microbial community has been associated with terrestrial plants with challenges to access nutrients, novel and often-stressful conditions, and pathogens. There are elements (including bacteria and fungi) of the phytomicrobiome associated with all major plant structures (flowers, fruits, stems, leaves, roots). The most comprehensible example is the nitrogen-fixing rhizobia associated with legumes (Gray and Smith 2005). Many members of the phytomicrobiome cannot be cultured, and it has only been since the advent of metagenomics (Hirsch and Mauchline 2012) and related methods that we are able to assess how membership is changed by conditions, plant genotype (Delaplace et al. 2015; Poli et al. 2016; Wintermans et al. 2016), and plant development. Current microbes differ in their efficiency, and are used for the production of crops and enhancement of soil fertility for sustainable agriculture. The stress-tolerant microbe's inoculation increases yield of pulse up to 20–30% over an uninoculated crop. The enhancement red gram yields up to 35% over nonnative rhizobium inoculation in an acidic condition (Table 1.2) This is due to the easy acclimatization of the inoculated native strain in comparison to non-native strains (Sethi 2019).

The yield and quality of green gram seed, black gram seed, okra fruit, finger millet grain, berseem grain (Pattanayak 2012, 2016) rice grain (Pattanayak et al. 2007, 2006a, b) increase due to the application of microbes in the plant rhizosphere, with integrated application nutrients and soil amendments for maintaining soil health.

1.6 POTENTIAL OF PLANT–MICROBE INTERACTION FOR DEVELOPING A LOW-INPUT SUSTAINABLE AGRICULTURE

Inorganic fertilizers are not cost-effective and, on the other hand, its injudicious use worsens soil health. The application of organic fertilizers or inoculation of plant growth-promoting rhizobacteria enhances root activity by producing different exopolysachharides, which create a suitable rhizosphere environment for microbe and host root proliferation.

As with diazotrophs, the production of growth hormones assists with amelioration of the root surface area, which may have indirect effects on the P-solubilizing ability of the bacteria (Sharma et al. 2016). It appears that the evaluation of ability to solubilize P in vitro does not participate as efficiently as under field conditions. Thus, it may be concluded that the biofertilizers with dual action might have been mediated by directly solubilizing the inorganic source of P, mineralizing organic P, and stimulating root growth or formation of mycorrhiza.

The significant quantities of N can be supplied by rhizobacteria in crop plants (Dobbelaere et al. 2003). However, inoculation of Azospirillum sp. in cereal (wheat, sorghum, and maize) contributed 5 kg N ha^{-1}year^{-1}. This quantity is extremely small in comparison to 150–200 kg N ha^{-1}year^{-1} (as is common practice in modern agriculture). Contribution of free-living rhizobacteria for crop plants in Australia is even less than 10 kg N ha^{-1}year^{-1} (Unkovich and Baldock 2008). Peoples et al. (2002) also pointed out that the range of N$_2$ fixation varied from 0–15 kg N ha^{-1}year^{-1}. Further, value of N$_2$ fixation ranged between 1–10 kg N ha^{-1}year^{-1} as suggested by Bottomley and Myrold (2007). Therefore, the ability of plant-growth-promoting bacteria to fix atmospheric N$_2$ is considered an important criterion when classifying them as a biofertilizer which can be supplemented for enhancing the bioavailability of nutrients to crops. Thus, it can be used as an input for sustainable agriculture.

1.7   CONCLUSION

Microbes of the phytomicrobiome form a holobiont in combination with the plant, and this plant–microbe association provides a wide range of resources and benefits to the plant, as well as the microbial community. The plant–microbe association is of great importance to agriculture, owing to the rich diversity of root exudates and plant cell debris that attracts diverse and unique patterns of microbial colonization. Microbes of the rhizomicrobiome play key roles in enhancing plant growth by nutrient acquisition and assimilation, improved soil structure, secreting and modulating extracellular molecules such as hormones, secondary metabolites, antibiotics, and various signal compounds. The plant–microbe intercommunication can improve the dynamics and efficiency of many nutrients, as well as sustained growth and yield of crop plants, due to interactive influences of the soil, plant, and environmental factors, as described in this chapter. Any one or more of these factors may adversely affect the availability of nutrients. Along with these, PGPR enhances plant growth and productivity by enhancing various signaling mechanisms (direct and indirect). Studies have been undertaken of the healthy plant–microbe association, whereby crops escape from various abiotic as well as biotic stresses, thereby providing optimum conditions for their growth throughout their life cycle. The role of plant–microbe interaction in phytoremediation, bioremediation, and rhizoremediation has also been thoroughly discussed in this chapter. Ultimately, the process leads to a sustainable agroecosystem. How the interaction between plant and microbes could help make steps towards a sustainable agroecosystem has been thoroughly discussed in this chapter. An understanding of plant–microbe interaction and intercommunication is of the utmost importance when making efforts towards a sustainable ecosystem – and even towards a very predictable agroecosystem. Plants in companionship with rhizobial microbes can lead to a significantly more productive and sustainable agroecosystem, which in turn will support the life of all living organisms.

REFERENCES

Ali, S.Z., Sandhya, V., Grover, M. et al. 2011. Effect of inoculation with a thermo tolerant plant growth promoting *Pseudomonas putida* strain AKMP7 on growth of wheat (*Triticums* pp.) under heat stress. *Journal of Plant Interactions* 6:239–246.

Almeida, H.J., Pancelli, M.A., Prado, R.M. et al. 2015. Effect of potassium on nutritional status and productivity of peanuts in succession with sugar cane. *Journal of Soil Science and Plant Nutrition* 15:1–10.

Ambrosio, R., Ortiz-Marquez, J.C., Curatti, L. 2017. Metabolic engineering of a diazotrophic bacterium improves ammonium release and biofertilization of plants and microalgae. *Metabolic Engineering* 40:59–68.

Arora, N.K., Mishra, J. 2016. Prospecting the roles of metabolites and additives in future bioformulations for sustainable agriculture. *Applied Soil Ecology* 107:405–407.

Baek, K.H., Skinner, D.Z. 2012. Production of reactive oxygen species by freezing stress and the protective roles of antioxidant enzymes in plants. *Journal of Agricultural Chemistry and Environment* 1:34–40.

Baetz, U., Martinoia, E. 2014. Root exudates: The hidden part of plant defense. *Trends in Plant Science* 19(2):90–98.

Berendsen, R.L., Pieterse, C.M., Bakker, P.A. 2012. The rhizosphere microbiome and plant health. *Trends in Plant Science* 17:478–486.

Bever, J.D., Dickie, I.A., Facelli, E. et al. 2010. Rooting theories of plant community ecology in microbial interactions. *Trends in Ecology & Evolution* 25:468–478.

Bottomley, P.J., Myrold, D.D. 2007. Biological N inputs. In: *Soil Microbiology, Ecology and Biochemistry* (ed. Paul, E.), 3658–4387. Oxford: Academic.

Brader, G., Compant, S., Mitter, B., Trognitz, F., Sessitsch, A. 2014. Metabolic potential of endophytic bacteria. *Current Opinion in Biotechnology* 27:30–37.

Bulgarelli, D., Garrido-Oter, R., Munch, P.C. et al. 2015. Structure and function of the bacterial root microbiota in wild and 887 domesticated barley. *Cell Host Microbe* 17:392–403.

Bulgarelli, D., Rott, M., Schlaeppi, K. et al. 2012. Revealing structure and assembly cues for Arabidopsis root-inhabiting bacterial microbiota. *Nature* 488:91–95.

Chagas, M.B.O., Prazeres Dos Santos, I., Nascimento, da Silva L.C., Correia, M.T.D.S., Magali de Araújo, J. et al. 2017. Antimicrobial activity of cultivable endophytic fungi associated with *Hancornia speciosa* gomes bark. *Open Microbiology Journal* 11:179–188.

Chesson, P. 2000. Mechanisms of maintenance of species diversity. *Annual Review of Ecology and Systematics* 31(1):343–366.

Ciccazzo, S., Esposito, A., Rolli, E. et al. 2014. Different pioneer plant species select specific rhizosphere bacterial communities in a high mountain environment. *Springer Plus* 3:391.

Compant, S., Clement, C., Sessitsch, A. 2010. Plant growth-promoting bacteria in the rhizo- and endosphere of plants: Their role, colonization, mechanisms involved and prospects for utilization. *Soil Biology and Biochemistry* 42:669–678.

Cunningham, S.D., Berti, W.R. 1993. Phytoremediation of contaminated soils: Progress and promise. *Proceedings of Symposium on Bioremediation and Bioprocessing – ACS Meeting 205*. Denver, CO: American Chemical Society, 265–268.

De, O.E.J., Rabinovitch, L., Monnerat, R.G. et al. 2004. Molecular characterization of *Brevibacillus laterosporus* and its potential use in biological control. *Applied and Environmental Microbiology* 70:6657–6664.

Delaplace, P., Delory, B.M., Baudson, C. et al. 2015. Influence of rhizobacterial volatiles on the root system architecture and the production and allocation of biomass in the model grass *Brachypodium distachyon* (L.) *P. Beauv. BMC Plant Biology* 15:195.

Dobbelaere, S., Vanderleyden, J., Okon, Y. 2003. Plant growth-promoting effects of diazotrophs in the rhizosphere. *Critical Review of Plant Science* 22:107–149.

Egamberdieva, D., Jabborova, D., Hashem, A. 2015. Pseudomonas induces salinity tolerance in cotton (*Gossypium hirsutum*) and resistance to Fusarium root rot through the modulation of indole-3-acetic acid. *Saudi Journal of Biological Sciences* 22:773–779.

Geddes, B.A., Ryu, M.H., Mus, F., Garcia Costas, A., Peters, J.W., Voigt, C.A. et al. 2015. Use of plant colonizing bacteria as chassis for transfer of N2-fixation to cereals. *Current Opinion of Biotechnology* 32:216–222.

Ghavami, N., Alikhani, H.A., Pourbabaee, A.A. 2016. Study the effects of siderophore-producing bacteria on zinc and phosphorous nutrition of Canola and Maize plants. *Communication of Soil Science and Plant* 47:1517–1527.

Glick, B.R. 1995. The enhancement of plant growth by free-living bacteria. *Canadian Journal of Microbiology* 41:109–117.

Glick, B.R. 2012. Using soil bacteria to facilitate phytoremediation, *Biotechnology Advances* 28:367–374.

Gray, E.J., Smith, D.L. 2005. Intracellular and extracellular PGPR: Commonalities and distinctions in the plant-bacterium signaling processes. *Soil Biology & Biochemistry* 37:395–412.

Grover, M., Ali, S.Z., Sandhya, V. et al. 2011. Role of microorganisms in adaptation of agriculture crops to abiotic stresses. *World Journal of Microbiology and Biotechnology* 27:1231–1240.

Hirsch, P.R., Mauchline, T.H. 2012. Who's who in the plant root microbiome. *Nature Biotechnology* 30:961–962.

Ho, Y.N., Chiang, H.M., Chao, C.P. et al. 2015. Plant biocontrol of soil borne Fusarium wilt of banana through a plant endophytic bacterium, *Burkholderia cenocepacia* 869 T2. *Plant and Soil* 387:295–306.

Hua, J. 2013. Modulation of plant immunity by light, circadian rhythm, and temperature. *Current Opinion in Plant Biology* 16:406–413.

Hussain, Z., Khattak, R.A., Irshad, M. 2016. Effect of saline irrigation water on the leachability of salts, growth and chemical composition of wheat (*Triticum aestivum* L.) in saline-sodic soil supplemented with phosphorus and potassium. *Journal of Soil Science and Plant Nutrition* 16:604–620.

Jensen, E.S., Peoples, M.B., Boddey, R.M. 2012. Legumes for mitigation of climate change and provision of feedstocks for biofuels and biorefineries. *Agronomy of Sustainable Development* 32:329–364.

Keel, C., Schnider, U., Maurhofer, M. et al. 1992. Suppression of root diseases by *Pseudomonas fluorescens* CHA0: Importance of the bacterial secondary metabolite 2, 4-diacetyl phloroglucinol. *Molecular Plant–Microbe Interactions* 5:4–13.

Kour, D., Rana, K.L., Yadav, N., Yadav, A.N., Rastegari, A.A., Singh, C., Negi, P., Singh, K., Saxena, A.K. 2019a. Technologies for biofuel production: Current development, challenges, and future prospects. In: *Prospects of Renewable Bioprocessing in Future Energy Systems* (eds. Rastegari, A.A. et al.), *Biofuel and Biorefinery Technologies*, Vol. 10, 1–50. Berlin: Springer.

Kuiper, I., Bloemberg, G.V., Lugtenberg, B.J.J. 2001. Selection of a plant-bacterium pair as a novel tool for rhizostimulation of polycyclic aromatic hydrocarbon-degrading bacteria. *Molecular Plant–Microbe Interaction* 14:1197–1205.

Lareen, A., Burton, F., Schäfer, P. 2016. Plant root-microbe communication in shaping root microbiomes. *Plant Molecular Biology* 90:575–587.

Lim, H.S., Kim, Y.S., Kim, S.D. 1991. *Pseudomonas stutzeri* YPL-1 genetic transformation and antifungal mechanism against *Fusarium solani*, an agent of plant root rot. *Applied and Environmental Microbiology* 57:510–516.

Lundberg, D.S., Lebeis, S.L., Paredes, S.H. et al. 2012. Defining the core *Arabidopsis thaliana* root microbiome. *Nature* 488:86.

Ma, Y., Prasad, M.N.V., Rajkumar, M. et al. 2011. Plant growth promoting rhizobacteria and endophytes accelerate phytoremediation of metalliferous soils. *Biotechnological Advances* 29:248–258.

Majeed, A., Abbasi, M.K., Hameed, S. 2015. Isolation and characterization of plant growth-promoting rhizobacteria from wheat rhizosphere and their effect on plant growth promotion. *Frontiers in Microbiology* 6:198.

Mathew, D.C., Ho, Y.N., Gicana, R.G. et al. 2015. Rhizosphere associated symbiont, *Photo bacteriums* pp. strain MELD1, and its targeted synergistic activity for phytoprotection against mercury. *PLoS One* 10:0121178.

Meena, V.S., Maurya, B.R., Verma, J.P. 2014. Does a rhizospheric microorganism enhance K+ availability in agricultural soils. *Microbiology Research* 169:337–347.

Meena, V.S., Maurya, B.R., Verma, J.P. et al. 2015. Potassium solubilizing rhizobacteria (KSR): Isolation, identification, and K-release dynamics from waste mica. *Ecological Engineering* 81:340–347.

Meena, V.S., Maurya, B.R., Verma, J.P. et al. 2016. *Potassium Solubilizing Microorganisms for Sustainable Agriculture*. Berlin: Springer.

Mendes, R., Garbeva, P., Raaijmakersm, J.M. 2013. The rhizosphere microbiome: Significance of plant beneficial, plant pathogenic, and human pathogenic microorganisms. *FEMS Microbiology Reviews* 37:634–663.

Nguema-Ona, E., Vicre-Gibouin, M., Cannesan, M.A., Driouich, A. 2013. Arabinogalactan proteins in root-microbe interactions. *Trends Plant Science* 18:440–449.

Novo, L., Castro, P., Alvarenga, P., Da, S.F. 2018. Plant growth promoting rhizobacteria assisted phytoremediation of mine soil. In: *Bio-Geotechnologies for Mine Site Rehabilitation*. doi:10.1016/B978-0-12-812986-9.00016-6.

Ormeño-Orrillo, E., Hungria, M., Martínez-Romero, E. 2013. Dinitrogen-fixing prokaryotes. In: *The Prokaryotes: Prokaryotic Physiology and Biochemistry* (eds. Rosenberg, E., de Long, E.F., Lory, S., Stackebrandt, E., Thompson, F.), 427–451. Berlin: Springer-Verlag. doi:10.1007/978-3-642-30141-4_72.

Padma, S.D., Sukumar, J. 2015. Response of mulberry to inoculation of potash mobilizing bacterial isolate and other bio-inoculants. *Global Journal of Bioscience and Biotechnology* 4:50–53.

Pattanayak, S.K. 2012. *QRT Report, 2007–2012, AINP on Soil Biodiversity Biofertilizers*. Submitted to Indian Council of Agricultural Research, New Delhi. Orissa University of Agriculture and Technology, Bhubaneswar.

Pattanayak, S.K. 2016. Biological nitrogen fixation – Status, potential and prospects in supplementing nitrogen needs of the crops. *Indian Journal of Fertilisers* 12:94–103.

Pattanayak, S.K., Das, P.K., Pany, B.K. 2006a. Nutrient management through biofertilizers. In: *Nutrient Management in Crops in Soils of Orissa* (ed. Mitra, G.N.), 162–184. Jodhpur, India: Scientific Publishers.

Pattanayak, S.K., Panda, D., Pradhan, N.K. 2006b. Nutrient management in cereal crops. In: *Nutrient Management in Crops in Soils of Orissa* (ed. Mitra, G.N.), 45–74. Jodhpur, India: Scientific Publishers.

Pattanayak, S.K., Rao, D.L.N., Mishra, K.N. 2007. Effect of biofertilisers on yield, nutrient uptake and N economy of rice peanut cropping sequence. *Journal of the Indian Society of Soil Science* 55:184–189.

Peoples, M.B., Brockwell, J., Herridge, D.F., Rochester, I.J., Alves, B.J.R., Urquiaga, S., Boddey, R.M., Dakora, F.D., Bhattarai, S., Maskey, S.L. et al. 2009. The contributions of nitrogen-fixing crop legumes to the productivity of agricultural systems. *Symbiosis* 48:1–17.

Peoples, M.B., Giller, K.E., Herridge, D.F., Vessey, J.K. 2002. Limitations to biological nitrogen fixation as a renewable source of nitrogen for agriculture. In: *Nitrogen Fixation: Global Perspectives* (eds. Finan, T.M., O'Brian, M.R., Layzell, D.B., Vessey, J.K., Newton, W.), 356–360. Wallingford: CAB International.

Poli, A., Lazzari, A., Prigione, V., Voyron, S., Spadaro, D., Varese, G.C. 2016. Influence of plant genotype on the cultivable fungi associated to tomato rhizosphere and roots in different soils. *Fungal Biology* 120:862–872.

Prashar, P., Kapoor, N., Sachdeva, S. 2013. Biocontrol of plant pathogens using plant growth promoting bacteria. In: *Sustainable Agriculture Reviews*, Vol. 12, 319–360. Dordrecht: Springer Netherlands.

Pratiwi, H., Aini, N., Soelistyono, R. 2016. Effects of *Pseudomonas fluorescens* and sulfur on nutrients uptake, growth and yield of groundnut in an alkaline soil. *Journal of Degraded Mining Lands Management* 3:507–516.

Richardson, A.E., Barea, J.M., Mcneill, A.M., Prigent-Combaret, C. 2009. Acquisition of phosphorus and nitrogen in the rhizosphere and plant growth promotion by microorganisms. *Plant and Soil* 321:305–339.

Rogers, C., Oldroyd, G.E.D. 2014. Synthetic biology approaches to engineering The nitrogen symbiosis in cereals. *Journal of Experimental Botany* 65:1939–1946.

Rubio, L.M., Ludden, P.W. 2008. Biosynthesis of the iron molybdenum cofactor of nitrogenase. *Annual Review on Microbiology* 62:93–111.

Saeidnejad, A.H., Pouramir, F., Naghizadeh, M. 2012. Improving chilling tolerance of maize seedlings under cold conditions by spermine application. *Notulae Scientia Biologicae* 4:110.

Saikia, R., Gogoi, D., Mazumder, S., Yadav, A., Sarma, R., Bora, T. et al. 2011. *Brevibacillus laterosporus* strain BPM3, a potential biocontrol agent isolated from a natural hot water spring of Assam, India. *Microbiological Research* 166:216–225.

Sethi, D. 2019. *Isolation and Characterization of Efficient Strains of Native Rhizobium and Evaluation of their Potency as Bio-fertilizer for Enhancing Productivity of Red Gram in Odisha*. Ph.D. Thesis submitted to Odisha University of Agriculture and Technology, Bhubaneswar.

Shahbaz, M., Ashraf, M. 2013. Improving salinity tolerance in cereals. *Critical Reviews in Plant Sciences* 32:237–249.

Sharma, M., Mishra, V., Rau, N., Sharma, R.S. 2015. Increased iron-stress resilience of maize through inoculation of siderophore producing *Arthrobacter globiformis* from mine. *Journal of Basic Microbiology* 56:719–735.

Sharma, P., Kumawat, K.C., Kaur, S. 2016. Plant growth promoting rhizobacteria in nutrient enrichment: Current perspectives. In: *Biofortification of Food Crops*. doi:10.1007/978-81-322-2716-8_20.

Singh, C., Tiwari, S., Boudh, S., Singh, J.S. 2017a. Biochar application in management of paddy crop production and methane mitigation. In: *Agro-Environmental Sustainability: Managing Environmental Pollution* (eds. Singh, J.S., Seneviratne, G.), second ed., 123–146. Switzerland: Springer.

Singh, C., Tiwari, S., Gupta, V.K., Singh, J.S. 2018. The effect of rice husk biochar on soil nutrient status, microbial biomass and paddy productivity of nutrient poor agriculture soils. *Catena* 171:485–493.

Singh, C., Tiwari, S., Singh, J.S. 2017b. Impact of rice husk biochar on nitrogen mineralization and methanotrophs community dynamics in paddy soil. *International Journal of Pure and Applied Bioscience* 5:428–435.

Singh, C., Tiwari, S., Singh, J.S. 2017c. Application of biochar in soil fertility and environmental management: A review. *Bulletin of Environment, Pharmacology and Life Sciences* 6:07–14.

Singh, C., Tiwari, S., Singh, J.S. 2019. Biochar: A sustainable tool in soil 2 pollutant bioremediation. In: *Bioremediation of Industrial Waste for Environmental Safety* (eds. Bharagava, R.N., Saxena, G.), 475–494. Berlin: Springer.

Smith, D.L., Gravel, V., Yergeau, E. 2017. Editorial: Signaling in the phytomicrobiome. *Frontiers in Plant Science* 8:611.

Stinson, K.A., Campbell, S.A., Powell, J.R., Wolfe, B.E., Callaway, R.M., Thelen, G.C. et al. 2006. Invasive plant suppresses the growth of native tree seedlings by disrupting belowground mutualisms. *PLoS Biology* 4:140.

Suzuki, N., Rivero, R.M., Shulaev, V., Blumwald, E., Mittler, R. 2014. Abiotic and biotic stress combinations. *New Phytologist* 203:32–43.

Tank, N., Saraf, M. 2010. Salinity-resistant plant growth promoting rhizobacteria ameliorates sodium chloride stress on tomato plants. *Journal of Plant Interactions* 5:51–58.

Thijs, S., Sillen, W., Weyens, N., Vangronsveld, J. 2017. Phytoremediation: State-of-the-art and a key role for the plant microbiome in future trends and research prospects. *International Journal of Phytoremediation* 19:23–38.

Tiwari, S., Singh, C., Boudh, S., Rai, P.K., Gupta, V.K., Singh, J.S. 2019a. Land use change: A key ecological disturbance declines soil microbial biomass in dry tropical uplands. *Journal of Environmental Management* 242:1–10.

Tiwari, S., Singh, C., Singh, J.S. 2018. Land use changes: A key ecological driver regulating methanotrophs abundance in upland soils. *Energy, Ecology, and the Environment* 3:355–371.

Tiwari, S., Singh, C., Singh, J.S. 2019b. Wetlands: A major natural source responsible for methane emission. In: *Restoration of Wetland Ecosystem: A Trajectory Towards a Sustainable Environment* (eds. Upadhyay, A.K. et al.), 59–74. Berlin: Springer.

Udvardi, M., Poole, P.S. 2013. Transport and metabolism in legume-rhizobia symbioses. *Annual Review on Plant Biology* 64:781–805.

Unkovich, M., Baldock, J. 2008. Measurement of a symbiotic N2 fixation in Australian agriculture. *Soil Biology Biochemistry* 40:2915–2921.

Verma, P., Yadav, A.N., Khannam, K.S., Panjiar, N., Kumar, S., Saxena, A.K. et al. 2015. Assessment of genetic diversity and plant growth promoting attributes of psychrotolerant bacteria allied with wheat (*Triticum aestivum*) from the northern hills zone of India. *Annals of Microbiology* 65:1885–1899.

Vogelsang, K.M., Bever, J.D. 2009. Mycorrhizal densities decline in association with non-native plants and contribute to plant invasion. *Ecology* 90:399–407.

Wintermans, P.C., Bakker, P.A., Pieterse, C.M. 2016. Natural genetic variation in Arabidopsis for responsiveness to plant growth-promoting rhizobacteria. *Plant Molecular Biology* 90:623–634.

Zhang, A., Zhao, G., Gao, T., Wang, W., Li, J., Zhang, S., Zhu, B. 2013. Solubilization of insoluble potassium and phosphate by *Paenibacillus kribensis* X-7: A soil microorganism with biological control potential. *African Journal of Microbiological Research* 7:41–47.

2 Soil Quality Monitoring in Selected Regions of Raebareli, Uttar Pradesh, and the Relationship of Microbes in Soil Fertility Management

Ajay Neeraj and R. Hiranmai Yadav

CONTENTS

2.1   INTRODUCTION

India is the second-most populous country after China, and most of the population depend on agriculture to make a living. Consequently, soil health and soil quality are of the utmost importance. Degradation and improper practice of agriculture techniques affect soil quality and environmental sustainability. A sustainable environment can be achieved by maintaining and enhancing soil quality. Degradation of surface soil and improper agriculture practices affect productivity and environmental sustainability. A sustainable environment can be achieved by maintaining and enhancing soil quality (Doran and Zeiss 2000). It is the component of the environment forming the uppermost layer of the Earth's crust and a basic source of nutrients for biodiversity.

The word "soil" originated from the Latin word "solum" which means "ground" or "floor." Degradation of forest and pollution soil becomes unfertile, harmful for the crop and human health. In plant growth, different types of minerals (N, P, K) are present in the soil, in addition to water, which is the carrier of minerals in the plant. The quality of soil and organic content also depend upon the place of soil, texture, and climate condition (Nath 2014). A sustainable environment can be achieved by maintaining and enhancing soil quality. Soil is the most important resource on the earth as well; it acts as a good environment filter because soil bears most of the minerals which are required for plant growth and soil quality.

Soil is mandatory for the plant, animal, and landform. It affects the organism habitat, and diversity consists of a wide range of different habitats. Soil controls and influences the exchange of water between the land and atmosphere, as well as acting as a sink for gases like N_2, CO_2, and O_2. Soil is a mixture of an organic compounds, living organisms, and minerals, which are continuously interacting with each other for balancing the environment. Soil takes part in the biogeochemical cycle. Nitrogen, carbon dioxide, and oxygen with the water cycle. Soil organism is an important part of soil because of agriculture, forest, and landform. Soil quality depends upon the physicochemical assets of the soil. Changes in the natural composition of soil have become unfavorable for humans as well as for the ecosystem. Therefore, soil microbes play a key role in soil quality and soil health.

2.2   SOIL FERTILITY AND NUTRIENT RECYCLING

In the field of agriculture, soil fertility and the nutrient content required for growing plants both play an important role. Sustainable environment, as well as sustainable agriculture productivity, depends upon the soil fertility because if the soil productivity is good then production or a plant grows well in the presence of nutrients. The uncontrolled population growth has resulted in demand for a huge amount of agriculture production, hence farmers are forced to use a high amount of fertilizers. This high amount of fertilizer is a solution for a short duration of time but the long-term impact of using fertilizers makes the soil unsuitable for agriculture, thereby reducing crop production. Different types of macro and micronutrients present in the soil help plant growth. Soil fertility regulated by the nutrient cycle maintains the nutrient balance. There are many nutrient cycles that regulate the nutrient cycle in the environment. In the environment there are 17 elements – without such elements, plants simply cannot grow. There are two types of nutrients present in the soil that are essential for plant growth: macronutrients and micronutrients, the difference being in their concentration.

1. *Macronutrients*: Carbon, Hydrogen, Nitrogen, Phosphorous, Sulphur, Calcium, Magnesium, Potassium, Chloride
2. *Micronutrients*: Ferrous, Manganese, Zinc, Copper, Nickel, Molybdenum

2.2.1 BIOGEOCHEMICAL CYCLE

Soil plays a vital role in the biogeochemical cycle and in the hosting of the largest diversity of organisms on land. The biogeochemical cycle has been divided into four cycles: one is in the biosphere, which the living organisms occupy; and the other three are in the lithosphere, atmosphere, and hydrosphere. However, where the biosphere overlaps the lithosphere, atmosphere, or hydrosphere, there is a zone occupied by living organisms.

The biogeochemical cycle can differ in two pathways:

• Sedimentary cycle
• Gaseous cycle (Figure 2.1)
    1. *Sedimentary cycle*: In the sedimentary cycle, the transport of matter through the ground involves water or the transportation of matter between the lithosphere and the hydrosphere.

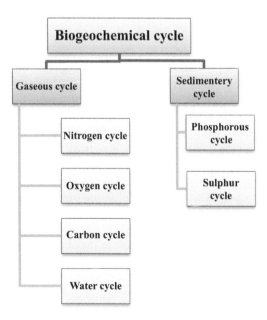

FIGURE 2.1 Different types of biogeochemical cycle.

- *Phosphorous cycle*: Phosphorous is usually found in the water, soil, and sediment. Phosphorus cannot be found in the air or gaseous state because phosphorous at the standard temperature and pressure, is found in liquid form. So the cycle of phosphorous is transported through the water, soil, and sediment. The phosphorous cycle is the slowest cycle in the sedimentary cycle.
- *Sulfur cycle*: It is conveyed by physical activity like run off water, wind, and activity. Sulfur in its natural form is found in solid form. Sulfur compounds like H_2SO_4, SO_X and other salts of sulfur or organic sulfur, travel from ocean to the atmosphere, to land, and then return in the ocean via rainfall.

2. *Gaseous cycle*: In this cycle, the transport of material occurs through the atmosphere.
   - *Carbon cycle*: Carbon is one of the most important elements that sustain life on earth. CO_2 and CH_4 in the earth's atmosphere have a substantial effect on the earth's heat balance. CO_2 absorbs the infrared radiation in the atmosphere.
   - *Nitrogen cycle*: Nitrogen is the most abounded gas in the atmosphere, at approximately 78%. The nitrogen cycle is most important for plant life. Nitrate, nitrite, and ammonia are compounds of nitrogen which help in the growth of plants or life.
   - *Oxygen cycle*: The oxygen cycle describes the movement of oxygen from the atmosphere to the lithosphere through the biosphere. Plants perform photosynthesis, which is the main producer of oxygen on the Earth. A plant takes CO_2 from the atmosphere and produces O_2.

The above biological cycle is explained as follows:

A) *Carbon cycle*: Carbon is the fundamental element of a living body. Most of the fuel used for energy is primarily made of carbon, including the food web. Carbon is present in the atmosphere, soil, ocean, and earth crust. Different forms of carbon such as CO_2 play an important role in the regulating climate. The biological cycle converts different carbon forms. CO_2 taken by the plant in photosynthesis creates new plant materials by converting C. Sugars, starches, fats, and proteins are necessary for any life which is made of carbon. The movement of carbon in different reservoirs like earth crust, ocean, soil, atmosphere, etc. is called carbon flux. All of these processes are fluxes that can cycle carbon among various pools within ecosystems, and eventually release it back to the atmosphere (Dioxide 1985). Carbons from the decaying of dead matter are converted into organic matter by microorganisms, and released back to the atmosphere. The Earth acts as a reservoir for the addition and removal of carbon from the atmosphere. CO_2 helps to maintain the Earth's average temperature. Dead plant material and animal waste through the microbes decompose in organic matter, so here soil microbes store the carbon on a temporary basis. (Figure 2.2).

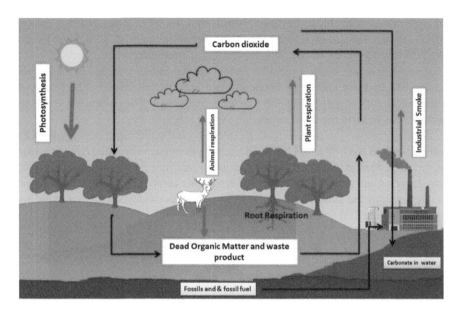

FIGURE 2.2 Different steps of the carbon cycle.

B) *Nitrogen cycle*: In the atmosphere, 78% of nitrogen is present. Nitrogen is required for the biological process. The abundance of N_2 in the atmosphere cannot be used directly by the plants. So the fixation of nitrogen helps to convert nitrogen into a usable form for plants. In the nitrogen cycle, different steps are utilized to fix the nitrogen.

C) *Nitrogen fixation*: Firstly, when the atmospheric nitrogen converts into ammonia NH_3 it is in an organic form of nitrogen, which a plant cannot take, and so NH_3 converts into ammonium (NH_4^+) that plants can use. This process is termed "nitrogen fixation". Some of these bacteria (*Rhizobium*) live in symbiosis with certain legume plants and others are free-living bacteria, such as cyanobacteria or *Azotobacter* (Shridhar 2012.)

Nitrification: Nitrification is a two-step process in which ammonia is firstly transformed into nitrites (NO_2^-) and then to nitrates (NO_3^-). Nitrification is the biological oxidation of ammonia with oxygen into nitrite, followed by the oxidation of these nitrites into nitrates, which is performed by two different bacteria (nitrifying bacteria). In the first step of the nitrification, *Nitrosominas* and *nitrosococcus* take part in the process; in the second step of the nitrification, *Nitrobacter* bacteria convert nitrite into nitrate. Nitrosifyers, bacterium like *Nitrosomonas* change NH_4^+ ions into nitrite ($NH_4^+ + O_2 \rightarrow NO_2^- + H_2O + H^+$). Then nitrifying bacteria like *Nitrobacter* convert nitrite in the nitrate by the oxidizing ($NO_2^- + O_2 \rightarrow NO_3^-$). Both the afore-mentioned bacteria are typically found in water and soil. They live particularly in regions where ammonia exists in high amounts, such as wastewater and manure. Nitrification does not provide significantly to agricultural land, so the nitrate is not always available for the plant. During the denitrification process, microbes quickly consume the nitrate.

In cooperation with ammonification, nitrification forms a mineralization activity which refers to the complete decomposition of organic material, with the release of available nitrogen compounds. This replenishes the nitrogen cycle.

Ammonification: When a plant or animal dies or excretes waste, so the decomposing-like bacteria and fungi break it down into organic matter. The release of ammonia occurs in a dissolved form in water in the soil. Then ammonia combines with H^+ ions to form the ammonium.

Denitrification: The transformations into the gaseous nitrogen compounds form nitrates such as N_2O, NO, and N_2 by different types of bacteria called denitrification, or nitrate reduction. Bacteria use nitrate as an alternate of O_2 during the respiration progress and convert into another nitrogenous compound. In the end, nitrogen is released into the atmosphere in the form of nitrous oxide, or nitrogen oxide and nitrogen (Figure 2.3).

$$NO_3^- \rightarrow NO_2^- \rightarrow NO \rightarrow N_2O \rightarrow N_2$$

D) *Phosphorous cycle*: The phosphorous cycle is the interchange of phosphorous elements through the lithosphere, hydrosphere, and biosphere, and living organisms requiring phosphorous for biological function, as it is required for the formation of nucleotides (Kumar et al. 2015). Phosphorous is the most important element providing nutrients for plants and animals. DNA is double helix is linked by the phosphorous. Phosphorus is an important nutrient for animals and plants (Childers et al. 2011). It is the main component of cell development and used to store energy in the form of ATP, DNA, and lipids. Deficiency of the phosphorus may decrease the crop

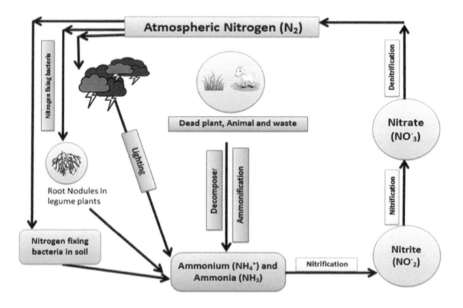

FIGURE 2.3   Different steps of the nitrogen cycle.

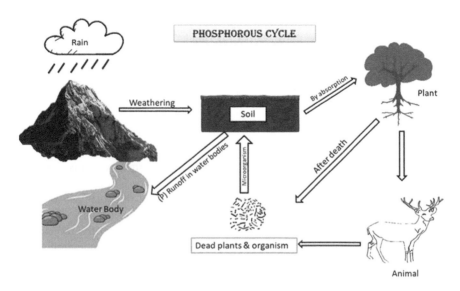

FIGURE 2.4 Different steps of the phosphorous cycle.

yield. The main source of P is found in rocks and the first step of the P cycle involves the extraction of P from the rocks by the process of weathering. Weathering event occurs with the help of rainfall, wind, temperature, and other environmental factors. Once the phosphorus is reached in the soil, plants and micro-organisms take phosphorus from the soil for growth. Plants can take phosphorus directly from the water, in the reservoirs with rainwater; phosphorus reaches the water, which can be taken by aquatic. Death and decay of plants and animals occurs, by microorganism of the soil, air returns P to the environment (Figure 2.4).

2.2.2 IMPORTANCE OF THE BIOGEOCHEMICAL CYCLE

The potential for life would not be possible without the biogeochemical cycle, as all elements like carbon, nitrogen, oxygen, and hydrogen would not be able to circulate without this (Redfield 1958). The ecosystem depends upon the biogeochemical cycle as it is between the living and non-living, so the flow of element is supplied all living organisms (Jacobson et al. 2000).

2.3 SIGNIFICANCE OF MICROBES IN SOIL FERTILITY MANAGEMENT

The soil ecosystem consists of a huge diversity of microbes such as protozoans, bacteria, actinomycetes, fungi, and algae. This diversity of microorganisms plays a vital role in the biogeochemical cycles of the environment, the formation of soils through complex physical, chemical and biological processes, and soil fertility management. The crucial role of microorganisms in the cycling of C, N, and P compound has been described in numerous studies of soil productivity

Some microbes play a vital role in improving plant nutrients with the help of the nitrogen-fixing bacteria or the mycorrhizal fungi. To access such support, plants depend upon soil microbes which possess the metabolic machinery to depolymerize and mineralize organic from the nitrogen, phosphorus, and sulfur. The substances of these microbial cells are subsequently released, either through turnover and cell lysis, or via protozoic predation (Bonkowski 2004; Richardson et al. 2009).

2.3.1 CARBON

The process of supply of nutrients to all living organisms from the soil is facilitated by a carbon exchange between the biosphere and atmosphere. During this exchange, the generated carbon is much greater than that generated by all human activity. This cycling of carbon is largely dependent upon the microorganisms which fix the atmospheric carbon into the soil and hence promote plant growth and the transformation of organic material in the soil. Different ecosystems (high-latitude permafrost, grassland soils, tropical forests, and others) have a large amount of carbon stored in them. Microbes play a vital role in the longevity and stability of this carbon, as well as preventing its emission to the atmosphere in the form of greenhouse gases. Unfortunately, this microbial-mediated carbon and other biogeochemical cycling are underestimated in various studies, hence resulting in limited predictive capabilities. If there were to be greater focus on these microbial-mediated activities, we would be able to further elaborate the conditions, and possibly ensure the regulation of carbon sequestration to a much greater extent.

Therefore, small changes in the rate and magnitude of biological carbon cycling may have a serious impact on the ecosystem; this may include the capture, store, or release of carbon, and vice versa. These changes in the chemical and physical properties of soil such as quality and amount of soil organic matter, pH, and redox conditions have an immense impact on the biological communities, dynamics, structure, and functions of the soil. (Lombard et al. 2011). Earlier studies reveal the biotic processes of decomposition at the molecular, organismal, and community level (Kelly and Tate 1998; Singh et al. 2018, 2017a, b, c, 2019; Tiwari et al. 2018, 2019a, b; Kour et al. 2019). This suggests the importance of microbes in the global cycle (Kandeler et al. 2001). The high amount of microbial biomass can be found in organic C rich-litter layers of forest and boreal grassland, where the fungi biomass dominates (Schimel et al. 1999). Dead and decaying plants and animals are the main source of energy for microorganisms. The organic matter is mineralized (decomposed completely) to CO_2, or it is humified. Where the mineralization and humification process goes simultaneously. These two processes are carried out by the microorganisms (fungi, actinomycetes, and bacteria) using enzymatic reactions. Most of the reservoir of the carbon store in the terrestrial ecosystem has great efficiency in containing carbon than the atmosphere. Environmental conditions and the microbial reactions control the transformation of carbon in the soil to a great capacity, allowing the sequestration of a huge amount of carbon to improve the soil quality, and be beneficial to the environment. The present day terrestrial ecosystem fix maximum CO_2 from the atmosphere then returns to the atmosphere through respiration, which is about 25% of global fossil fuels. (Le Quere et al. 2009). So the net

TABLE 2.1
Different Types of Microbes Involved in a Carbon Cycle

Bacillus spp.
Clostridium Thermocellum (B)
Penicillium (F)
Aspergillus (f)
Streptomyces (B)
Pseudomonas (B)
Trichoderma (F)
Phanerochaete chrysosporium (F)
Cellulomonas (B)
Verticillium (F)
Nocardia (F)
Phanerochaete chrysosporium (F)

carbon sequestration differs in various locations because of varying land management practices. Microorganism uses the original carbon after the transformation of carbon complex organic compounds remain in the soil. Humus is a combination of the many compound sands, and is important in a water and nutrient-holding capacity. Humus is composed of humic acid and fulvic acid. The two main methods of humus formation are modification of the lignin and the degradation of the plant residues. Plant residues return the soil to various organic compounds by microbial degradation. Decomposition is the biological process where a complex organic compound converts into simple organic and inorganic forms. In the carbon cycle transfer of the organic and inorganic compound by the plant and microorganism from soil to air and then atmosphere (Brussaard, Breen land 1994). In the process, if the decomposition... different products are released, such as carbon dioxide, energy, water and plant nutrient, and organic compound resynthesized. Dead organic matter is converted into a complex organic compound by the successive decomposition (Juma 1999). Due to slow decomposition, soil aggregation, and stability increase, there is an ability to retain soil nutrients (Table 2.1).

2.3.2 NITROGEN

Nitrogen is an essential part of our life form. It found in the amino acid, protein, and many organic compounds which result from activity during nitrogen fixation (Egamberdieva and Kucharova 2008). Biological nitrogen is fixed by the prokaryotes. It may be symbiotic or free-living in nature. In the soil, biologically nitrogen fixation depends upon the nitrogenase enzyme. The nitrogen cycle is the most important nutrient cycle in the terrestrial ecosystem. The nitrogen cycle involves several different processes (nitrogen fixation, mineralization, nitrification, denitrification) due to the large input of the nitrogen fertilizer to agriculture field influence the process of the nitrification and denitrification, and N_2O production increases due to the resultant denitrification and nitrification (Akiyama et al. 2006) Denitrifying fungi play an important role in the N_2O production under the aerobic and anaerobic condition by denitrifying fungi.

2.3.2.1   Nitrogen Fixation

Nitrogen gas is made of two atoms of nitrogen allied by a very strong triple bond so the molecules of nitrogen required very high energy for the breakdown of the triple bond so it is chemically unreactive. So some amount of nitrogen is fixed by the atmospheric lighting and maximum nitrogen fixation achieved by the bacteria, which is about 60% of the total nitrogen gas in the atmosphere.

The conversion of nitrogen gas into ammonia in soil and legume plants occurs by the action of microorganisms, and it requires more energy. The 16 moles of ATP and nitrogenase enzymes that easily attack bonds; therefore, nitrogen atoms are combined with hydrogen. It can be expressed as:

$$N_2 + 3H_2\ 2NH_3$$

Nitrogen-fixing bacteria are responsible for carrying out this reaction. A required amount of nitrogen is available for plants. Nitrogen-fixing bacteria show a symbiotic relationship with legume plants and provides nitrogen fertilizer. Nitrogen is a primary and limiting factor for agriculture that enhances the food production and crop growth by the use of a huge amount of N-fertilizer in agriculture.

2.3.2.2   Types of Nitrogen-Fixing Bacteria

In soil, certain free-living nitrogen-fixing bacteria are present that independently fix nitrogen, e.g. *Azotobacter* (aerobic in nature) and *Clostridium* (anaerobic in nature). Some other nitrogen-fixing bacteria exhibit symbiotic association with legume plants that form root nodules containing *Rhizobium* (e.g. peas and beans). The free-living rhizobia are infecting through the root, and form a nodule in 1 plants. A nodule plant tissue is twisted within the root while the rhizobia bacterium are inserted through legume plants via an infection thread. The infection thread passes through the immature portion of root and bacterium easily enters the roots' hair. Rhizobia are then involved in the root cortex. The legume plants provide nutrients to the bacteroids which synthesize the ATP which are present in large amounts. It is used for the conversion of nitrogen into ammonia in the presence of enzymes with metal molybdenum. It provides an oxygen-free environment.

Small nodules are easily visible within a week of infection. The group of *Rhizobium* bacteria are packed in each root nodule. Such non-legume plants are the root-nodulated group of woody species, e.g. *Frankia. Actinorhizal* root nodule is formed by the infection of filamentous rhizobium bacteria.

2.3.2.3   Adapting to the Environment

The conversion of nitrogen gas into ammonia occurs by the nitrogen-fixing bacteria which contain a complex enzyme called a "nitrogenase enzyme". It supplies the H_2 ion and energy from the ATP. Nitrogenase enzyme is very sensitive to O_2. This is not a problem with free-living anaerobic bacteria such as Clostridium. Free-living anaerobic bacteria have a variety of mechanisms to protect the nitrogenase complex. *Azotobacter* affects this problem by having the highest rate of respiration of any organism, thus maintaining a low level of oxygen in their cells.

Rhizobium bacteria contain *leghaemoglobin*. It is mainly nitrogen and oxygen carrier hemopro which are present in nitrogen-fixing root nodules of leguminous plants. It works similarly to hemoglobin, i.e. it binds to oxygen. A pink pigment is produced by legumes, which is responsible for colonizing nitrogen-fixing bacteria, i.e. rhizobia, which resembles the symbiotic association between plants and bacterium. It delivers adequate oxygen for the metabolic functions of the bacteroids, but prevents the accumulation of free oxygen that would finish the action of nitrogenase. *Frankia* and *Anabaena* are able to eliminate oxygen by carrying out the fixation in specialized structures, known as a vesicle and heterocyst. The thick walls of the vesicle and heterocyst function as an oxygen dispersion barrier.

2.3.2.4 Nitrification

This is the process of oxidation that converts ammonium compounds into nitrites or nitrate by the nitrifying bacteria *Nitrosomonas* and *Nitrococcus* respectively. In the nitrification process, the nitrite is converted to nitrate by the soil bacteria *Nitrobacter*. Energy gets released during the oxidation process. The nitrifying bacteria are chemoautotrophs and easily use the source of energy to convert organic complexes into inorganic (photo-autotrophs use light energy to produce organic compounds from inorganic ones).

Nitrification is a two-step process:

1. (NO_2^-). Ammonium ions convert into the nitrite by *Nitrosomonas* (Nitrite is toxic to plants in higher concentration).
2. Then the nitrite is converted into nitrate by *Nitrobacter* that can be taken up by the plant.

Denitrification: Denitrifying bacteria convert the nitrate primarily into nitrogen gas, e.g. *Pseudomonas*. The soil-like wetland and swampy ground contain denitrifying bacteria that transform the nitrate. For the breakdown of the nitrite bacteria, they need the respiration to get oxygen.

Ammonification (decay): Ammonification is the process of the conversion of organic nitrogen into inorganic N2. The N2-fixing bacteria are free-living in the soil and are classified into various classes based on different criteria as some are oxygen-dependent, e.g. *Azotobacter* (aerobic) whereas some are anaerobic, e.g. *Clostridium* (anaerobic).

Some microbes live in the symbiotic association with plant-like root-modulated legumes, e.g. beans, peas with Rhizobium then again root-nodulated non-legumes, a varied group of woody species such as alder, with, e.g., *Frankia*. These filamentous bacteria infect the roots of plants forming actinorhizal root nodules (Tables 2.2 and 2.3).

TABLE 2.2

Nutrient Bioavailability and Aggregate Formation in Bacteria and Fungi

Reaction	Micro-Organism	Conditions	Process
Nitrogen fixation	Nitrogen-fixing bacteria, e.g. *Rhizobium*	Aerobic/ anaerobic	The first step in the synthesis of almost all nitrogenous compounds is fixation of nitrogen gas to other forms so that organisms can use it.
Ammonification (decay)	Ammonifying bacteria (decomposers)	Aerobic/ anaerobic	In the decomposition, several bacteria and fungi break the complex matter like protein and animal waste, releasing the ammonium ions then converting into the compound.
Nitrification	Nitrifying bacteria, e.g. *Nitrosomonas, Nitrobacter*	Aerobic	Nitrification occurs in two steps. In the first step ammonia or ammonium ions are oxidized to nitrate; in the second step they are converted into nitrite, which is the most usable form of plant.
Denitrification	Denitrifying bacteria	Anaerobic	NO_3^- is converted into N_2 gas, returning nitrogen to the air and completing the cycle.

Source: https://microbiologysociety.org

TABLE 2.3

Different Types of Microbes in the Nitrogen Cycle

Nitrosococcus oceani

Nitrobacter winogradskyi

Fusarium oxysporum

Fusarium solani

Cylindrocarpon tonkinense

Trichosporon cutaneum

Gibberella fujikuroii

Gluconacetobacter Diazotrophicus

Acetobacter nitrogenifigens

Gluconacetobacter kombuchae

Gluconacetobacter johannae

Gluconacetobacter azotocaptans

Swaminathania salitolerans

Acetobacter peroxydans

Nitrosopumilus maritimus

2.3.3 PHOSPHORUS

Phosphorus plays a vital role as nutrition for plants, following nitrogen (N) and also helps to various physiological changes in plants such as photosynthesis, energy movement, macromolecular biosynthesis, and respiration (Khan et al. 2010). Phosphorous directly affects the plant growth in both organic and inorganic forms. Generally, phosphorous is found in the form of insoluble minerals complexes by the regular use of phosphoric fertilizer (Rengel and Marschner 2005). Organic matter is one of the best reservoirs of immobilizing phosphorous present in the range of 20 to 80% (Richardson 2001). There is a very small amount existing in a soluble form for the plant (Zou et al. 1992). Tropical land soil contains the highest phosphorous capacity. Supplements due to the soil phosphorous and available phosphorous are added to the agriculture soil by the excess use of phosphorous fertilizer, which impacts on the soil health as well as contributes to the degradation of terrestrial and aquatic resources (Tilman et al. 2001). The excess and non-stop application of phospho-fertilizers directly influences the soil fertility (Gyaneshwar et al. 2002) by disturbing the microbial diversity which leads to fertility loss and reduced crop yield. Soil microorganisms act as sources as well as sink of the nutrient but also affects the speciation (Oberson 2005). However, there are very limited effects of phosphate nutrients on crop fields. The buffer capacity of Cl-phosphate rich soil can explain the limited effects of phosphate nutrients leading to the retreat of protons released by microorganisms (Jones and Darrah 1994). The growing condition of rhizospheric microorganisms is totally different than laboratory conditions to assess the phosphate solubilization rate (Table 2.4).

2.3.4 MICRONUTRIENTS

Soil health is totally dependent upon the physical, chemical, and biological properties of the soil, and also a good indicator of the health of the soil. Due to the continuous degradation of the soil quality, it becomes unfertile. Soil organic matter helps to sustain the soil quality. A close link is generally found between ecosystem metabolism and terrestrially derived nutrients in temperate ecosystems.

For good yield, there is a need for at least 16 macro and micronutrients in the soil. Knowing the effect on the yield of tree plants is due to the lack of nutrients in the soil. Nutrients are is one of the most important factors affecting the development of plants. Microbial bacteria in the soil are economically and economically beneficial in addition to increasing the optimum level of plant and increasing yield (Richardson et al. 2009).

It indicates relations between the soil's inner and outside components for the viable food production system. The effective soil microbes play an important role, they are responsible for several biological transformations and the various pool of the carbon and macro, micronutrients, which aid the successive establishment of soil-plant-microbe interaction. Biological activity in soil concentrated on the top soil as organic residues convert biomass into carbon dioxide and water (Singh et al. 2016) These efficient microbes are further associated with the transformation and degradation of waste materials and synthetic organic compounds (Sindhu et al. 2016). The roles of these efficient rhizospheric microbes possess…. The soil organic matter (SOM)

TABLE 2.4

Microbes Involved in the Phosphorous Cycle

Gordonia sp. fungus	Hoberg et al. 2005
Pseudomonas (fluorescens) bacteria	
Chlamydomonas reinhardtii (algae)	Wykoff et al. 1999
Pseudomonas sp.	Jorquera et al. 2008
Enterobacter sp.	
Pantoea sp.	
Bacillus megaterium	
Bacillus circulans	
Bacillus subtilis	
Bacillus polymyxa	
Bacillus sircalmous	
Pseudomonas striata	
Sacchromonospora viridis	
Actinomodura citrea	
Micromonospora echinospora	
Saccharopolyspora hirsute	
Streptoverticillium album	
Streptomyces albus	
Streptomyces cyaneus	
Thermonospora mesophila	
Gigaspora, Scutellospora	
Entrophospora,Pacispora	
Pseudomonas cepacia	

helps in retaining the soil health as well as its quality, inactive toxic compounds, suppress pathogens, and protect environmental sustainability. In order to achieve this, it is necessary to have interaction among the soil's internal and external components for better food production. The potential soil microorganisms play vital roles, as they are responsible for various biological transformations and different pools of carbons (C) and macro and micronutrients, which facilitate the subsequent establishment of soil plant–microbe interaction. But due to the alarmingly growing worldwide population, this biogeochemical cycling of the nutrients has become incapable of fixing a good amount of nutrients into the soil to produce a sufficient yield.

In this way, the organic residues are converted into biomass to CO_2, H_2O, and the nutrients. These efficient microbes are further associated with the transformation and degradation of waste materials and synthetic organic compounds (Sindhu et al. 2016).

2.4   SOIL QUALITY MONITORING IN RAEBARELI

Raebareli district situated on the Indo-Gangetic plain contains fertile alluvial soil which is good for crop production. The land of Raebareli district is gently flat and undulating and shows characteristics of six physiographic systems:

1. Ganga khadars
2. Ganga recent alluviams
3. Ganga flats
4. Sai uplands
5. Sai low lands
6. Sai flats

The average rainfall in the district is 932 mm with a maximum and minimum temperature of 44.2°C and 2.3°C. Loamy sandy, sandy loam, clay loam, and silt are found in the district. The district is irregular in shape but fairly compact. It is a part of the Lucknow division and lies between latitude 25° 49' north and 26° 36' north and longitude 80° 41' east and 81° 50' east. In the north the district is bounded by the tehsil Mohanlalganj of the district Lucknow and the tehsil Haidargarh of the district Barabanki, while on the east it is bounded by the tehsil of Salon district Amethi, and in the south is the Kunda tehsil of the district Pratapgarh. On the west lies the Bighapur tehsil of district Unnao. According to the census of 2011, the geographical area of the district is 3286 km².

2.4.1 LAND USE AND LAND COVER IN THE STUDY AREA

Knowledge of land use and land cover is important for many planning and management activities concerned with the surface of the earth. The term "land cover" relates to the type of feature present on the surface of the earth, wheat fields, roads, rail tracks, lakes, etc. The term "land use" relates to the human activity or economic function associated with a specific piece of land. As an example, a tract of land on the fringe of an urban area may be used for family housing. Land use is the most common cause of the loss or degradation of natural habitat due to multiple interacting with the socio-economy caused by the loss of biota, and the loss of natural ecosystem and habitat (Singh and Pandey 2019). The study area consists of different land use/land cover classes. Due to continuous growth in the human population, degradation of land increase, and the extinction of various flora and fauna, the land needs to be protected through a policy for land use and management. The land change is promoted by the disordered and uncontrolled settlement. Excessive agriculture practices, the construction of private infrastructure, and the excavation of soil resources affect land. The impacts of land use range widely from disturbance of territory, isolation of habitat, and reduction of forest cover. Agricultural land, wasteland, water, and urban or built-up land (Singh and Pandey 2019) have been studied, showing that land use and land cover have undergone change; such transformations are caused by the population, technology, and the agriculture sector of Raebareli district. It is essential that the existing policy focuses on the protection of agriculture and promotes the use of sustainable agriculture. In the Raebareli district, the majority of the land is used for agriculture and a very small urban area exists. According to the environmental statistics of India 2019, the total area of Uttar Pradesh is 2,40,92800 ha, where there is degradation of land 337524 ha, which indicates that about 1.41% of land degradation has occurred and the salinity of land in 2003–2005 is 636202 ha. From a total built up area of 1141.41 km², the agricultural area was about 191084.27 km² and forest area 12554.58 km² (Table 2.5).

TABLE 2.5
Different Land Use and Land Pattern Cover in Uttar Pradesh

S. No.	Land Type	Area (hectare or square kilometer)
1.	Total Area of Uttar Pradesh	24092800 ha
2.	Degradation of Land	337524 ha
3.	Total Agriculture Area	191084.27 km^2
4.	Total Forest Area	12554.58 km^2
5.	Wet Lands / Water Bodies	11173.45 km^2

In the Raebareli district, agricultural land constitutes the largest area. Soil is a basic source of plant nutrient, as well as for humans, and play an important role in balancing the environment through the biogeochemical cycle. Due to the increasing population, it is required to increase the food crop, as well as infrastructure builds by the human which increase the soil degradation (Figures 2.5 and 2.6).

According to the NRSC (National Remote Sensing Centre) 2011–2012 report on land use and land cover information, the total geographical area of Raebareli district is 4609.0 km^2. Here, the majority of the area is agricultural crop land which is 3069.16 km^2 of the total land (66.59%), and then agricultural plantation land is 382.84 km^2 (8.31%). The built-up urban land and rural land are 2.69% and 3.04%, water bodies contribute 115.07 km^2 (0.11%), then salt-affected and barren uncultivable land contributes 350.52 km^2 (2.79%) to the Raebareli geographic area. Degradation of soil affects the crop production and also results in the loss of macro and micronutrients. Physical attributes of the soil can be estimated by the soil bulk density, soil porosity, and water holding capacity. All the afore-mentioned properties are improved, resulting in soil sustainability for plant growth and function of the biological activity. Hoyt et al. (1986) observed that the transformation of micronutrients from non-available to

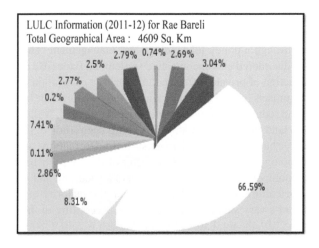

FIGURE 2.5   Different land use and land pattern cover of Raebareli district in Uttar Pradesh.
Source: https://bhuvan.nrsc.gov.in.

LULC Class		Area (Sq.Km)	LULC Class		Area (Sq.Km)
	Builtup,Urban	124.12		Builtup,Rural	140.29
	Agriculture,Crop land	3069.16		Agriculture,Plantation	382.84
	Agriculture,Fallow	131.99		Forest,Deciduous	4.89
	Barren/unculturable/ Wastelands, Salt Affected land	341.46		Barren/unculturable/ Wastelands, Gullied/Ravinous Land	9.06
	Barren/unculturable/ Wastelands, Scrub land	127.5		Wetlands/Water Bodies, Inland Wetland	115.07
	Wetlands/Water Bodies, River/Stream/canals	128.74		Wetlands/Water Bodies, Reservoir/Lakes/Ponds	33.88
Total					4609.00

FIGURE 2.6 Different land use types and degraded land covers of Raebareli district in Uttar Pradesh. Source: https://bhuvan.nrsc.gov.in.

available is influenced by the soil pH. Reeves (1997) documented that a continuous cropping system causes soil degradation, which reduces the soil organic carbon so tillage practice is the best method to conserve the soil organic carbon and maintain the soil health. Soil consists of solid, liquid, and gaseous phases which are more complex than air and water. Earlier studies (Wei et al. 2014; Guo and Gifford 2002; Don et al. 2011) show that natural systems lose carbon when converted to agriculture, with the exception of the conversion of forest into grazing land, where some studies indicated carbon gain (Guo and Gifford 2002) and some noted carbon loss (Don et al. 2011). The conversion of a natural system in the agricultural field adversely affects soil carbon, organic matter, water, and other factors. So the protection of the natural ecosystem is important for the soil carbon, organic matter, water, and biota (Don et al. 2011). The decrease of soil organic carbon due to the land use affects the temperate regions, which is followed by the tropical and then by the boreal regions. Wei et al. (2014) and Chimdi et al. (2012) demonstrated that change in land use patterns influence soil quality. There is a positive relationship between soil pH and exchangeable Mg_2^+ and Ca_2^+ ions. Land use changes also caused a decline in cation exchange capacity, present base saturation, exchangeable bases, and increased bulk density, which affect the soil and interrupt the root growth. Further detailed investigations are required to find out the difference between different land use types on soil moisture content, pH, cation exchange capacity (CEC), organic carbon, total nitrogen (TN), and available phosphorus. Analysis of soil pH also shows a significant positive relationship between soil pH and extractable soil Fe_3^+ and Mn_2^+ ions among the land uses. It is observed that the biological parameters of soil such as soil organic content, microbial biomass, and soil nitrogen status play an important role in soil health and soil function, which are also related to the water-holding capacity and soil degradation. The soil organic carbon may enhance the availability of zinc while an increase in soil $CaCO_3$ may decrease the availability of soil zinc. Purswani (2018) reported that in Gandhinagar, the land pattern contains most of the fertile soil found in the vegetation class, while urban class shows the least amount of organic carbon. Maikhuri and Rao (2012) studied soil health; soil quality can be defined as the ability of soil to act as the biological system within the land use boundaries. Nutrient imbalance affects the soil health and soil quality. Soil functions so that it may sustain the productivity of soil as well as maintain the surrounding environment. The application of manure with soil improves the nutrients as well as organic matter and lower bulk density which increase the soil quality. (www.fao.org) According to the FAO, soil fertility is defined as the ability of soil to supply essential plant nutrient and soil water in the adequate amount and proportion for the plant growth and reproduction in the absence of toxic substance which may inhibit the plant growth.

2.5 MICROBES AND ITS ROLE IN SOIL QUALITY MANAGEMENT

All the soil microbes play an important role in soil fertility and the production of crops. Microbes manage the soil air, water, and the soil nutrients. Soil microbes play a major role in the ecosystem, since the microbes manage all nutrient cycles of the Earth.

2.5.1 MICROBES IN DIFFERENT LAND USES

Human activity like the construction of buildings and other development projects strongly affect the soil ability in which microbes play a key role. Microorganisms can help create a maximum biomass and play a significant role in the soil ecosystem function of the soil formation process. Microbial activity was lowest in the arable field and high in grassland in top soil. It means the availability of organic matter is higher in the grass, but land-specific bacterial activity is higher in the arable field than forest and grassland. Plowing accelerates the rate of decomposition. Land use affects the microbial community in the rooted zone in the upper soil. Maximum biomass and the activity are found below the 30 cm depth. This means the C horizon should not be neglected when we are studying the nutrient cycling and carbon sequestration (Van leeuwen et al. 2017). Land use affects the microbial community; it may also affect soil structure and texture, ultimately being harmful for the fertile soil.

2.5.2 MICROBES AND SOIL QUALITY

Soil microbes, bacteria, archaea, and fungi play diverse and often critical roles in these ecosystem services. All major elemental cycles like carbon, nitrogen, and phosphorous soil microbes contribute a vast metabolic diversity of microbes. These elemental cycles affect the soil ecosystem as well as the soil properties which improve the soil quality. The status of healthy and good quality soil can be indicated by the microbial biomass of carbon (MCB). Soil microorganisms are collective agents of transformation in soil organic matter and nutrients, and serve as the sink of soil organic matter to the atmospheric CO_2 (Lal et al. 1997) that increase soil carbon content, which in turn indicate a higher microbial mass in the soil (Sparling et al. 2003).

2.5.3 MICROBES IN NUTRIENT RECYCLING

A variety of soil factors are known to increase nutrient availability and plant productivity. In soils, microbes play a pivotal role in cycling nutrients essential for life (Table 2.6).

TABLE 2.6

Role of Soil Microbes and Environmental Factors in Nutrient Availability and Plant Productivity

S. No.	Nutrient/Structure	Bacteria	Fungi
1.	Nitrogen	Diazotrophs fix N2 as ammonia through their metabolic process, bacteria from family Rhizobiaceace living in the soil infect plant root from nodules and fix N in structure through a complex enzyme system.	Fungi do not fix N but provide growth-limiting nutrients to bacteria for N fixation.
2.	Phosphorous	P solubilization or availability is enhanced by P mineralization as well as production of siderophores and organic acids in the soil.	Increase P bioavailability through mineralization in soil mycelial transport, P solubilization by siderophores, N assimilation, and CO_2 release.
3.	Potassium	Potassium bacteria release various processes such as acidolysis, chelation and complexolysis, and exchange reaction.	Influence the K mobilization through mycelial transport as well as by the K solubilization process that involves the release of H^+, CO_2 organic acid such as citrate, malate, and oxalate.
4.	Iron	Produces sidephores which have affinity to chelate and solubilization iron form minerals or organic compound.	Transport of Fe from minerals to organic soil horizon for decomposition and mineralization, and release chelator for Fe translocation in soil.
5.	Aggregate formation	Produced peripheral slime polymer and decompose organic materials to from organo-materials' product that are associated with soil to form aggregates.	Hyphal network entrap soil particle and forces them together.

These microbes increase organic matter that boosts the availability of N, P, K, and Fe in soil (Stevens et al. 2014). Soil microbes play a major role as saprotrophs in completing the carbon cycle, and converting organic material formed by primary producers back to carbon dioxide during respiration. They are sometimes aided in this process by higher animals (herbivores and carnivores) that digest particulate organic material with the help of microbes residing in their intestinal tracts. This process is known as decomposition and involves the degradation of non-living organic material to obtain energy for growth. Mineralization of the organic compound occurs when it is degraded completely into inorganic products such as carbon dioxide, ammonia, and water.

2.6 CONCLUSION

Soil microbe is of significance in soil fertility in different land uses and different biogeochemical cycles in the environment. Microbes play a major role in the survival of both plants and animals with the provision of nutrients. Decomposition is a process in which microbes convert dead matter (fine roots, litter, etc.) into valuable organic matter; a different type of microbe species has a different type of characteristic in the soil. Microbes of the soil also play an important role in the returning of nutrients to their minerals form, so plants take nutrients again from soil. This process is known as mineralization. Microbes break the complex compounds of organic matter and convert them into a simple organic matter, a jelly-like substance called "humus". Soil microorganisms maintain the inputs of organic matter which is an important source of carbon, energy, and nutrients for soil. Anthropogenic activity like erosion, chemical pollution, land use, etc. accelerates degradation of the soil microbes. Climate change, temperature, and precipitation also influence the soil microbes. Thus for the conservation of soil microbes and its diversity for the environment, balance is required in organic farming, as well as the use of organic fertilizer which enhances the soil attributes and quality of soil. This means the soil can be sustained for more prolonged durations in the future.

REFERENCES

Akiyama, H., Yan, X., Yagi, K. 2006. Estimations of emission factors for fertilizer-induced direct N_2O emissions from agricultural soils in Japan: Summary of available data. *Soil Science and Plant Nutrition* 52:774–787.

Bonkowski, M. 2004. Protozoa and plant growth: The microbial loop in soil revisited. *New Phytologist* 162:617–631.

Borron, S. 2006. *Building Resilience for an Unpredictable Future: How Organic Agriculture Can Help Farmers Adapt to Climate Change.* Food and Agriculture Organization of the United Nations, Rome.

Brussaard, L. 1994. Interrelationships between biological activities, soil properties and soil management. In: *Soil Resilience and Sustainable Land Use.* CAB International, Wallingford, Oxfordshire, pp. 309–329.

Childers, D.L., Corman, J., Edwards, M., Elser, J.J. 2011. Sustainability challenges of phosphorus and food: Solutions from closing the human phosphorus cycle. *Bioscience* 612:117–124.

Chimdi, A., Gebrekidan, H., Kibret, K., Tadesse, A. 2012. Status of selected physicochemical properties of soils under different land use systems of Western Oromia, Ethiopia. *Journal of Biodiversity and Environmental Sciences* 23:57–71.

Dioxide, A.C. 1985. *The Global Carbon Cycle.* Department of Energy ER-0239, Washington, DC.

Don, A., Schumacher, J., Freibauer, A. 2011. Impact of tropical land use change on soil organic carbon stocks – A meta-analysis. *Global Change Biology* 174:1658–1670.

Doran, J.W., Zeiss, M.R. 2000. Soil health and sustainability: Managing the biotic component of soil quality. *Applied Soil Ecology* 151:3–11.

Egamberdieva, D., Kucharova, Z. 2008. Cropping effects on microbial population and nitrogenase activity in saline arid soil. *Turkish Journal of Biology* 322:85–90.

FAO Inter-Departmental. *Working Group on Biological Diversity.* FAO, Rome, Italy, 312 pp.

Guo, L.B., Gifford, R.M. 2002. Soil carbon stocks and land use change: A metaanalysis. *Global Change Biology* 84:345–360.

Gyaneshwar, P., Naresh, K.G., Parekh, L.J., Poole, P.S. 2002. Role of soil microorganisms in improving P nutrition of plants. *Plant Soil* 245:83–93.

Hoberg, E., Marschner, P., Lieberei, R. 2005. Organic acid exudation and pH changes by *Gordonia* sp. and *Pseudomonas fluorescens* grown with P adsorbed to goethite. *Microbiological Research* 160:177–187.

Hoyt, P., Mackenzie, A.F., Neilsen, D. 1986. Distribution of soil Zn fractions in British Columbia interior orchard soils. *Canadian Journal of Soil Science* 66:445–454.

Jacobson, M., Charlson, R.J., Rodhe, H., Orians, G.H. 2000. *Earth System Science: From Biogeochemical Cycles to Global Changes*, Vol. 72. Academic Press, Cambridge, MA.

Jones, D.L., Darrah, P.R. 1994. Role of root derived organic acids in the mobilization of nutrients from the rhizosphere. *Plant and Soil* 166:247–257.

Jorquera, M.A., Hernández, M.T., Rengel, Z., Marschner, P., de la Luz Mora, M. 2008. Isolation of culturable phosphobacteria with both phytate-mineralization and phosphate-solubilization activity from the rhizosphere of plants grown in a volcanic soil. *Biology and Fertility of Soils* 44:1025.

Joseph, G. *Program Manager* 301-903-1239. joseph.graber@science.doe.gov.

Juma, N.G. 1999. *The Pedosphere and Its Dynamics. A Systems Approach to Soil Science. Volume 1: Introduction to Soil Science and Soil Resources*. Salman Productions, University of Alberta, Edmonton.

Kandeler, E., Stemmer, M., Gerzahek, H. 2001. Organic matter and soil microorganisms – Investigations from the micro- to the macro-scale. *Die Bodenkultur* 117:2.

Kelly, J.J., Tate, R.L. 1998. Effects of heavy metal contamination and remediation on soil microbial communities in the vicinity of a zinc smelter. *Journal of Environmental Quality* 27:609–617.

Khan, M.S., Zaidi, A., Ahemad, M., Oves, M., Wani, P.A. 2010. Plant growth promotion by phosphate solubilizing fungi – Current perspective. *Archives of Agronomy and Soil Science* 56:73–98.

Kour, D., Rana, K.L., Yadav, N., Yadav, A.N., Rastegari, A.A., Singh, C., Negi, P., Singh, K., Saxena, A.K. 2019. Technologies for biofuel production: Current development, challenges, and future prospects. In: Rastegari, A.A. et al. (Eds.), *Prospects of Renewable Bioprocessing in Future Energy Systems, Biofuel and Biorefinery Technologies*, Vol. 10. Springer, Berlin, pp. 1–50.

Kumar, G., Singh, G., Tripathi, B.N., Kumar, R., Tiwari, U.S., Kumar, A. 2015. Status of available micronutrients in soils of Rae Bareli, Uttar Pradesh. *The Ecoscan* 4:289–297.

Lal, R., Kimble, J.M., Follett, R.F., Stewart, B.A. (Eds.) 1997. *Soil Processes and the Carbon Cycle*, Vol. 11. CRC Press, Boca Raton, FL.

Le Quéré, C., Raupach, M.R., Canadell, J.G., Marland, G. et al. 2009. Trends in the sources and sinks of carbon dioxide. *Nature Geoscience* 2:831–836.

Lombard, N., Prestat, E., van Elsas, J.D., Simonet, P. 2011. Soil-specific limitations for access and analysis of soil microbial communities by metagenomics. *FEMS Microbiology Ecology* 78:31–49.

Maikhuri, R.K., Rao, K.S. 2012. Soil quality and soil health: A review. *International Journal of Ecology and Environmental Sciences* 381:19–37.

Nath, T.N. 2014. Soil texture and total organic matter content and its influences on soil water holding capacity of some selected tea growing soils in Sivasagar district of Assam, India. *International Journal of Chemical Sciences* 12:1419–1429.

Oberson, A. 2005. Microbial turnover of phosphorus in soil. In: *Organic Phosphorus in the Environment*. CAB International, Wallingford, pp. 133–164.

PCGCC Pew Center on Global Climate Change. 2004. *Coping with Global Climate Change: The Role of Adaptation in the United States*. PCGCC, Arlington, TX.

Purswani, E. 2018. Assessment of soil characteristics in different land-use systems in Gandhinagar, Gujarat. *Proceedings of the International Academy of Ecology and Environmental Sciences* 83:162.

Redfield, A.C. 1958. The biological control of chemical factors in the environment. *American Scientist* 46:230A–221A.

Reeves, D.W. 1997. The role of soil organic matter in maintaining soil quality in continuous cropping systems. *Soil Tillage Research* 43:131–167.

Rengel, Z., Marschner, P. 2005. Nutrient availability and management in the rhizosphere: Exploiting genotypic differences. *New Phytologist* 168:305–312.

Richardson, A.E. 2001. Prospects for using soil microorganisms to improve the acquisition of phosphorus by plants. *Functional Plant Biology* 28:897–906.

Richardson, A.E., Barea, J.M., McNeill, A.M., Prigent-Combaret, C. 2009. Acquisition of phosphorus and nitrogen in the rhizosphere and plant growth promotion by microorganisms. *Plant and Soil* 321:305–339.

Schimel, J.P., Gulledge, J.M., Clein-Curley, J.S., Lindstrom, J.E., Braddock, J.F. 1999. Moisture effects on microbial activity and community structure in decomposing birch litter in the Alaskan taiga. *Soil Biology and Biochemistry* 31:831–838.

Shridhar, B.S. 2012. Review: Nitrogen fixing microorganisms. *International Journal of Microbiological Research* 3:46–52.

Sindhu, S.S., Parmar, P., Phour, M., Sehrawat, A. 2016. Potassium-solubilizing microorganisms KSMs and its effect on plant growth improvement. In: *Potassium Solubilizing Microorganisms for Sustainable Agriculture*. Springer, New Delhi, pp. 171–185.

Singh, C., Chowdhary, P., Singh, J.S., Chandra, R. 2016. Pulp and paper mill wastewater and coliform as health hazards: A review. *Microbiology Research International* 4:28–39.

Singh, C., Tiwari, S., Boudh, S., Singh, J.S. 2017a. Biochar application in management of paddy crop production and methane mitigation. In: Singh, J.S., Seneviratne, G. (Eds.), *Agro-Environmental Sustainability: Managing Environmental Pollution*, 2nd Edition. Springer, Switzerland, pp. 123–146.

Singh, C., Tiwari, S., Gupta, V.K., Singh, J.S. 2018. The effect of rice husk biochar on soil nutrient status, microbial biomass and paddy productivity of nutrient poor agriculture soils. *Catena* 171:485–493.

Singh, C., Tiwari, S., Singh, J.S. 2017b. Impact of rice husk biochar on nitrogen mineralization and methanotrophs community dynamics in paddy soil. *International Journal of Pure and Applied Bioscience* 5:428–435.

Singh, C., Tiwari, S., Singh, J.S. 2017c. Application of biochar in soil fertility and environmental management: A review. *Bulletin of Environment, Pharmacology and Life Sciences* 6:07–14.

Singh, C., Tiwari, S., Singh, J.S. 2019. Biochar: A sustainable tool in soil 2 pollutant bioremediation. In: Bharagava, R.N., Saxena, G. (Eds.), *Bioremediation of Industrial Waste for Environmental Safety*. Springer, Berlin, pp. 475–494.

Singh, S., Pandey, B.V. 2019. Agricultural land-use pattern in Raebareli District, Uttar Pradesh: A geographical appraisal. *International Journal of Research and Analytical Reviews* 61: 1269–2348.

Sparling, G., Parfitt, R.L., Hewitt, A.E., Schipper, L.A. 2003. Three approaches to define desired soil organic matter contents. *Journal of Environmental Quality* 32:760–766.

Stevens, W.B., Sainju, U.M., Caesar, A.J., West, M., Gaskin, J.F. 2014. Soil-aggregating bacterial community as affected by irrigation, tillage, and cropping system in the northern great plains. *Soil Science* 179:11–20.

Tilman, D., Fargione, J., Wolff, B., D'antonio, C., Dobson, A., Howarth, R., Swackhamer, D. 2001. Forecasting agriculturally driven global environmental change. *Science* 292:281–284.

Tiwari, S., Singh, C., Boudh, S., Rai, P.K., Gupta, V.K., Singh, J.S. 2019a. Land use change: A key ecological disturbance declines soil microbial biomass in dry tropical uplands. *Journal of Environmental Management* 242:1–10.

Tiwari, S., Singh, C., Singh, J.S. 2018. Land use changes: A key ecological driver regulating methanotrophs abundance in upland soils. *Energy, Ecology, and the Environment* 3:355–371.

Tiwari, S., Singh, C., Singh, J.S. 2019b. Wetlands: A major natural source responsible for methane emission. In: Upadhyay, A.K. et al. (Eds.), *Restoration of Wetland Ecosystem: A Trajectory Towards a Sustainable Environment.* Springer, Berlin, pp. 59–74.

Van Leeuwen, J.P., Djukic, I., Bloem, J., Lehtinen, T., Hemerik, L., de Ruiter, P.C., Lair, G.J. 2017. Effects of land use on soil microbial biomass, activity and community structure at different soil depths in the Danube floodplain. *European Journal of Soil Biology* 79:14–20.

Wei, X., Shao, M., Gale, W., Li, L. 2014. Global pattern of soil carbon losses due to the conversion of forests to agricultural land. *Scientific Reports* 4:4062.

Wykoff, D.D., Grossman, A.R., Weeks, D.P., Usuda, H., Shimogawara, K. 1999. Psr1, a nuclear localized protein that regulates phosphorus metabolism in Chlamydomonas. *Proceedings of the National Academy of Sciences* 96:15336–15341.

Zou, X., Binkley, D., Doxtader, K.G. 1992. A new method for estimating gross phosphorus mineralization and immobilization rates in soils. *Plant and Soil* 147:243–250.

3 Biofertilizers in Boosting Agricultural Production

Subhrajyoti Mishra, K. M. Karetha,
Rajeswari Das, and D. K. Dash

CONTENTS

3.1   INTRODUCTION

Agriculture is a branch of science, the art and practice of dealing with the cultivation of soil, growing of crops and rearing of farm animals. It is the culture of soil which sustains our beloved culture and civilization. The term "agriculture" comes from the Latin word "agricultura" where "ager" means "a field" and "cultura" means "cultivation", as in the strict sense whereby we can say it is the "tillage of the soil". Thus, a literal meaning of this word yields "tillage of a field or fields". Agriculture is a very primitive operation and started in an organized way in around 8000 B.C., contributing food from nearly 11% (1.5 billion ha) of the globe's land surface (13.4 billion ha) and around 2.7 billion ha with crop production potential. According to data presented by the World Bank in 2016, around 37.431% of the world's total area is under agriculture, which provides more than 50% of employment. The history of agriculture dates back 7,000 to 10,000 years to the Neolithic era or the New Stone Age where there existed seven major Neolithic crops *viz.*, emmer wheat, einkorn wheat, peas, lentils, bitter vetch, hulled barley, chickpeas and flax. As the years passed, various developments in agriculture proceeded, and achieved a remarkable momentum, culminating in the Green Revolution era, which occurred around the 1940s in the world, and in the 1960s in India. Though we have witnessed a drastic enhancement in food grain production concurrent with the increased population, the underlying problems are many. Reduction in soil fertility, ground water pollution, destruction of soil flora and fauna, land degradation, environmental pollution, heavy metal toxicity, etc. have a serious impact on conventional high input agriculture systems. Improper use of chemicals and pesticide, with the frequent removal of ground water, poses a serious threat to the agriculture, which leads to the search for alternative ways in reducing the hazardous effect of those intensive agricultural inputs. Various new methods and improved technologies helped growers to achieve better production in the past, yet we are unable to achieve the optimum. Organic farming, sustainable agriculture, zero budget farming, no tillage operation, biodynamic farming and permaculture were developed owing to health consciousness among consumers. Nevertheless, each of these practices needs standardization. Advanced methods, production tools and technology are still developing to enhance agricultural production, among which the use of biofertilizers in invigorating the crop production is one.

If we take the example of the preparation of curd, we may notice that we need to add a small amount of previously made curd (*Juana* in Odia; *Melvana* in Gujarati; Jaman in Hindi; and *Starter* in English) to the bowl of milk for rapid fermentation to curd. In this process, through the addition of a little curd, we add the useful *Lactobacillus* bacteria to grow and multiply in the milk, thus producing curd sooner. Even in some cases, we may add lime juice in order to create a favorable condition for the growth of the *Lactobacillus* bacteria, which needs more time for curd setting as compared to the previous sort. Similarly, soil is the storehouse of all the essential and beneficial nutrients necessary for the plant growth but they are not instantly available for the root to uptake, due to overuse of chemical fertilizers and pesticides. In the past, farmers used much bulky organic matter like decomposed cow dung and farm residue, which are the storehouse of numerous beneficial microorganisms. In recent times, in order to achieve greater production, the use of the chemicals *viz.*, inorganic fertilizers, pesticides, growth retardants in the field, kills the soil-hibernating organisms who helped to mobilize the fixed nutrient to the plant-available form. Thus, the nutrient cycle gets hampered. Biofertilizers are, in this way, microbial inoculants or strains which are added to soil, seed or root to create a favorable condition for the absorption and growth of the plant root in the rhizosphere. They are not nutrient sources like chemical fertilizers but instead they convert the unavailable form of nutrient to an available form. The renowned natural farming practitioner and receiver of the fourth highest civilian award of India, "The Padma Shri" during the 2016 Subhash Palekar of Maharashtra, correctly mentioned that our soils are deficient in the active microbial population, due to the improper use of agrochemicals, thereby reducing productivity. To sustain the agroecosystem, we should return to zero budget farming or natural farming.

Biofertilizer is a self-explanatory term. Its literal meaning is the use of living microorganisms as fertilizer. It is the use of single or multiple types of microorganism, along with a suitable carrier to enhance the availability of a specific nutrient in the rhizosphere when added, thus improving the plant growth, yield, produce quality and stress resistance. According to Vessey (2003), biofertilizers are commonly referred to as the fertilizer that contains living soil microorganisms to enhance the availability and uptake of mineral nutrients for plants. Mazid et al. (2011) described them as the substance which contains living microorganisms which when applied to seed, plant surfaces or soil colonizes the rhizosphere or the interior of the plant and promotes growth by increasing the availability of primary nutrients to the host plant. These are the artificially multiplied microbial inoculants of specific strains of either bacteria or fungus to improve plant growth and productivity (Mazid and Khan 2014). With an increase in awareness among consumers regarding health, physiological and mental wellbeing, the demand for organic produce is becoming a huge agricultural issue.

Fertilizers are the sources of nutrients for plants, which may be of organic fertilizers, including FYM, compost, vermicompost, green manure, etc.; and inorganic fertilizers, including urea, MOP, SSP and many more. Biofertilizers are not direct sources of nutrients, rather they help the plant root to avail of more nutrients simultaneously by increasing the nutrient availability in the rhizosphere. They colonize in the soil or plant or both ecosystems, thus helping the plant throughout the life cycle, whilst also counteracting the negative impact of agrochemicals. Chemical fertilizers give more yields in less time, whereas biofertilizers need more time to colonize and act, but they are persistent in soil for longer periods.

3.2   USE OF MICROORGANISMS AS BIOFERTILIZERS

Microorganisms are the primary sources of biofertilizers. The era of biofertilizers started around 1891, when laboratory culture of *Rhizobia* was developed under the name "Nitragi" by Nobbe and Hiltner for the first time (Nobbe et al. 1891; Nobbe and Hiltner 1893). They also explained that plants attract a suitable and specific microbe based on its root exudates. Starting with *Rhizobia*, a vigorous and expanded research for other N-fixing microorganisms began and soon it was found that there were other non-symbiotic bacteria which could fix atmospheric nitrogen. In India, systematic research on biofertilizers started with the first study of N. V. Joshi (1920) at Pusa, Bihar. He attempted to discover the relation between microbes and the organic matter decomposition in the soil, nitrification, nitrogen fixation and losses from the soil. Subsequently, extensive research was completed by other pioneer scientists like N. R. Dhar and co-workers at Allahabad (1932–36), P. E. Lander in Punjab (1925–27) and B. Viswanath in Coimbatore (1925).

Albert Howard at the Imperial Agricultural Research Institute, Pusa was the first to realize the importance of humus and mycorrhizae in crop plants and perennial trees, helping their growth by the efficient removal of nutrients by the root from the soil. Subsequently, the following were all examined by various eminent scientists: various methods of composting; the recommendation of nitrogen-fixing bacteria in cereal crops to increase productivity; the mechanism of microbial activities; the decomposition of cellulosic materials; the relation of soil fertility and microbes; studies on actinomycetes; the presence and use of cyanobacteria in rice ecosystem; the isolation, characterization and microbial multiplication under laboratory conditions. Hence, this has provided additional literature to major fields of research on the subject of biofertilizers in the agroecosystem. Legume (berseem) and cereal (wheat) crop rotation enrich the soil with nitrogen when fed with heavy phosphorous, due to enhanced activity like nitrogen fixation, ammonification and nitrification (Acharya et al. 1953). Early developments were seen with phosphorous-solubilizing bacteria i.e. *Bacillus* strain were procured from Russia and in the latter half of the 20th century, the vast majority of biofertilizer strains were prepared in India. In the early part of the 21st century, the molecular characterization of microorganisms, their use in stress management and the dissemination of microbes and related technology is of much concern (Rao and Patra 2009). The relevant institutions, researchers and extension agents are engaged in soon achieving beneficial results in a wide range of crop plants.

3.3   PRODUCTION AND DEMAND SCENARIO OF BIOFERTILIZERS

Since time immemorial, production and demand have been very closely connected. With the increasing demand, production increases likewise, but the opposite may or may not happen in reality. As biofertilizers are living organisms, any time passing following their preparation inevitably reduces their viability, thus the optimum benefits cannot be obtained. Therefore, biofertilizers are prepared based on demand and hence supplied to the grower after packaging. In recent years, health consciousness among consumers has created a favorable atmosphere in which to proceed towards an organic or low chemical input farming system. Though there has been a huge demand for more than two decades, still both the production and marketing sectors of biofertilizers are fragmented and locality-specific. This accounts for the small percentage overall of chemical fertilizer production in the market. A number of chemical fertilizer plants also produce biofertilizers, albeit in small quantities. At a national (Indian) level, apart from the fertilizer plants, M/s. Agro Evo Limited, a joint venture of Hoechst and Schering, is one of the major units. In terms of product, *Rhizobium* accounts for the bulk of biofertilizer production. In Gujarat, fertilizer companies like GSFC, KRIBCO and GUJCOMASOL produce biofertilizers in small quantities. Some of the small-scale units in production are as follows:

- M/s. Agriland Biotech
- M/s. Ocean Agro Industries
- M/s. Gujarat Life Science Pvt. Ltd.

All of these are located in Baroda, Gujarat, India. The demand estimation for biofertilizer is quite difficult to assess due to unorganized data, as well as the underdevelopment of marketing channels. It is estimated that there is a lot of hidden potential, provided systematic production, research and marketing are carried out.

In a recent study (the Global Industry Report 2012–2022), it can be observed that the global biofertilizer market size was estimated at USD 787.8 million in 2016 and it will exponentially increase thereafter till 2022 (Market Research Report 2018). The major driving force in this regard is awareness among consumers as to the residual toxicity of agrochemicals. Among different biofertilizers, it has been observed that demand was higher for nitrogen-fixing biofertilizers, followed by the phosphorous solubilizing-biofertilizers (Figure 3.1). Weaker demand was seen in the case of the K, Zn and other micronutrient-and silicate-solubilizing biofertilizers. The compound annual growth rate (CAGR) of phosphate solubilizers was 13.9% from 2015 to 2022 accounting for the biofertilizer with the highest demand.

Among the different methods of application, seed treatment (more than 60%) is mostly adopted due to easiness and rapidity of the method, followed by soil application.

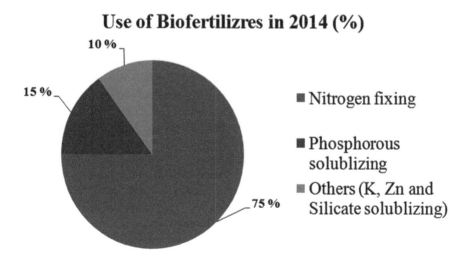

FIGURE 3.1   Global use of different biofertilizers in 2014 (modified from Market Research Report 2018).

3.4   SOIL MICROBIAL LOAD AND THEIR
FUNCTIONS IN AGRICULTURAL FIELDS

Soil is the storehouse of various microbes which may be harmful or beneficial to the agricultural ecosystem. The parent rock of a soil determines the nutrient content of the soil. The microbes help to mobilize the nutrients. They also interact with the phyllosphere, in a variety of ways from decomposition of the organic matter or during the biogeochemical cycle in order to activate disease in plants. As we descend in the soil, so the microbial population decreases. The upper 10 cm of soil constitutes the maximum microbial population. The range of bacterial, fungal and actinomycetes load in the upper 10 cm of soil ranged from 141×10^5 to 271×10^5 CFU/g, 124×10^3 to 27×10^4 CFU/g and 129×10^3 to 60×10^4 CFU/g respectively [CFU=Colony Forming Unit] (Krishna et al. 2012). That suggests the major microbes in the soil are bacteria followed by actinomycetes and fungus. In most cases, the microbial number was highest where organics and nutrients were available in abundance. The presence of the maximum number or amount of plant roots in the top 10–15 cm of soil has a strong influence on increasing the microbial load. Diverse plant species are known to make and discharge dissimilar types of organic carbon compounds (Lynch and Whipps 1990), and those root exudates are the main carbon and energy source. They are also a major factor in determining the composition of soil microbial communities in the neighborhood of plant roots (Bertin et al. 2003). Most microorganisms are heterotrophs; they need a specific host for their food and at the proximal of zone soil (10–15 cm) have a greater carbon source (Trumbore 2000), favoring occurrence of their maximum numbers.

3.5   IMPORTANCE OF BIOFERTILIZER FOR
SUSTAINABLE AGRICULTURE

Sustainable agriculture is the practice of cultivation which maintains the farm ecosystem in the long term, while producing maximum healthy food grains. The health of plant, soil and farm animals are maintained in this system. Soil and its properties have a major impact in this system. For example, the soil with no use of agrochemicals produced better-tasting foods as compared to soil with a high and frequent use of agrochemicals. The soil of both macroorganisms and microorganisms play a very crucial role in food grain production in a sustainable agricultural ecosystem. The presence of microorganisms modifies the habitat and makes it possible for other life forms to survive and function in a desirable way. Soil organisms contribute to the critical soil functions, by acting as the primary driving agents of nutrient recycling, regulating the dynamics of soil organic matter, soil carbon sequestration, greenhouse gas emission, modifying soil physical structure and water regimes, enhancing the amount and efficiency of nutrient acquisition by vegetation and enhancing plant health. They also reduce the activity of harmful microbes.

3.6   TYPES OF BIOFERTILIZERS

Biofertilizers are essentially crop-specific. Even though they are plenty in the organic cultivation system, they have to be added to the soil for enhancing the various bio-physicochemical properties of the soil with regards to the nutrient. For instance, for a pulse crop, the various strains of *Rhizobium*, for N_2 fixation; for rice the Azolla-anabaena; Azotobacter for N_2 fixation; for fruit plants like litchi, the Vascular Arbuscular Mycorrhiza (VAM) are being used. The in vitro culture of some previously identified and specific strain of microbes is used for the preparation of biofertilizers. They are broadly categorized into five different types based on their activity which are discussed in the following subheadings.

3.6.1 NITROGEN-FIXING BIOFERTILIZERS (NFBs)

Nitrogen is the most abundant element in the earth, comprising 78.09 and 79.2% of the atmosphere (Figure 3.2) and soil respectively; sill is the limited element for the plant due to difficulty with its fixation in soil and uptake by the plant roots. Some microbes, either in the associated or non-associated form, help to fix the atmospheric nitrogen in the plant-available form, thus reducing the problem of fixation of N_2, denitrification and leaching loss of this element.

The nitrogen-fixing biofertilizers which depend on their association with plants are broadly classified into two types: the symbiotic nitrogen fixers, which can fix the nitrogen when in association with the plant root system; and asymbiotic nitrogen fixer, which can fix nitrogen in soil in a free living state, i.e. they don't require the plant association. The former group includes *Rhizobium* species, *Frankia* species, Cyanobacteria etc., whereas *Azotobacter chroococcum*, *Medicago sativa*, *Beijerinckia mobilis* and *Clostridium* species, free-living cyanobacteria (blue-green algae) etc., are placed in the latter group.

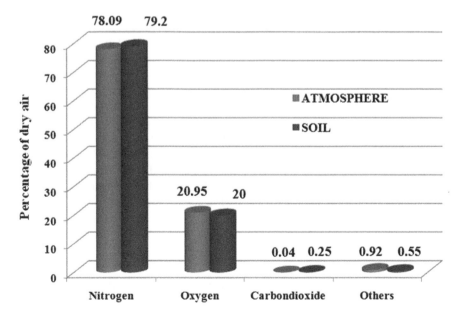

FIGURE 3.2   Gaseous composition of air (atmosphere) and soil.

3.6.1.1   Symbiotic Nitrogen Fixating Biofertilizers and Their Interaction in Rhizo-Phyllosphere

This group has close interaction with the plant system. It includes *Rhizobia* (*Rhizobium, Bradyrhizobium, Sinorhizobium, Azorhizobium Mesorhizobium, Allorhizobium*), *Frankia*, Cyanobacteria and Trichodesmium. The *Rhizobia* strains may fix the atmospheric N_2 up to 450 kg/ha depending upon the species and host (Stamford et al. 1997; Unkovich et al. 1997; Spaink et al. 1998; Vance 1998; Graham and Vance 2000; Unkovich and Pate 2000), where they form the nodules (root nodules) (Thomas and Singh 2019).

3.6.1.2   Asymbiotic/Free-Living Nitrogen-Fixating Biofertilizers and Their Interaction in Rhizosphere

Organisms which could fix the atmospheric nitrogen, with or without association with the plant or its part are known as asymbiotic or free-living nitrogen-fixing biofertilizers. It is difficult to estimate the N_2 fixation rate of free-living nitrogen-fixing biofertilizers. Some eminent scientists have recorded the efficiency of this group *viz.*, 3–10 kg N ha^{-1} in *Medicago sativa* plant (Roper et al. 1995), 2–15 mg N g^{-1} of carbon source in culture media by *Azotobacter chroococcum* in arable soils, which also produces abundant slime to aggregate the soil, 20–30 kg N ha^{-1} by free-living cyanobacteria (blue-green algae) in rice cultivation in India (Kannaiyan 2002). Smolander and Sarsa (1990) observed the fixation of N_2 by free-living strain of nodulating bacteria i.e. Frankia in its host as well as non-host species. Polyanskaya et al. (2002) recorded significant improvement of N_2 fixation by *Beijerinckia mobilis* and *Clostridium* species in cucumber plant when the inoculants were sprayed on a leaf and soaked with the seeds.

Another group classified as nitrogen-fixing biofertilizers is that of associative nitrogen-fixing organisms. They have less intimate association with the plant root. These include *Acetobacter diazotrophicus* and *Herbaspirillum* species in sugarcane, sorghum and maize (Triplett 1996; James et al. 1997; Boddey et al. 2000); *Azoarcus* species in *Leptochloa fusca* grass (Malik et al. 1997); different species of *Alcaligenes, Azospirillum, Bacillus, Enterobacter, Herbaspirillum, Klebsiella* and *Pseudomonas* in rice and maize (James 2000); and *Azospirillum* with great host specificity, comprising a variety of annual and perennial plants (Bashan and Holguin 1997). Several studies have shown that due to nitrogen fixation and production of growth-promoting substances, *Azospirillum* enhanced the growth and yield of wheat, rice, sunflower, carrot, oak, sugar beet, tomato, brinjal, pepper and cotton (Okon 1985; Bashan et al. 1989; Okon and Labandera-Gonzalez 1994). The biofertilizer of *Acetobacter diazotrophicus* was found to fix and make available up to 70% of sugarcane crop's N_2 requirement, which accounts for 150 kg N ha^{-1} annually (Boddey et al. 1995). In both of these groups (i.e. free-living and associative nitrogen-fixing biofertilizers), the plant is able to obtain N_2 only after the death of the microorganisms, which can fix the atmospheric nitrogen.

3.6.2   Phosphate-Solubilizing Biofertilizers (PSBs) and Mycorrhizae

Phosphorous is the second most limiting plant element after nitrogen (Schachtman et al. 1998). In degraded soil like alkaline soils, they are attached to the clay

TABLE 3.1

List of Phosphate-Solubilizing and Phosphate-Mobilizing Microbes

Phosphate-Solubilizing Microbes	Phosphate-Mobilizing Microbes
Bacillus megaterium var. phosphaticum, *B. subtilis, B. circulans, B. polymyxa, Pseudomonas striata, Penicillium* spp., *Aspergillus awamori, Trichoderma, Rhizoctonia solani, Rhizobium, Burkholderia, Achromobacter, Agrobacterium, Microccocus, Aereobacter, Flavobacterium* and *Erwinia*	Arbuscular mycorrhiza (*Glomus* sp., *Gigaspora* sp., *Acaulospora* sp., *Scutellospora* sp., *Sclerocystis* sp.), ectomycorrhiza (*Laccaria* spp., *Pisolithus* spp., *Boletus* spp., *Amanita* spp.), ericoid mycorrhizae (*Pezizella ericae*) and *mycorrhiza* (*Rhizoctonia solani*)

Source: modified from Thomas and Singh 2019

materials and become unavailable to the plant roots. Some microbes, either freely or in association with the plant root, help in solubilizing the fixed P. The PSB like *Bacillus* and *Pseudomonas* can increase phosphorus availability to plants by mobilizing it from the unavailable forms in the soil (Richardson et al. 2009). These bacteria and certain soil fungi such as *Penicillium* and *Aspergillus* bring about the dissolution of bound phosphates in soil through the secretion of organic acids characterized by lower pH in the phyllosphere. The application rock phosphate, along with a PSB (*Bacillus megaterium* var. phosphaticum) to sugarcane, was found to increase sugar yield and juice quality by 12.6% and it reduced the phosphorus requirement by 25%, thereby further causing a 50% reduction of the costly superphosphate usage. The phosphate-solubilizing and phosphate-mobilizing microbes are listed in Table 3.1.

3.6.2.1 Mycorrhizal Associations and Role in Plant Phosphorous Absorption

Certain higher plants in their root have the association of mycorrhiza, which helps the plant to avail certain important benefits *viz.*, improved plant growth, yield, adoptability, root absorption of nutrients (Sylvia et al. 2005). As this is a symbiotic association, both the plant and fungus are benefited. For example, in a litchi fruit orchard, with the founding of a new plantation, one has to add some soil from an already established plantation to each pit. This is because the mycorrhiza of established plantation can easily be made available for young plants, favoring their easy uptake of nutrients and better growth.

3.6.2.2 Role of Phosphorous-Solubilizing Microbes

Some rhizospheric microorganisms (except mycorrhizal fungi), particularly P-solubilizing bacteria (PSB) and fungi (PSF), can enhance plant P acquisition by directly increasing solubilization of P or by indirect hormone-induced stimulation of plant growth (Richardson et al. 2009). The PSB or PSF may mobilize soil P by the acidification of soil, through the release of enzymes (such as phosphatases and phytases) or the production of carboxylates such as gluconate, citrate and oxalate.

3.6.3   K-Solubilizing Biofertilizers

Some rhizobacteria can solubilize the insoluble forms of potassium, which is another essential nutrient necessary for plant growth (Jakobsen et al. 2005). The higher biomass yields due to increased potassium uptake have been observed with *Bacillus edaphicus* (for wheat), *Paenibacillus glucanolyticus* (for black pepper) and *Bacillus mucilaginosus* in coinoculation with the phosphate-solubilizing *Bacillus megaterium* (for eggplant, pepper and cucumber) (Meena et al. 2014; Etesami et al. 2017).

3.6.4   Biofertilizers for Micronutrients

Micronutrients are required in a significantly lower concentration; without them the plant life cycle cannot be completed. Again, these elements are present in the earth crust in a minute concentration; they mostly put forth deficiency symptoms in the plant. Some microorganism helps in solubilizing the insoluble form and makes it available for plant uptake. *Bacillus subtilis*, *Thiobacillus thiooxidans* and *Saccharomyces* spp. can solubilize insoluble zinc compounds (Ansori and Gholami 2015). Certain microorganisms like *Bacillus* sp. can hydrolyze beneficial element *viz.*, silicates and aluminum silicates and make those available for plants, which helps in increasing plant growth, resistance to stress and yield (Cakmakci et al. 2007).

3.6.5   Plant Growth-Promoting Rhizobacteria

Besides the nitrogen-fixing PSBs and K-solubilizing biofertilizers, there are certain other soil bacteria which help in plant growth by the synthesizing of growth-promoting chemicals (Bashan 1998). They are known as plant growth-promoting rhizobacteria (PGPRs). Some PGPRs, like *Bacillus pumilus*, *Bacillus licheniformis* and some species of *Azospirillum*, produce substantial quantities of plant hormone (Bashan et al. 1990; Bashan and Holguin 1997; Gutierez-Manero et al. 2001). Others like *Paenibacillus polymyxa* show beneficial properties, including nitrogen fixation, phosphorus solubilization, production of antibiotics, cytokinins, chitinase and other hydrolytic enzymes, in addition to increased soil porosity (Timmusk et al. 1999). These indicate that there is potential scope for diverse microbes as biofertilizers.

   The antagonistic mechanisms of PGPRs against phytopathogenic microorganisms include the production of antimicrobial metabolites like siderophores and antibiotics, gaseous products like ammonia and fungal cell wall-degrading enzymes which cause cytolysis, leakage of ions, membrane disruption, inhibition of mycelial growth and protein biosynthesis (Idris et al. 2007; Lugtenberg and Kamilova 2009). *Pseudomonas* strains can produce antifungal metabolites *viz.*, phenazines, pyrrolnitrin, pyoluteorin and cyclic lipopeptides of viscosinamide, which can prevent *Pythium ultimum* infection in the case of sugar beet. *Pseudomonas fluorescens* produces siderophores preventing the growth and proliferation of *Pythium ultimum*, *Rhizoctonia batatticola* and *Fusarium oxysporum* (Cox and Adams 1985; Leeman et al. 1996; Hultberg et al. 2000). *Pseudomonas aeruginosa* produces the siderophores and induces resistance against *Botrytis cinerea* (on bean and tomato) and *Colletotrichum lindemuthianum* (on bean) (De-Meyer and Hofte 1997; Audenaert et al. 2002). Some species of *Pseudomonas* produce chitinase and laminase, which can lyse *Fusarium solani* mycelia. In addition, PGPRs provide protection against some soil-borne diseases, insect pests and plant diseases (Wani et al. 2013).

3.7 IMPACT OF BIOFERTILIZERS IN CROP GROWTH AND PRODUCTION

Enhanced yield, quality and resistance/tolerance to stresses has been noticed with the utilization of biofertilizers, along with maintenance of the soil's bio-physico-chemical properties, like soil structure, porosity, pH, CEC. When these are added to the soil they multiply and inhabit the rhizosphere, thus providing a better condition for crop growth, as compared to a conventional cultivation system with the use of high value inputs. They participate in nutrient recycling and organic matter decomposition (Sinha et al. 2014; Sivakumar et al. 2013). Biofertilizers, along with the chemical fertilizers, help to improve crop growth and yield. Previously, these were only used in the legume and paddy crops but more recently, they have been used in various other crops including wheat, maize, sugarcane, vegetables, flower and fruits.

3.7.1 Biofertilizer-Mediated Induced Stress Resistance in Crop Plants

In the current agricultural context, stress is the most important factor for the reduction in crop productivity. Many tools and techniques of modern science help to overcome plant stress in adverse conditions. Certain plant growth-promoting rhizobacteria (PGPRs) help the plant to cope with stress. They act as a bioprotectant, possibly as an alternative to highly poisonous agrochemicals (Yang et al. 2009).

3.7.1.1 Biofertilizers Improving Resistance/Tolerance to Abiotic Stress

Heat, wind, soil salinity, alkalinity, radiation and heavy metal are major types of abiotic stress. Certain PGPRs help to overcome the negative impact of stress through various defense techniques. *Pseudomonas aeruginosa* has been shown to withstand biotic and abiotic stress (Pandey et al. 2012). *Rhizobium trifolii* inoculated with *Trifolium alexandrinum* were observed to have higher plant biomass and nodulation in salinity stress conditions (Hussain et al. 2002; Antoun and Prevost 2005). *P. fluorescens* MSP-393 produces osmolytes and salt stress-induced proteins to overcome the negative effects of salt (Paul and Nair 2008). *P. putida* Rs-198 enhanced the germination rate and several growth parameters of cotton under alkaline and high salt conditions by increasing the rate of uptake of K^+, Mg^{2+} and Ca^{2+} and by decreasing the absorption of Na+ (Yao et al. 2010). A few strains of *Pseudomonas* have plant tolerance with 2,4-diacetylphloroglucinol (DAPG) (Schnider-Keel et al. 2000). Systemic response was found to be induced against *P. syringae* in *Arabidopsis thaliana* by *P. fluorescens* DAPG (Weller et al. 2012). Calcisol produced by PGPRs *viz.*, *P. alcaligenes* PsA15, *Bacillus polymyxa* BcP26 and *Mycobacterium phlei* MbP18 provides tolerance to high temperatures and salinity stress (Egamberdiyeva 2007). It has been demonstrated that inoculation of plants with AM fungi also improves plant growth under salt stress (Ansari et al. 2013). *Achromobacter piechaudii* increased the biomass of tomato and pepper plants under 172 mM NaCl and water stress (Alavi et al. 2013).

Root endophytic fungus *Piriformospora indica* was found to defend host plants against salt stress. Inoculation of PGPR alone or along with AM like *Glomus intraradices* or *G. mosseae* resulted in better nutrient uptake and improvement in normal physiological processes in *Lactuca sativa* under stress conditions, whereas *P. mendocina* increased shoot biomass under salt stress (Kohler and Caravaca 2010). Mechanisms involved in osmotic stress tolerance employing transcriptomic and microscopic strategies revealed a considerable change in response to salt stress (Gao et al. 2012). A combination of AM fungi and N_2-fixing bacteria helped the legume plants in overcoming drought stress (Aliasgharzad et al. 2006). Better results of *A. brasilense* along with AM can be seen in other crops such as tomato, maize and cassava (German et al. 2000; Casanovas et al. 2002; Creus et al. 2005). *A. brasilense* and AM combination improved plant tolerance to various abiotic stresses (Joe et al. 2009). The additive effect of *Pseudomonas putida* or *Bacillus megaterium* and AM fungi was effective in alleviating drought stress (Marulanda et al. 2009; Singh et al. 2018, 2017a, b, c, 2019; Tiwari et al. 2018, 2019a, b; Kour et al. 2019). Application of *Pseudomonades* sp. under water stress improved the antioxidant and photosynthetic pigments in basil plants. Interestingly, a combination of three bacterial species caused the highest CAT, GPX and APX activity and chlorophyll content in leaves under water stress (Heidari and Golpayegani 2012). *Pseudomonas* spp. was found to cause a positive effect on the seedling growth and seed germination of *A. officinalis* L. under water stress (Liddycoat et al. 2009). Photosynthetic efficiency and the antioxidative response of rice plants subjected to drought stress were observed to increase after inoculation of arbuscular mycorrhiza (Ruiz-Sanchez et al. 2010). The beneficial effects of mycorrhizae have also been reported under both the drought and saline conditions (Aroca et al. 2013).

Heavy metals such as cadmium, lead, mercury from hospital, industrial and factory waste accumulate in the soil and enter plants through the root system (Gill et al. 2012). *Azospirillium* spp, *Phosphobacteria* spp and *Glucanacetobacter* spp. isolated from the rhizosphere of a rice field and mangroves were found to be more tolerant to heavy metals, especially iron (Samuel and Muthukkaruppan 2011). *P. potida* strain 11 (P.p.11), *P. potida* strain 4 (P.p.4) and *P. fluorescens* strain 169 (P. f.169) can protect canola and barley plants from the inhibitory effects of cadmium via IAA, siderophore and 1-aminocyclopropane-1-carboxylate deaminase (ACCD) (Baharlouei et al. 2011). It was reported that rhizoremediation of petroleum-contaminated soil can be expedited by adding microbes in the form of an effective microbial agent (EMA) to the different plant species such as cotton, ryegrass, tall fescue and alfalfa (Tang et al. 2010).

3.7.1.2 Biofertilizers Improving Resistance/Tolerance to Biotic Stress

PGPRs acting as biological agents were observed to be one of the alternatives of chemical agents to provide resistance to various pathogen attacks (Murphy et al. 2000). Apart from acting as growth-promoting agents, they also produce resistance against pathogens by producing metabolites (Backman and Sikora 2008). *Bacillus*

subtilis GBO3 can induce defense-related pathways *viz.*, salicylic acid (SA) and jasmonic acid (JA) (Ryu et al. 2004). Application of PGPR isolates *viz.*, *B. amyloliquefaciens* 937b and *B. pumilus* SE-34 provide immunity against tomato mottle virus
(Murphy et al. 2003). *B. megaterium* IISRBP 17, characterized from the stem of
black pepper, acts against *Phytophthora*. Inoculation of the Washington navel orange
with *Pseudomonas fluorescens* strain 843 was not only highly effective in increasing
the production and quality of the fruit but also in inhibiting the nematode (Abdelaal
et al. 2010).

3.8 PROS AND CONS OF BIOFERTILIZERS

Numerous benefits can be obtained from biofertilizers some of which are listed
below:

- Assistance in increased recycling, mineralization and uptake of mineral
  nutrients.
- Synthesis of vitamins, amino acids, auxins, gibberellins and plant growth-
  regulating substances.
- Bioremediation of heavy metals in contaminated soils.
- Antagonism with potential plant pathogens through the competition and
  development of an amensal relationship, based on production of antibiotics,
  siderophores and/or hydrolytic enzymes.
- Maintenance of the soil structure and buffering capacity.
- Provision of a better plant root environment.

Though there are many advantages to using microbial fertilizers, they still have some
disadvantages. The major constraints in biofertilizer use are listed below:

- Use of less efficient microbial strain for large scale production.
- Shorter shelf life of microbes.
- Lack of technically efficient personnel in the production unit.
- Non-availability of suitable production facility *viz.*, equipment, space, stor-
  age, efficient packaging, etc.
- Non-availability of sufficient fund, awareness and retail outlet.
- Inefficient production system to meet the seasonal biofertilizer requirement.
- Soil and after-care are not properly observed after biofertilizer inoculation.
- Unfamiliarity with biofertilizers, as well as ignorance of environmental
  problems, is a major issue with their use.

3.9 TECHNIQUES INVOLVED IN BIOFERTILIZERS

3.9.1 FILLER MATERIALS/CARRIER OF BIOFERTILIZERS

Conventionally, biofertilizers are not the sole element in their final prepared form.
They are inoculated with a suitable properly sterilized carrier or filler material before
supplying to the grower. For the preparation of seed inoculant, the carrier material is

milled to fine powder with a particle size of 10–40 μm. According to the *Handbook for Rhizobia* (Somasegaran and Hoben 1994), the properties of a good carrier material for seed inoculation are as follows:

- Non-toxicity to inoculants' bacterial strain.
- Good moisture absorption capacity.
- Easy to process and free of lump-forming materials.
- Easy to sterilize by autoclaving or gamma-irradiation.
- Availability in adequate amounts.
- Inexpensive.
- Good adhesion to seeds.
- Good pH buffering capacity.
- Non-toxicity to plants (Senoo 2006).

Depending upon the demand, microorganisms can multiply in a very short span of time because of their large surface area compared to the volume size. These filler materials are mostly used to provide a base for microbial activity. These should be easily available, inexpensive, have a high water-holding capacity, free from contamination, support a longer life of microbes, retain a higher microbial count and ultimately be a good source of organic matter. Some commonly used carrier materials in the production of good-quality biofertilizers are neutralized peat soil, lignite, vermiculite, charcoal, press mud, farmyard manure, soil mixture, talc, rice barn, paddy straw compost, rock phosphate pellets, clay, etc. However, due to certain disadvantages of possessing a lower shelf-life, temperature sensitivity, being prone to contamination and the retention of low cell counts, liquid formulations have been developed for *Rhizobium, Azospirillum, Azotobacter* and *Acetobacter*. Despite its higher cost, it has the advantages of easier production, higher cell counts, longer shelf-life, no contamination, storage up to 45° C and greater competence in soil (Ngampimol and Kunathigan 2008), as compared to conventional carrier materials.

3.9.2 METHODS OF APPLICATION

Biofertilizers can be added to the rhizosphere by seed treatment, root treatment, and through soil application (Figure 3.3). Seed treatment is the most effective, economical, easy and rapid method of microbial inoculation to the seeds (Sethi et al. 2014). In this method, the seeds are mixed with the biofertilizer slurry (a mixture of biofertilizers and water) evenly. For achieving a good and tight coating of inoculate on the seed surface use of adhesive, such as gum arabic, methy lethyl cellulose, sucrose solutions and vegetable oils is recommended (Senoo 2006). Then the seeds are dried under shade and must be sown within 24 hours of treatment. The liquid biofertilizers can be treated to seed by placing the seeds in a polythene bag or buckets. Two or more biofertilizers can be treated to the plant at an optimum dose without any antagonistic effect (Chen 2006). It is the most commonly used method.

Root treatment is generally a common approach for transplanted crops *viz.*, sugarcane, cereals, vegetables, fruits, flowers, plantation crops, cotton, etc., where the seedling roots are dipped in the suspension (water suspension) of biofertilizers for

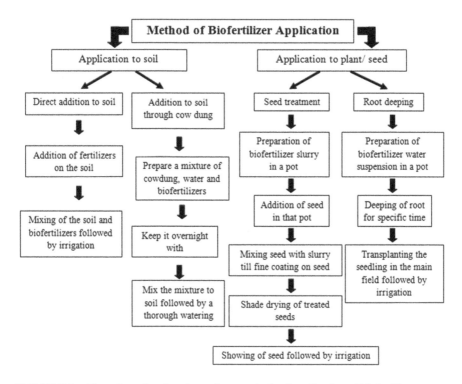

FIGURE 3.3   Flow chart showing the various methods of application of biofertilizers.

a sufficient period of time before transplanting. The time depends on the type of crop used.

In soil application, the single (or combination of) biofertilizer(s) is/are directly added to soil and mixed thoroughly or added in furrow (granular biofertilizer) under or alongside the seeds (Senoo 2006). The biofertilizers can also be applied to soil by adding cow dung. In this method, the biofertilizers are added to cow dung, mixed and kept overnight. Precautions should be taken to maintain the sufficient moisture content of either the dung or the soil in the soil application method. Pindi and Satyanarayana (2012) suggested that phosphate-solubilizing microbial biofertilizer can be applied to soil by mixing with cow dung and rock phosphate after procuring them overnight with maintenance of a 50% moisture content.

3.9.3   Time of Application

Biofertilizers are very effective when added to the seed soon before sowing, and to the seedling before transplanting. When added to soil, these are drenched either at the time of sowing or at different crop growth stages.

3.9.4   Dose for Application

A specific dose for the different methods used in biofertilizers' application:

- For seed treatment, one packet (200 g) is sufficient to treat 10–12 kg.
- For seedling root, dip the suspension prepared by mixing 1 kg (5 packets) biofertilizer culture in 10–15 liters of water.
- For soil application in short duration (less than 6 months) crop, 10–15 packets (each of 200 g) of biofertilizers are mixed with 40–60 kg of well-decomposed cattle manure or with 40–60 kg soil for one acre of land.
- For soil application in a long duration crop (perennial crop) 20–30 packets of biofertilizer (each containing 200 g) are mixed with 80–120 kg cattle manure or soil per acre.
- For application in a standing crop, perennial plants are pruned once in a year. After pruning, the soil in the bed is dug up with a fork with due care to avoid any damage to the roots. A mixture of biofertilizer and FYM/soil is then applied by incorporating it into the soil, followed by irrigation.

3.10   CAN BIOFERTILIZERS BE ENOUGH FOR CONVERSION TO ORGANIC FARMING PRACTICES?

Global food production recorded an exponential growth rate from 1970 to 1995, due to the Green Revolution, which uses high-yielding inputs such as improved and high-yielding varieties (HYVs), irrigation, chemical fertilizers and synthetic pesticides. In the 21st century, consumers are more focused on food quality, more specifically the nutritional value of each and every particle of food consumed. Hence, organic farming is a current necessity in order to fulfill the demand of consumers and maintain a healthy society. Due to the heavy use of agrochemicals, productive soil is not only degraded but also unhealthy. If we consider one example: nowadays, minor pests *viz.*, thrips, mites, mealy bugs, etc., are considered more of a threat, hence the need for frequent spraying of systematic insecticides. Such practices leave a chemical residue in the final product. This reduces the nutritive value, as well as slowly adding poison to the human body.

Biofertilizers are a very good source of nutrition for a plant, whilst also building the plant's resistance to certain types of pests and pathogens. But as we know, for the most part, biofertilizers help to solubilize the soil-bounded nutrient and make these available for plant uptake. Thus biofertilizers alone may provide the solution. In addition, there is a need for adding organic manure in order to fulfill the crop nutrition demand.

3.11   INTERVENTION NECESSARY AND FUTURE PROSPECTS

Due to the low cost of the biofertilizers in comparison with chemical fertilizers, thus far some farmers are reluctant to use the organic option. Little interest is shown on their part; when the management of soil health comes to the fore, they think they might have crop loss, eventually helping the soil to degrade more. Even now, no farmers ever ask for the soil microbial count while submitting the sample for analysis. Soil parameters like soil pH, EC, organic carbon, N, P, K, the micronutrients, etc. have been measured and suitable recommendations noted, but in no instance has there been a clear recommendation with regards to the use of biofertilizers. From the above discussion, it is certain that awareness among farmers as to the use of biofertilizers is still lacking. Such a paucity in knowledge is the main reason for poor demand. Hence it is of prime concern to enhance understanding among growers, thus helping them produce the best quality strain. A modern and straightforward marketing strategy in regards to biofertilizers must be implemented.

3.12   CONCLUSION

In modern times, conventional agriculture tends towards sustainable and organic agriculture, thus biofertilizers are an important component in nutrition management. They provide a very good alternative to agrochemicals *viz.*, fertilizers and pesticides. In addition, they can help in fixing or making available nitrogen to the plant root, solubilize phosphorous, mobilize the potassium and make it available for plant uptake, produce hormones, antimetabolites to enhance plant growth and decompose the organic matters. This helps to enhance the crop yield and soil bio-physicochemical properties, finally assisting with the crop plant's ability to overcome biotic and abiotic stress. Nevertheless, increased awareness is vital, from its production until its use in the grower's field. It also creates a path for entrepreneurship in the manufacturing, transportation and distribution sectors in relation to biofertilizers.

REFERENCES

Abdelaal, S., El-Sheikh, M.H., Hassan, H.A.S., Kabeil, S.S. 2010. Microbial bio-fertilization approaches to improve yield and quality of Washington navel Orange and reducing the survival of nematode in the soil. *Journal of American Science* 6:264–271.

Acharya, C.N., Jain, S.P., Jha, J. 1953. Studies on the building up of soil fertility by the phosphatic fertilization of legumes. Influence of growing berseem on the nitrogen content of the soil. *Journal of the Indian Society of Soil Science* 1:55–64.

Alavi, P., Starcher, M.R., Zachow, C., Muller, H., Berg, G. 2013. Root-microbe systems: The effect and mode of interaction of stress protecting agent (SPA) *Stenotrophomonas rhizophila* DSM14405T. *Frontier Plant Science* 4:141.

Aliasgharzad, N., Reza, M., Salimi, G.N. 2006. Effects of arbuscular mycorrhizal fungi and *Bradyrhizobium japonicum* on drought stress of soybean. *Biologia* 19:324–328.

Ansari, M.W., Trivedi, D.K., Sahoo, R.K., Gill, S.S., Tuteja, N. 2013. A critical review on fungi mediated plant responses with special emphasis to *Piriformospora indica* on improved production and protection of crops. *Plant Physiology and Biochemistry* 70:403–410.

Ansori, A., Gholami, A. 2015. Improved nutrient uptake and growth of maize in response to inoculation with *Thiobacillus* and mycorrhiza on an alkaline soil. *Communication Soil Science Plant Annals* 46:2111–2126.

Antoun, H., Prevost, D. 2005. Ecology of plant growth promoting rhizobacteria. In: *PGPR: Biocontrol and Biofertilization*. Edited by Z.A. Siddiqui. Dordrecht: Springer, 1–38.

Aroca, R., Ruiz-Lozano, J.M., Zamarreno, A.M., Paz, J.A., García-Mina, J.M., Pozo, M.J., Lopez-Raez, J.A. 2013. Arbuscular mycorrhizal symbiosis influences strigolactone production under salinity and alleviates salt stress in lettuce plants. *Journal of Plant Physiology* 170:47–55.

Audenaert, K., Pattery, T., Cornelis, P., Hofte, M. 2002. Induction of systemic resistance to *Botrytis cinerea* in tomato by *Pseudomonas aeruginosa* 7NSK2: Role of salicylic acid, pyochelin and pyocyanin. *Molecular Plant–Microbe Interactions* 15:1147–1156.

Backman, P.A., Sikora, R.A. 2008. Endophytes: An emerging tool for biological control. *Biology Control* 46:1–3.

Baharlouei, K., Pazira, E., Solhi, M. 2011. Evaluation of inoculation of plant growth-promoting rhizobacteria on cadmium. In: International Conference on Environmental Science and Technology IPCBEE, Vol. 6. Singapore: IACSIT Press.

Bashan, Y. 1998. Inoculants of plant growth-promoting bacteria for use in agriculture. *Biotechnology Advance* 16:729–770.

Bashan, Y., Harrison, S.K., Whitmoyer, R.E. 1990. Enhanced growth of wheat and soybean plant inoculated with *Azospirillum brasilense* is not necessary due to general enhancement of mineral uptake. *Applied and Environment Microbiology* 56:769–775.

Bashan, Y., Holguin, G. 1997. Azospirillum-plant relationships: Environmental and physiological advances (1990–1996). *Canadian Journal of Microbiology* 43:103–121.

Bashan, Y., Ream, Y., Levanony, H., Sade, A. 1989. Non-specific responses in plant growth, yield, and root colonization of noncereal crop plants to inoculation with *Azospirillum brasilense* Cd. *Canadian Journal of Botany* 67:1317–1324.

Bertin, C., Yang, X., Weston, L.A. 2003. The role of root exudates and allelochemicals in the rhizosphere. *Plant and Soil* 256:67–83.

Boddey, R.M., Da-Silva, L.G., Reis, V., Alves, B.J.R., Urquiaga, S. 2000. Assessment of bacterial nitrogen fixation in grass species. In: *Prokaryotic Nitrogen Fixation: A Model System for Analysis of a Biological Process*. Edited by E.W. Triplett. Wymondham: Horizon Scientific Press, 705–726.

Boddey, R.M., de-Oliveira, O.C., Urquiaga, S., Reis, V.M., Olivares, F.L., Baldani, V.L.D., Dobereiner, J. 1995. Biological nitrogen fixation associated with sugar cane and rice: Contributions and prospects for improvement. *Plant Soil* 174:195–209.

Cakmakci, R., Donmez, M.F., Erdogan, U. 2007. The effect of plant growth promoting rhizobacteria on barley seedling growth, nutrient uptake, some soil properties and bacterial counts. *Turkish Journal of Agriculture* 31:189–199.

Casanovas, E.M., Barassi, C.A., Sueldo, R.J. 2002. Azospirillum inoculation mitigates water stress effects in maize seedlings. *Current Research in Communication* 30:343–350.

Chen, J.H. 2006. The combined use of chemical and organic fertilizers and/or biofertilizer for crop growth and soil fertility. In: International Workshop on Sustained Management of the Soil Rhizosphere System for Efficient crop Production and Fertilizer Use. Bangkok, Thailand: Land Development Department, 16–20.

Cox, C.D., Adams, P.A. 1985. Siderophore activity of pyoverdin for *Pseudomonas aeruginosa*. *Infection and Immunology* 48:130–138.

Creus, C.M., Graziano, M., Casanovas, E.M., Pereyra, M.A., Simontacchi, M., Puntarulo, S. 2005. Nitric oxide is involved in the *Azospirillum brasilense*-induced lateral root formation in tomato. *Planta* 221:297–303.

De-Meyer, G., Hofte, M. 1997. Salicylic acid produced by the rhizobacterium *Pseudomonas aeruginosa* 7NSK2 induces resistance to leaf infection by *Botrytis cinerea* on bean. *Phytopathology* 87:588–593.

Egamberdiyeva, D. 2007. The effect of plant growth promoting bacteria on growth and nutrient uptake of maize in two different soils. Applied Soil Ecology 36:184–189.

Etesami, H., Emami, S., Alikhani, H.A. 2017. Potassium solubilizing bacteria (KSB): Mechanisms, promotion of plant growth, and future prospects, a review. *Journal of Soil Science and Plant Nutrition* 17:897–911.

Gao, X., Lu, X., Wu, M., Zhang, H., Pan, R., Tian, J., Li, S., Liao, H. 2012. Co-Inoculation with rhizobia and AMF inhibited soybean red crown rot: From field study to plant defense-related gene expression analysis. *PLoS One* 7:33977.

German, M.A., Burdman, S., Okon, Y., Kigel, J. 2000. Effects of *Azospirillum brasilense* on root morphology of common bean (*Phaseolus vulgaris* L.) under different water regimes. *Biology and Fertility of Soils* 32:259–264.

Gill, S.S., Khan, N.A., Tuteja, N. 2012. Cadmium at high dose perturbs growth, photosynthesis and nitrogen metabolism while at low dose it up regulates sulfur assimilation and antioxidant machinery in garden cress (*Lepidium sativum* L.). *Plant Science* 182:112–120.

Graham, P.H., Vance, C.P. 2000. Nitrogen fixation in perspective: An overview of research and extension needs. *Field Crops Research* 65:93–106.

Gutierez-Manero, F.J., Ramos-Solano, B., Probanza, A., Mehouachi, J., Tadeo, F.R., Talon, M. 2001. The plant-growth-promoting rhizobacteria *Bacillus pumilus* and *Bacillus licheniformis* produce high amounts of physiologically active gibberellins. *Physiology of Plant* 111:206–211.

Heidari, M., Golpayegani, A. 2012. Effects of water stress and inoculation with plant growth promoting rhizobacteria (PGPR) on antioxidant status and photosynthetic pigments in basil (*Ocimum basilicum* L.). *Journal of Saudi Society Agriculture Science* 11:57–61.

Hultberg, M., Alsanius, B., Sundin, P. 2000. In vivo and in vitro interactions between *Pseudomonas fluorescens* and *Pythium ultimum* in the suppression of damping-off in tomato seedlings. *Biology Control* 19:1–8.

Hussain, N., Mujeeb, F., Tahir, M., Khan, G.D., Hassan, N.M., Bari, A. 2002. Effectiveness of *Rhizobium* under salinity stress. *Asian Journal of Plant Science* 1:12–14.

Idris, E.E., Iglesias, D.J., Talon, M., Borriss, R. 2007. Tryptophan-dependent production of indole-3-acetic acid (IAA) affects level of plant growth promotion by *Bacillus amyloliquefaciens* FZB42. *Molecular Plant–Microbe Interactions* 20:619–626.

Jakobsen, I., Leggett, M.E., Richardson, A.E. 2005. Rhizosphere microorganisms and plant phosphorus uptake. In: *Phosphorus, Agriculture and the Environment*. Edited by J.T. Sims, A.N. Sharpley. Madison, WI: Am. Soc. Agronomy, 437–494.

James, E.K. 2000. Nitrogen fixation in endophytic and associative symbiosis. *Field Crops Research* 65:197–209.

James, E.K., Olivares, F.L., Baldani, J.I., Dobereiner, J. 1997. *Herbaspirillum*, an endophytic diazotroph colonizing vascular tissue in leaves of *Sorghum bicolor* L. *Moench Journal of Experimental Botany* 48:785–797.

Joe, M.M., Jaleel, C.A., Sivakumar, P.K., Zhao, C.X., Karthikeyan, B. 2009. Co-aggregation in *Azospirillum brasilensense* MTCC-125 with other PGPR strains: Effect of physical and chemical factors and stress endurance ability. *Journal of Taiwan Institute Chemical Engineering* 40:491–499.

Joshi, N.V. 1920. *Memoirs of the Department of Agriculture in India: Bacteriological Series*, Vol. 1. Pusa, Delhi, India: Dept. of Agriculture, Indian Agricultural Research Institute, 246.

Kannaiyan, S. 2002. *Biotechnology of Biofertilizers*. New Delhi: Narosa Publishing House, Alpha Science Int'l Ltd.

Kohler, J., Caravaca, F. 2010. An AM fungus and a PGPR intensify the adverse effects of salinity on the stability of rhizosphere soil aggregates of *Lactuca sativa* Roldan. *Soil Biology and Biochemistry* 42:429–434.

Kour, D., Rana, K.L., Yadav, N., Yadav, A.N., Rastegari, A.A., Singh, C., Negi, P., Singh, K., Saxena, A.K. 2019. Technologies for biofuel production: Current development, challenges, and future prospects. In: *Prospects of Renewable Bioprocessing in Future Energy Systems, Biofuel and Biorefinery Technologies*. Edited by A.A. Rastegari et al., Vol. 10. Berrlin: Springer, 1–50.

Krishna, M.P., Rinoy, V., Hatha, A.A.M. 2012. Depth wise variation of microbial load in the soils of midland region of Kerala: A function of important soil physicochemical characteristics and nutrients. *Indian Journal of Education Information Management* 1(3):126–129.

Leeman, M., Den-Ouden, F.M., Van-Pelt, J.A., Dirkx, F.P.M., Steijl, H., Bakker, P.A.H.M., Schippers, B. 1996. Iron availability affects induction of systemic resistance to *Fusarium* wilt of radish by *Pseudomonas fluorescens*. *Phytopathology* 86:149–155.

Liddycoat, S.M., Greenberg, B.M., Wolyn, D.J. 2009. The effect of plant growth promoting rhizobacteria on asparagus seedlings and germinating seeds subjected to water stress under greenhouse conditions. *Canadian Journal of Microbiology* 55:388–394.

Lugtenberg, B., Kamilova, F. 2009. Plant-growth-promoting rhizobacteria. *Annual Review of Microbiology* 63:541–556.

Lynch, J.M., Whipps, J.M. 1990. Substrate flow in the rhizosphere. In: *Plant and Soil*, Vol. 129, 1–10, and *The Rhizosphere and Plant Growth*. Edited by D.L. Keister, P.B. Cregan. Dordecht: Kluwer, 15–24.

Malik, K.A., Bilal, R., Mehnaz, S., Rasul, G., Mirza, M.S., Ali, S. 1997. Association of nitrogen-fixing, plant growth-promoting rhizobacteria (PGPR) with kallar grass and rice. *Plant Soil* 194:37–44.

Market Research Report. 2018. *Biofertilizers Market Size, Share and Trends Analysis Report by Product (Nitrogen Fixing, Phosphate Solubilizing), by Application (Seed Treatment, Soil Treatment) and Segment Forecasts, 2012–2022*. https://www.grandviewresearch.com/industry-analysis/biofertilizers-industry.

Marulanda, A., Barea, J.M., Azcon, R. 2009. Stimulation of plant growth and drought tolerance by native microorganisms (AM) from dry environments: Mechanisms related to bacterial effectiveness. *Journal of Plant Growth Regulator* 28:115–124.

Mazid, M., Khan, T.A. 2014. Future of bio-fertilizers in Indian agriculture: An overview. *International Journal of Agricultural and Food Research* 3(3):10–23.

Mazid, M., Khan, T.A., Mohammad, F. 2011. Potential of NO and H_2O_2 as signalling molecules in tolerance to abiotic stress in plants. *Journal of Industrial Research and Technology* 1(1):56–68.

Meena, V.S., Maurya, B.R., Verma, J.P. 2014. Does a rhizospheric microorganism enhance K+ availability in agricultural soils. *Microbiology Research* 169:337–347.

Murphy, J.F., Reddy, M.S., Ryu, C.M., Kloepper, J.W., Li, R. 2003. Rhizobacteria mediated growth promotion of tomato leads to protection against cucumber mosaic virus. *Phytopathology* 93:1301–1307.

Murphy, J.F., Zehnder, G.W., Schuster, D.J., Sikora, E.J., Polstan, J.E., Kloepper, J.W. 2000. Plant growth promoting rhizobacteria mediated protection in tomato against tomato mottle virus. *Plant Disease* 84:79–84.

Ngampimol, H., Kunathigan, V. 2008. The study of shelf life for liquid biofertilizer from vegetable waste. *AU Journal of Technology* 11:204–208.

Nobbe, F., Hiltner, L. 1893. Impfet den Boden!. *Sachsische Landwirtschaftliche Zeitschrift* 16:1–5.

Nobbe, F., Schmid, E., Hiltner, L., Hotter, E. 1891. Versuche uber die Stickstoff – Assimilation von Leguminosen. *Landwirtsch Vers-Stn* 39:327–359.

Okon, Y. 1985. *Azospirillum* as a potential inoculant for agriculture. *Trends in Biotechnology* 3:223–228.

Okon, Y., Labandera-Gonzalez, C.A. 1994. Agronomic applications of *Azospirillum*: An evaluation of 20 years worldwide field inoculation. *Soil Biology and Biochemistry* 26:1591–1601.

Pandey, P.K., Yadav, S.K., Singh, A., Sarma, B.K., Mishra, A., Singh, H.B. 2012. Cross-species alleviation of biotic and abiotic stresses by the endophyte *Pseudomonas aeruginosa* PW09. *Journal of Phytopathology* 160:532–539.

Paul, D., Nair, S. 2008. Stress adaptations in a plant growth promoting Rhizobacterium (PGPR) with increasing salinity in the coastal agricultural soils. *Journal of Basic Microbiology* 48:1–7.

Pindi, P.K., Satyanarayana, S.D.V. 2012. Liquid microbial consortium – A potential tool for sustainable soil health. *Journal of Biofertilizer Biopesticide* 3:124.

Polyanskaya, L.M., Vedina, O.T., Lysak, L.V., Zvyagintsev, D.G. 2002. The growth-promoting effects of *Beijerinckia mobilis* and *Clostridium* sp. cultures on some agricultural crops. *Microbiology* 71:109–115.

Rao, D.L.N., Patra, A.K. 2009. Soil microbial diversity and sustainable agriculture. *Journal of the Indian Society of Soil Science* 57:513–530.

Richardson, A.E., Barea, J.M., Mcneill, A.M., Prigent-Combaret, C. 2009. Acquisition of phosphorus and nitrogen in the rhizosphere and plant growth promotion by microorganisms. *Plant and Soil* 321:305–339.

Roper, M.M., Gault, R.R., Smith, N.A. 1995. Contribution to the N status of soil by free-living N_2-fixing bacteria in a Lucerne stand. *Soil Biology Biochemistry* 27:467–471.

Ruiz-Sanchez, M., Aroca, R., Munoz, Y., Polon, R., Ruiz-Lozano, J.M. 2010. The arbuscular mycorrhizal symbiosis enhances the photosynthetic efficiency and the antioxidative response of rice plants subjected to drought stress. *Journal of Plant Physiology* 167:862–869.

Ryu, C.M., Farag, M.A., Hu, C.H., Reddy, M.S., Kloepper, J.W., Pare, P.W. 2004. Bacterial volatiles induce systemic resistance in *Arabidopsis*. *Plant Physiology* 134:1017–1026.

Samuel, S., Muthukkaruppan, S.M. 2011. Characterization of plant growth promoting rhizobacteria and fungi associated with rice, mangrove and effluent contaminated soil. *Current of Botany* 2:22–25.

Schachtman, D.P., Reid, R.J., Ayling, S.M. 1998. Phosphorus uptake by plants: From soil to cell. *Plant Physiology* 116:447–453.

Schnider-Keel, U., Seematter, A., Maurhofer, M., Blumer, C., Duffy, B., Gigot-Bonnefoy, C., Reimmann, C., Notz, R., Defago, G., Haas, D., Keel, C. 2000. Autoinduction of 2,4-diacetylphloroglucinol biosynthesis in the biocontrol agent *Pseudomonas fluorescens* CHA0 and repression by the bacterial metabolites salicylate and pyoluteorin. *Journal of Bacteriology* 182:1215–1225.

Senoo, K. 2006. *Carrier Materials, Carriers for Biofertilizers, Biofertilizer Manual*. Japan: FNCA Biofertilizer Project Group, Japan Atomic Industrial Forum, 41–44.

Sethi, S.K., Sahu, J.K., Adhikary, S.P. 2014. Microbial biofertilizers and their pilot-scale production. In: *Microbial Biotechnology and Progress Trends*. Boca Raton, FL: CRC Press, 297.

Singh, C., Tiwari, S., Boudh, S., Singh, J.S. 2017a. Biochar application in management of paddy crop production and methane mitigation. In: *Agro-Environmental Sustainability: Managing Environmental Pollution*. Edited by J.S. Singh, G. Seneviratne, 2nd ed. Switzerland: Springer, 123–146.

Singh, C., Tiwari, S., Gupta, V.K., Singh, J.S. 2018. The effect of rice husk biochar on soil nutrient status, microbial biomass and paddy productivity of nutrient poor agriculture soils. *Catena* 171:485–493.

Singh, C., Tiwari, S., Singh, J.S. 2017b. Impact of rice husk biochar on nitrogen mineralization and methanotrophs community dynamics in paddy soil. *International Journal of Pure and Applied Bioscience* 5:428–435.

Singh, C., Tiwari, S., Singh, J.S. 2017c. Application of biochar in soil fertility and environmental management: A review. *Bulletin of Environment, Pharmacology and Life Sciences* 6:07–14.

Singh, C., Tiwari, S., Singh, J.S. 2019. Biochar: A sustainable tool in soil 2 pollutant bioremediation. In: *Bioremediation of Industrial Waste for Environmental Safety*. Edited by R.N. Bharagava, G. Saxena. Berlin: Springer, 475–494.

Sinha, R.K., Valani, D., Chauhan, K., Agarwal, S. 2014. Embarking on a second green revolution for sustainable agriculture by vermiculture biotechnology using earthworms: Reviving the dreams of Sir Charles Darwin. *International Journal of Agriculture and Biology* 1:50–64.

Sivakumar, T., Ravikumar, M., Prakash, M., Thamizhmani, R. 2013. comparative effect on bacterial biofertilizers on growth and yield of green gram (*Phaseolus radiate* L.) and cowpea (*Vigna siensis* Edhl.). *International Journal of Current Research and Academic Review* 1:20–28.

Smolander, A., Sarsa, M.L. 1990. Frankia strains of soil under *Betula pendula*: Behaviour in soil and in pure culture. *Plant Soil* 122:129–136.

Somasegaran, P., Hoben, H.J. 1994. *Handbook for Rhizobia: Methods in Legumes-Rhizobium Technology*. New York, NY: Springer-Verlag Inc., 450.

Spaink, H.P., Kondorosi, A., Hooykaas, P.J.J. (Eds.) 1998. *The Rhizobiaceae*. Dordrecht: Kluwer Academic Publishers.

Stamford, N.P., Ortega, A.D., Temprano, F., Santos, D.R. 1997. Effects of phosphorus fertilization and inoculation of *Bradyrhizobium* and mycorrhizal fungi on growth of *Mimosa caesalpiniaefolia* in an acid soil. *Soil Biology Biochemistry* 29:959–964.

Sylvia, D., Fuhrmann, J., Hartel, P., Zuberer, D. 2005. *Principles and Applications of Soil Microbiology*. Upper Saddle River, NJ: Pearson.

Tang, J., Wang, R., Niu, X., Wang, M., Zhou, Q. 2010. Characterization on the rhizoremediation of petroleum contaminated soil as affected by different influencing factors. *Biogeosciences Discussions* 7:4665–4688.

Thomas, L., Singh, I. 2019. Microbial biofertilizers: Types and applications. In: *Biofertilizers for Sustainable Agriculture and Environment. Soil Biology*. Edited by B. Giri, R. Prasad, Q.S. Wu, A. Varma, Vol. 55. Cham: Springer.

Timmusk, S., Nicander, B., Granhall, U., Tillberg, E. 1999. Cytokinin production by *Paenobacillus polymyza*. *Soil Biology and Biochemistry* 31:1847–1852.

Tiwari, S., Singh, C., Boudh, S., Rai, P.K., Gupta, V.K., Singh, J.S. 2019a. Land use change: A key ecological disturbance declines soil microbial biomass in dry tropical uplands. *Journal of Environmental Management* 242:1–10.

Tiwari, S., Singh, C., Singh, J.S. 2018. Land use changes: A key ecological driver regulating methanotrophs abundance in upland soils. *Energy, Ecology, and the Environment* 3:355–371.

Tiwari, S., Singh, C., Singh, J.S. 2019b. Wetlands: A major natural source responsible for methane emission. In: *Restoration of Wetland Ecosystem: A Trajectory Towards a Sustainable Environment*. Edited by A.K. Upadhyay et al. Berlin: Springer, 59–74.

Triplett, E. 1996. Diazotrophic endophytes: Progress and prospects for nitrogen fixation in monocots. *Plant Soil* 186:29–38.

Trumbore, S. 2000 Age of soil organic matter and soil respiration: Radiocarbon constraints on belowground C dynamics. *Ecological Applications* 10:399–411.

Unkovich, M.J., Pate, J.S. 2000. An appraisal of recent field measurements of symbiotic N_2 fixation by annual legumes. *Field Crops Research* 65:211–228.

Unkovich, M.J., Pate, J.S., Sanford, P. 1997. Nitrogen fixation by annual legumes in Australian Mediterranean agriculture. *Australian Journal of Agriculture Research* 48:267–293.

Vance, C.P. 1998. Legume symbiotic nitrogen fixation: Agronomic aspects. In: *The Rhizobiaceae*. Edited by H.P. Spaink. Dordrecht: Kluwer Academic, 509–530.

Vessey, J.K. 2003. Plant growth promoting rhizobacteria as biofertilizers. *Plant Soil* 255:571–586.

Wani, S.A., Chand, S., Ali, T. 2013. Potential use of *Azotobacter chroococcum* in crop production: An overview. *Currant Agriculture Research Journal* 1:35–38.

Weller, D.M., Mavrodi, D.V., van-Pelt, J.A., Pieterse, C.M., van-Loon, L.C., Bakker, P.A. 2012. Induced systemic resistance in *Arabidopsis thaliana* against *Pseudomonas syringae* pv. tomato by 2,4-diacetylphloroglucinol-producing *Pseudomonas fluorescens*. *Phytopathology* 102:403–412.

Yang, J.W., Kloepper, J.W., Ryu, C.M. 2009. Rhizosphere bacteria help plants tolerate abiotic stress. *Trends Plant Science* 14:1–4.

Yao, L., Wu, Z., Zheng, Y., Kaleem, I., Li, C. 2010. Growth promotion and protection against salt stress by *Pseudomonas putida* Rs-198 on cotton. *European Journal of Soil Biology* 46:49–54.

4 Bio-Fertilizers for Management of Soil, Crop, and Human Health

Umesh Pankaj

CONTENTS

4.1 INTRODUCTION

Today's increasing population is a big concern for all countries. To ensure food security for such a large population is a daunting task for a governing body of any country. The quality and quantity of food are going to be important challenges in the coming years, as continuous population growth requires the production of more agricultural products and increased production per unit area. To meet the ever-increasing global food demand, synthetic chemical fertilizers are used in excess quantities to improve agricultural productivity. Meanwhile, agricultural land has been continuously depleting and losing fertility due to such overuse of chemical fertilizers. Inorganic chemicals (e.g. synthetic fertilizers, pesticides, herbicides etc.) are being used by the growers for the fast growth of plants and enhanced productivity. However, several side effects of long-term use of chemical fertilizer have been reported (Figure 4.1), i.e. decrease in the soil organic carbon (SOC), thus leading to acidification or salinity, waterway eutrophication, depleted soil structure and soil productivity (Ge et al. 2008). Rapid industrial development and the increasing adaptation of agrochemical-based crop production practices since the Green Revolution have increased the persistent organic adulterations in the food chain. Agrochemicals equally cause harm to the nutritional

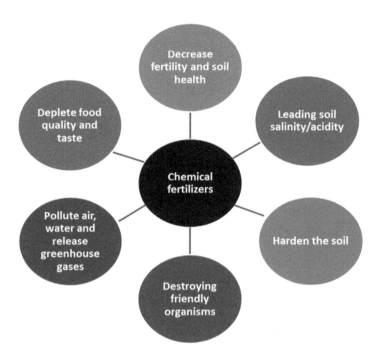

FIGURE 4.1   Harmful effects of excessive use of chemical fertilizers in the long term.

value of an agricultural crop seed product, and to the health of farmers, as well that of the users. Extreme and haphazard use of these agrochemicals has resulted in food contamination, weed and disease resistance and adverse ecological consequences, which together have a significant impact on human health. The long-term use of these chemicals promotes the magnification of toxic compounds in the ecosystem because chemicals are absorbed by most of the crops from the soil. Numerous synthetic fertilizers contain acid radicals, such as hydrochloride and sulfuric radicals, and therefore raise the soil acidity, undesirably affecting the soil and plant health. Highly recalcitrant chemical compounds can also be absorbed by some plants. The continuous consumption of such crops can lead to systematic disorders in humans. Quite a number of pesticides and herbicides have potential carcinogenicity. The increasing awareness of health challenges as a result of the consumption of poor-quality crops has led to a quest for new and improved technologies for the amelioration of both the quantity and quality of crops without jeopardizing human health. A reliable alternative to the use of chemical inputs is that of microbial inoculants that can act as bio-fertilizers, bioherbicide, biopesticides and biocontrol agents (Alori and Babalola 2018).

Fertilizer management is considered one of the main factors of sustainable agriculture; gradual replacement of chemical fertilizers with biological fertilizers is unavoidable due to their advantages and cost-effectiveness. The use of agrochemicals in soil is highly expensive and also produces a substantial amount of chemical residue. Both developing and developed countries have immediately stopped the overuse of chemical fertilizer, and proposed the plan "zero increment in chemical fertilizer until 2020" to reduce the consumption of chemical fertilizer. Therefore, the high effectiveness of fertilizer and alternate of chemical fertilizer is much sought after (Singh et al. 2019; Singh 2019; Singh and Singh 2019; Vimal and Singh 2019). So, there is a need to find a replacement or substitute for chemical fertilizer, which is cost-effective, sustainable and with a wide range of applicability. Microorganisms are able to carry out plant growth promotion, along with pest, disease and weed control. Microbial inoculants are beneficiary microorganisms, applied to either the soil or the plant in order to improve productivity and crop health. Microbial inoculants are natural-based products being widely used to control pests and improve the quality of the soil and crop, and hence human health.

An alternative to the increase of agricultural productivity in a sustainable manner, there is an increasing reliance on the manipulation of microorganisms that benefit soil and plant health. The knowledge of applied microbial inoculum is a long history which passes from generation to generation of farmers. It started with the culture of small-scale compost production that has evidently proved the ability of bio-fertilizer (Khosro and Yousef 2012). Bio-fertilizers are best defined as biologically active products or microbial inoculants viz., formulations containing one or more beneficial bacteria or fungal strains in easy to use and economical carrier materials which add, conserve and mobilize crop nutrients in the soil. In other words, bio-fertilizer is a substance that contains living microorganisms which when applied to seed, plant surfaces or soil colonizes the rhizosphere or the interior of the plant and promotes growth by increasing the availability of primary nutrients to the host plant (Mazid et al. 2011). Bio-fertilizers have emerged as a highly potent alternative to chemical fertilizers due to their being eco-friendly, easy to apply, non-toxic and having a cost-effective nature. Also, they make nutrients that are naturally abundant in soil or atmosphere, usable for plants and act as supplements to agrochemicals. Various beneficial aspects of bio-fertilizes are demonstrated in Figure 4.2. In the late 19th century, the first license for producing a biological fertilizer known as Nitragin was issued for the production of Rhizobium inoculants and after that, the inoculation of legumes started to be practiced in many countries using rhizobium fertilizers (Bagnasco et al. 1998). Liquid bio-fertilizers are special liquid formulations containing not only the desired microorganisms and their nutrients but also special cell protectants or chemicals that promote the formation of resting spores or cysts for longer shelf life and tolerance to adverse conditions.

4.2 EFFECTS OF CHEMICAL FERTILIZERS AND PESTICIDES ON SOIL AND HUMAN HEALTH

In modern agriculture, due to the heavy usage of chemical fertilizers and harmful pesticides on the crops, the sustainability of the agriculture systems collapsed, so the cost of cultivation soared at a high rate, resulting in a stagnant income for farmers stagnated. Thus, food security and safety have become a daunting challenge. The indiscriminate and imbalanced use of chemical fertilizers, especially urea, along with chemical pesticides and the unavailability of organic manures, has led to a considerable reduction in soil health. This has a heavy impact on the natural environment, as well as on human health, through the pollution of soils, waters and the whole food supply chain (Mazid and Khan 2014). Agricultural chemical inputs gain access

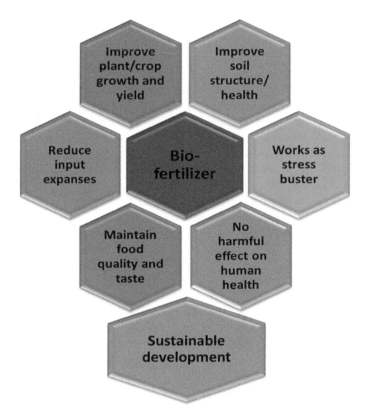

FIGURE 4.2 Multi-benefits of bio-fertilizers considered as nature's greatest gift.

into human body systems through three major means: (i) oral ingestion, (ii) infiltration through the skin and (iii) breathing (Roychowdhury et al. 2014). Pesticides have shown long-term resistance in food including vegetables, meat and fruits and in the human body (Battu et al. 2005). Quite a number of people are negatively affected by long-term exposure to agrochemicals, even at low levels (Kirkhorn and Schenker 2001). The illnesses range from respiratory disorders and musculoskeletal illnesses to dermal and cardiac-related diseases. These illnesses are encountered by farm owners, operators, family members and employees (Magauzi et al. 2011). In developing countries where less than 20% of the world agrochemical production is consumed, agrochemicals have been reported to account for 70% of acute poisoning among the working population (United States Environmental Protection Agency [USEPA] 2016). For example, in Nigeria, Ojo (2016) observed the factors that intensify health hazards from pesticide use; the use of low-cost pesticides but which proved to be the deadliest types (in terms of persistence and toxicity); poor pesticide education leading to extensive misuse; pesticide residue on locally consumed products; poor legislation and lack of enforcement of available legislation; inadequate information, awareness and knowledge of the inherent dangers of pesticides and inadequacies in medical recognition; responses to pesticide poisoning and the failure of regulatory systems (Alori and Babalola 2018).

4.3 PLANT GROWTH-PROMOTING BACTERIA (PGPR) SCREENING AND THEIR EFFECTIVENESS

For the isolation of efficient plant growth-promoting microbes, the plant rhizosphere is always a positive choice. The rhizosphere is a thin zone of soil surrounding the root zone that is immensely influenced by the root system (Hartmann et al. 2008). Compared to the neighboring bulk soil, this zone is rich in nutrients, due to the accumulation of a variety of organic compounds released by the roots through exudation, secretion and rhizodeposition. These organic compounds can be used as carbon and energy sources by microorganisms and microbial activity is particularly intense in the rhizosphere. The rhizosphere is therefore home to a variety of root-associated bacteria, commonly referred to as rhizobacteria. These rhizobacteria are screened for their potent activity involved in the improvement of nutrient availability to the plant by the fixation of atmospheric nitrogen, production of iron-chelating siderophores, organic matter mineralization (thereby meeting the nitrogen, sulfur, phosphorus nutrition of plants) and solubilization of insoluble phosphates. Another important activity involves the production of plant growth hormones and the stress-regulating hormone 1-aminocyclopropane -1-carboxylate (ACC) deaminase. However, the organism shows other indirect mechanisms in favor of plant growth and development including the inhibition of microorganisms that have a negative effect on the plant viz. hydrolysis of molecules released by pathogens, synthesis of enzymes that hydrolyze fungal cell walls, synthesis of HCN, improvement of symbiotic relationships with rhizobia and mycorrhizal fungi and insect pest control (Das et al. 2013). The rhizobacteria bacterial species is further identified using molecular tools and genome sequencing. Before going to the field, screened PGP microbes must be tested for their toxicity with neighboring micro-flora and environmental prospects. It must be shown that there is higher activity in the presence of wild microorganisms than in the soil (Khavazi et al. 2007).

4.3.1 EFFICACY OF BIO-FERTILIZERS

Several studies are available on laboratory, greenhouse and field screening and the utilization of plant growth-promoting bacteria (PGPB) for plant growth. Hence, a number of PGPR strains have been commercialized in the worldwide market. The commercially utilized plant growth-promoting and biocontrol microbes include the species of *Agro-bacterium, Azospirillum, Azotobacter, Rhizobium, Serratia, Delftia, Paenibacillus macerans, Bacillus, Burkholderia, Pseudomonas* and *Pantoea agglomerans* (Glick 2012). However, PGPR-inoculated crops represent only a small segment of worldwide agricultural practice. Another issue that needs to be considered here is that many highly efficient strains reported in literature have remained as artifacts of academic value only, and have not metamorphosed into commercial products (Bashan et al. 2014). This is due to inconsistent and varied responses obtained in field trials, which are largely influenced by the growing conditions and crop in which they were inoculated. The successful establishment of an introduced bacterial inoculant therefore depends on its survival in soil and the compatibility with the crop on which it is inoculated. This is besides its interaction with indigenous microflora;

furthermore, several other environmental factors also play an important role in determining the final outcome of the inoculation (Martínez-Viveros et al. 2010). Glick (2012) listed some important aspects to be considered for the extensive commercialization of PGPR which include: (i) determination of the traits that are most important for efficacious functioning and the subsequent selection of PGPR strains; (ii) consistency among regulatory agencies in different countries regarding release in environmental and safety issues; (iii) better understanding of the advantages and disadvantages of using rhizospheric/endophytic bacteria; (iv) selection of the strain that works well in a specific environment i.e. those that work in warm and sandy soil versus those that work well in a cold and wet environment; (v) development of a more effective means of application in different settings e.g. nursery versus field; (vi) a better understanding of the possible interaction between PGPR and soil fungi and host. An ideal PGPR should possess high rhizosphere competence, enhanced plant growth capabilities, ease of mass multiplication, a broad spectrum of action, excellent and reliable biological control activity (wherever applicable), should be safe for the environment, should be compatible with other rhizobacteria, and must tolerate desiccation, heat, oxidizing agents and UV radiations (Nakkeeran et al. 2005).

4.3.2 SHELF LIFE OF BIO-FERTILIZERS AND CARRIER MATERIAL

The inoculant formulation should have a sufficient shelf life at room temperature (Bashan 1998), due to the time required for the storage and field application. A variety of materials used as carriers have been shown to improve the survival for long periods, and the biological effectiveness of inoculants by protecting bacteria from biotic and abiotic stresses (Veen et al. 1997). A suitable carrier should be cheap, easy to use, mixable, packageable and available. Also, the carrier must permit gas exchange, particularly oxygen, have a high organic matter content and a high water-holding capacity (Bashan 1998; Ben Rebah et al. 2002). According to Somasegaran and Hoben (1994), the good carrier material must be non-toxic either to the bacterial inoculants or to the plant itself. Furthermore, Stephens and Rask (2000) and Ferreira and Castro (2005) stated that the carriers should have near neutral or readily adjustable pH, be abundant locally at a reasonable cost and be able to sterilize. These properties only indicate the potential for a good carrier, while the final selection of a carrier must be based on microbial multiplication and survival during storage, the general method of planting, equipment used for planting and the acceptable cost. The release of bio-inoculants in an entrapped formulation should not be too fast or too slow. Solid formulations include granules, micro-granules, wettable powders (talcum, biochar), wettable/water-dispersible granules (peat soil, vermiculite) and dusts used for live cell blending (Abadias et al. 2005; Guijarro et al. 2007).

Commonly, the liquid bio-formulations made using most additives are polymeric in nature, with high molecular weight known for long self-life of inoculants. These include gum arabic, glycerol and polyvinylpyrrolidone (PVP), etc. The major differences among solid and liquid bio-formulations are presented in Table 4.1. Arabic gum is a complex carbohydrate extracted from acacia and a commonly used adhesive for rhizobia. It protects the bacteria against desiccation and improves survival for a long time. PVP is a synthetic vinyl polymer that aids survival of *Bradyrhizobium*

TABLE 4.1

Differences between Solid-Based and Liquid Bio-Fertilizer

S. No.	Key Feature	Solid Carrier-Based Bio-Fertilizer	Liquid Bio-Fertilizer
1.	Health hazard	Lignite used as carrier in manufacturing which is hazardous to the production workers	Lignite is not required and no health hazard
2.	Shelf life	3–6 months	At least 1 year
3.	Viability of organisms	Estimation of cell viability is difficult	Can be easily estimated
4.	Packing & transportation of product	Difficult	Easy
5.	Application	Only seed, seedlings and soil application	Can be used in drip irrigation and foliar application
6.	Contamination	More chances	Fewer chances
7.	Compatibility with modern agricultural techniques	Traditionally used	Can be used in modern agriculture such as hydroponics
8.	Cost-effectiveness	Little bit costly due to solid carrier used and labor requirement	Solid carrier not required hence lower cost of production

japonicum. Similar to gum arabic and PVP, glycerol provides some protection from desiccation and additional protection from inhibitory seed coat exudates that are detrimental to inoculated rhizobia. The addition of glycerol to the culture medium preserved the viability of *Pseudomonas fluorescens* cells in liquid formulation for storage lasting 6 months. Glycerol as an amendment is used because it holds a considerable amount of water and protects cells from desiccation by slowing the drying rate.

4.3.3 BIO-FERTILIZER FORMULATION

After successful screening of potent microbes (bacteria, fungi and actinomycetes) having excellent plant growth-promoting attributes, biocontrol activity and better shelf life are subjected to bio-formulation. Formulation refers to the laboratory or industrial process of unifying the carrier with the bacterial strain. The success of bio-inoculant technology depends on two factors: microbial strain and inoculant formulation. Hitbold et al. (1980) and Lupwayi et al. (2000) showed that the quality of microbial inoculants depends primarily on the number of viable cells present in the inocula. Thus, the formulation step is very crucial to developing a successful bio-fertilizer. The formulation must be easy to handle, easy to apply so that it is delivered to the target in the most appropriate manner, able to protect bacteria from harmful environmental factors and also maintain or enhance the activity of the organisms

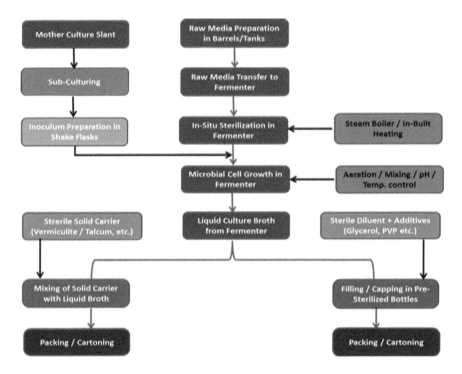

FIGURE 4.3   Steps involved in bio-formulation of solid and liquid bio-fertilizers.

in the field. For application, these bioinoculants have to be put in a carrier, either liquid or solid based, along with osmoprotectant, sticking agents, nutrients, etc.; the complete assembly thus prepared is called a bio-formulation. Bio-fertilizers are the preparations containing cells of micro-organisms which may be nitrogen fixers, phosphate solubilizers, sulfur oxidizers or organic matter decomposers; a formulation enables easy handling, long-term storage and effectiveness of bio-fertilizers (Figure 4.3).

4.3.4   CONTRIBUTION OF BIO-FERTILIZERS IN SOIL AND CROP HEALTH MAINTENANCE

Bio-fertilizers are one of the greatest gifts from nature in our agricultural science as a replacement to chemical fertilizers. The global market for bio-fertilizers is expected to exceed a market worth of USD 10.2 billion by 2018. Bio-fertilizers are usually amended with carrier material to increase effectiveness of the bioinoculants. It also increases water ration capacity (Ritika and Uptal 2014). On the basis of specific potent plant growth-promoting activity, bio-fertilizers are categorized into different types (Table 4.2). Bio-fertilizers keep the soil environment rich in all kinds of macro- and micronutrients via nitrogen fixation, phosphate and potassium solubilization or mineralization, release of plant growth-regulating substances, production of antibiotics and biodegradation of organic matter in the soil (Sinha et al. 2014). Bio-fertilizers, when applied as seed or soil inoculants, multiply and participate in

TABLE 4.2

Types of Bio-Fertilizers Being Used in Agriculture Crops for Improving the Growth and Productivity in a Sustainable Manner

S. No.	Types of Bio-Fertilizers on the Basis of their Activity	Micro-Organism Used
1.	Nitrogen fixating microorganisms • Fix atmospheric/elemental/ gaseous nitrogen (N_2) to available form (NH_4, NO_3)	**i) Rhizobium (symbiotic with legumes)** a) *R. leguminosarum* b) *R. japonicum* c) *R. lupine orinthopus* d) *R. melliloti* e) *R. phaseoli* f) *R. trifoli* **ii) Azospirillum** a) *A. lipoferum* b) *A. brasilense* c) *A. amazonense* **iii) Azotobacter (free living)** a) *A. chroococcum* b) *A. beijerinchii* c) *A. vinelandii* d) *A. paspali* e) *A. macrocytogenes* f) *A. insignis* g) *A. agilies* **iv) Acetobacter for sugarcane** a) *A. diazotrophicus* b) *A. pasturianus* **v) Blue-green algae (limit to paddy crop)** a) *Nostoc* b) *Anabaena*
2.	Phosphorous-solubilizing microorganisms • Bring insoluble phosphates into soluble forms by secreting organic acids	**i) *Pseudomonas*** a) *P. striata* b) *P. rathonis* c) *P. putida* **ii) *Micrococcus*** **iii) Bacillus** a) *B. polymyxa* b) *B. megaterium var. phosphaticum* c) *B. subtilis* d) *B. circulans* **iv) Flavobacterium sp.** **v) Penicillium sp.** **vi) Fusarium sp.** **vii) Sclerotium sp.** **viii) Aspergillus sp.**

(Continued)

TABLE 4.2 (CONTINUED)

Types of Bio-Fertilizers Being Used in Agriculture Crops for Improving the Growth and Productivity in a Sustainable Manner

S. No.	Types of Bio-Fertilizers on the Basis of their Activity	Micro-Organism Used
3.	Potash-mobilizing microorganisms • Releases K from the non-exchangeable reserve	*Frateuria aurentia*
4.	Zinc- and Sulphur-solubilizing microorganisms • Solubilize zinc and sulfur from the insoluble form by secretion of some organic acids	*Thiobacillus ferrooxidans* *Thiobacillus acidophilus*
5.	Biocontrol agents (biopesticides) • Indirectly benefited the crop by protection from pests; pesticides with synthetic chemicals have adverse toxicological effects	i) *Trichoderma* sp. ii) *Bacillus thuringiensis* iii) *Pseudomonas* sp. iv) *Streptomyces* sp. v) *Beauveria* sp. vi) *Paecilomyces* sp. vii) *Lecanicillium* sp. viii) *Metarhizium* sp. ix) *Verticillium* sp.
6.	Arbuscular Mycorrhizal Fungi Bio-fertilizers • Phosphorus acquisition and transfer to plant	a) *Funneliformis mosseae* b) *Rhizophagus intraradices* c) *Glomus aggregatum* d) *Rhizophagus fasciculatus*
7.	Plant growth hormone producer • Growth hormone like Auxin (Indol 3 AA)	a) *Enterobacter ludwigii* b) *Pseudomonas fragi* c) *Bacillus cereus*

Source: Mosa et al. 2016; https://www.iffco.in/index.php/ourproducts/index/bio-fertilizer

nutrient cycling and lead to crop productivity. Generally, 60% to 90% of the total applied fertilizer is lost and the remaining 10 to 40% is taken up by plants. Hence, bio-fertilizers can be an important component of integrated nutrient management systems for the sustaining of agricultural productivity and a healthy environment (Adesemoye and Kloepper 2009).

4.4 SUSTAINABLE CROP PRODUCTION FOR BETTER HUMAN HEALTH

Sustainable agriculture is proven to be one of the most demanding jobs in the present day. The need to increase agricultural production to meet the food requirements of the ever-increasing world population makes consistent maintenance of soil fertility essential. However, bio-fertilizers are a contribution of modern agricultural sciences which retards the nitrification for a sufficiently longer time and increases the soil fertility. Bio-fertilizers are important components of integrated nutrient management. They play a key role in the productivity and sustainability of soil, while protecting the environment, being cost-effective, eco-friendly and a renewable source of plant nutrients to supplement chemical fertilizers in a sustainable agricultural system. Unlike inorganic fertilizers, bio-fertilizers do not supply nutrients directly to plants. These are the microbial inoculants containing the living or latent cells of efficient strains used for application to seeds, soil or composting areas with the purpose of accelerating the microbial process. This procedure is used to augment the availability of nutrients that can easily be assimilated by plants, consequently colonizing the rhizosphere or the interior of the plant and promoting growth by converting nutritionally important elements into an available form. This is achieved through a biological process such as nitrogen fixation and solubilization of rock phosphate. Beneficial microorganisms in bio-fertilizers improve the plant growth, maintain the soil fertility and protect the plants from pests and diseases. Bio-fertilizers are promoted to harvest the naturally available, biological system of nutrient mobilization.

The substantial contribution of bio-fertilizer to the sustainable maintenance of human health has been reported. Bio-fertilizers improve the nutritious properties of fresh vegetables by increasing the antioxidant activity, the total phenolic compounds and chlorophyll (Khalid et al. 2017). Spinach inoculated with different bio-fertilizers was found to have 58.72 and 51.43% higher total phenolic content than the uninoculated control (Khalid et al. 2017). These secondary metabolites play preventive roles in cancer, neurodegenerative and cardiovascular disorders (Rodríguez-Morató et al. 2015). Inoculation of lettuce with *Azotobacter chroococcum* and *Glomus fasciculatum* also increased the total phenolic compounds, anthocyanins and carotenoids content of the vegetable (Baslam et al. 2011). Higher (48.02 and 40.46%) flavonoid content (antioxidant) was observed in lettuce co-inoculated with *G. fasciculatum* and *Glomus mosseae* (Baslam et al. 2011). Antioxidant biosynthesis by Arbuscular Mycorrhizal Fungi (AMF) has been reported (Carlsen et al. 2008; Nisha and Rajesh Kumar 2010; Eftekhari et al. 2012). Taie et al. (2008) documented a 75% increase of phenolic acid biosynthesis in soybean seedlings inoculated with Rhizobacteria. Karthikeyan et al. (2010) found that inoculating *Pseudomonas fluorescens* and *Bacillus megaterium* into *Catharanthus roseus* significantly increased the alkaloid content of the crop. Other reports suggest that the combination of PGPR and AMF is beneficial for growth promotion and secondary metabolite production of medicinal and aromatic plants under salt-affected soil and/or normal soil (Pankaj et al. 2019a, b, c, 2017; Trivedi et al. 2017; Verma et al. 2016; Khan et al. 2015; Singh et al. 2018, 2017a, b, c, 2019; Tiwari et al. 2018, 2019a, b; Kour et al. 2019).

4.5 CONCLUSION AND FUTURE PROSPECTS

From the conclusion of this chapter, government bodies, NGOs and scientific societies are highly encouraged to both educate and increase awareness in farmers about the ill-effects of agrochemical overuse. Bio-fertilizers have undoubtedly huge potential for future agriculture production in a sustainable manner. However, it is important to ensure that they are successfully applied in order to fulfill their role in a more sustainable agriculture. Food production by the use of bio-fertilizer is a viable alternative to the destructive health effects caused by the consumption of food produce by the use of agrochemicals such as pesticides, inorganic fertilizers, herbicides, etc. The knowledge of the mechanisms of actions employed by microbial inoculants will play a vital role in their use in sustainable agriculture. Excessive use of chemicals, as is often given as the recommended dose in agriculture, can be avoided and thus, can be removed from the human diet. It is confirmed from reviewing the research thus far that the use of bio-fertilizers will ensure healthy food safety for the future population.

REFERENCES

Abadias, M., Teixido, N., Usall, J., Solsona, C., Vinas, I. 2005. Survival of the postharvest biocontrol yeast Candida sake CPA-1 after dehydration by spray-drying. *Biocontrol Science and Technology* 15:835–846.

Adesemoye, A.O., Kloepper, J.W. 2009. Plant-microbes interactions in enhanced fertilizer use efficiency. *Applied Microbiology Biotechnology* 85(1):1–12.

Alori, E.T., Babalola, O.O. 2018. Microbial inoculants for improving crop quality and human health in Africa. *Frontiers in Microbiology* 9:2213.

Bagnasco, P., De La Fuente, L., Gaultieri, G., Noya, F., Arias, A. 1998. Fluorescent *Pseudomonas* spp. as biocontrol agents against forage legume root pathogenic fungi. *Soil Biology & Biochemistry* 30:1317–1323.

Bashan, Y. 1998. Inoculants of plant growth promoting bacteria for use in agriculture. *Biotechnology Advances* 16(2):729–770.

Bashan, Y., de-Bashan, L.E., Prabhu, S.R., Hernandez, J.P. 2014. Advances in plant growth-promoting bacterial inoculants technology: Formulations and practical perspectives 1998–2013. *Plant and Soil* 378:1–33.

Baslam, M., Garmendia, I., Goicoechea, N. 2011. Arbuscular mycorrhizal fungi (AMF) improved growth and nutritional quality of greenhouse-grown lettuce. *Agricultural and Biological Chemistry* 59:5504–5515.

Battu, R.S., Singh, B., Kang, B.K., Joia, B.S. 2005. Risk assessment through dietary intake of total diet contaminated with pesticide residues in Punjab, India. *Ecotoxicology and Environmental Safety* 62:132–139.

Ben Rebah, F.B., Tyagi, R.D., Prevost, D. 2002. Wastewater sludge as a substrate for growth and carrier for rhizobia: The effect of storage conditions on survival of *Sinorhizobium meliloti*. *Bioresource Technology* 83:145–151.

Carlsen, S., Understrup, A., Fomsgaard, I., Mortensen, A., Ravnskov, S. 2008. Flavonoids in roots of white clover: Interaction of arbuscular mycorrhizal fungi and a pathogenic fungus. *Plant and Soil* 302:33–43.

Das, A.J., Kumar, M., Kumar, R. 2013. Plant growth promoting PGPR: An alternative of chemical fertilizer for sustainable environment friendly agriculture. *Research Journal of Agriculture and Forestry Sciences* 1:21–23.

Eftekhari, M., Alizadeh, M., Ebrahimi, P. 2012. Evaluation of the total phenolics and quercetin content of foliage in mycorrhizal grape (*Vitis vinifera* L.) varieties and effect of postharvest drying on quercetin yield. *Industrial Crop and Product* 38:160–165.

Ferreira, E.M., Castro, I.V. 2005. Residues of the cork industry as carriers for the production of legumes inoculants. *Silva Lusitana* 13(2):159–167.

Ge, Y., Zhang, J.B., Zhang, L.M., Yang, M., He, J.Z. 2008. Long-term fertilization regimes affect bacterial community structure and diversity of an agricultural soil in northern China. *Journal of Soils Sediments* 8:43–50.

Glick, B.R. 2012. Plant growth promoting bacteria: Mechanisms and applications. *Scientifica* 2012:1–15.

Guijarro, B., Melgarejo, P., De Cal, A. 2007. Effect of stabilizers on the shelf-life of *Penicillium frequentans* conidia and their efficacy as a biological agent against peach brown rot. *Journal of Food Microbiology* 113:117–124.

Hartmann, A., Rothballer, M., Schmid, M. 2008. Lorenz Hiltner, a pioneer in rhizosphere microbial ecology and soil bacteriology research. *Plant and Soil* 312:7–14.

Hitbold, A.E., Thurlow, N., Skipper, H.D. 1980. Evaluation of commercial soybean inoculants by various techniques. *Agronomy Journal* 72:675–681.

Karthikeyan, B., Joe, M.M., Jaleel, C.A., Deiveekasundaram, M. 2010. Effect of root inoculation with plant growth promoting rhizobacteria (PGPR) on plant growth, alkaloid content and nutrient control of *Catharanthus roseus* (L.) G. Don. *Natura Croatica* 1:205–212.

Khalid, M., Hassani, D., Bilal, M., Asad, F., Huang, D. 2017. Influence of bio-fertilizer containing beneficial fungi and rhizospheric bacteria on health promoting compounds and antioxidant activity of *Spinacia oleracea* L. *Botanical Studies* 58:35.

Khan, K., Pankaj, U., Verma, S.K., Gupta, A.K., Singh, R.P., Verma, R.K. 2015. Bio-inoculants and vermicompost influence on yield, quality of *Andrographis paniculata*, and soil properties. *Industrial Crops and Products* 70:404–409.

Khavazi, K., Rejali, F., Seguin, P., Miransari, M. 2007. Effects of carrier sterilisation method and incubation on survival of *Bradyrhizobium japonicum* in soybean (*Glycine max* L.) inoculants. *Enzyme and Microbial Technology* 41:780–784.

Khosro, M., Yousef, S. 2012. Bacterial bio-fertilizers for sustainable crop production: A review. *APRN Journal of Agricultural and Biological Science* 7(5):237–308.

Kirkhorn, S., Schenker, M.B. 2001. *Human Health Effects of Agriculture: Physical Diseases and Illnesses*. N.A.S. Database, AHS-NET. Available at: http://nasdonline.org/1827/d001772/human-health-effects-of-agriculture-physical-diseases-and.html.

Kour, D., Rana, K.L., Yadav, N., Yadav, A.N., Rastegari, A.A., Singh, C., Negi, P., Singh, K., Saxena, A.K. 2019. Technologies for biofuel production: Current development, challenges, and future prospects. In: A. A. Rastegari et al. (Eds.), *Prospects of Renewable Bioprocessing in Future Energy Systems, Biofuel and Biorefinery Technologies*, Vol. 10. Berlin: Springer, pp. 1–50.

Lupwayi, N.Z., Olsen, P.E., Sonde, E.S. et al. 2000. Inoculant quallity and its evaluation. *Field and Crops Research* 65:259–270.

Magauzi, R., Mabaera, B., Rusakaniko, S. et al. 2011. Health effects of agrochemicals among farm workers in commercial farms of Kwekwe district, Zimbabwe. *Pan African Medical Journal* 9:26.

Martínez-Viveros, O., Jorquera, M.A., Crowley, D.E., Gajardo, G., Mora, M.L. 2010. Mechanisms and practical considerations involved in plant growth promotion by rhizobacteria. *Journal of Soil Science and Plant Nutrition* 10:293–319.

Mazid, M., Khan, T.A. 2014. Future of bio-fertilizers in Indian agriculture: An overview. *International Journal of Agricultural and Food Research* 3(3):10–23.

Mazid, M., Khan, T.A., Mohammad, F. 2011. Potential of NO and H_2O_2 as signalling molecules in tolerance to abiotic stress in plants. *Journal of Industrial Research & Technology* 1(1):56–68.

Mosa, W.F.A.E.-G., Pasztp, L.S., Frac, M., Trzcinski, P. 2016. Microbial products and biofertilizers in improving growth and productivity of apple – A review. *Polish Journal of Microbiology* 65(3):243–251.

Nakkeeran, S., Dilantha Fernado, W.G., Siddiqui, Z.A. 2005. Plant growth promoting rhizobacteria formulations and its scope in commercialization for the management of pests and diseases. In: Z.A. Siddiqui (Ed.), *PGPR: Biocontrol & Biofertilization*. Dordrecht, The Netherlands: Springer, pp. 257–296.

Nisha, M.C., Rajesh Kumar, S. 2010. Influence of arbuscular mycorrhizal fungi on biochemical changes in *Wedilla chinensis* (Osbeck) Merril. *Ancient Science of Life* 29:26.

Ojo, J. 2016. Pesticides use and health in Nigeria. *IFE Journal of Science* 8:981–991.

Pankaj, U., Singh, D.N., Singh, G., Verma, R.K. 2019a. Microbial inoculants assisted growth of *Chrysopogon zizanioides* promotes phytoremediation of salt affected soil. *Indian Journal of Microbiology* 59(2):137–146.

Pankaj, U., Singh, G., Verma, R.K. 2019b. Microbial approaches in management and restoration of marginal lands. In: Singh, J.S. (Eds.), *New and Future Developments in Microbial Biotechnology and Bioengineering: Microbes in Soil, Crop and Environmental Sustainability*. USA: Elsevier, pp. 295–305.

Pankaj, U., Verma, R.S., Yadav, A., Verma, R.K. 2019c. Effect of arbuscular mycorrhizae species on essential oil yield and chemical composition of commercially grown palmarosa (*Cymbopogon martinii*) varieties in salinity stress soil. *Journal of Essential Oil Research* 31(2):145–153.

Pankaj, U., Verma, S.K., Semwal, S., Verma, R.K. 2017. Assessment of natural mycorrhizal colonization and soil fertility status of lemongrass [(*Cymbopogon flexuosus*, Nees ex Steud) W. Watson] crop in subtropical India. *Journal of Applied Research on Medicinal and Aromatic Plants* 5:41–46.

Ritika, B., Uptal, D. 2014. Bio-fertilizer a way towards organic agriculture: A Review. *Academic Journals* 8(24):2332–2342.

Rodríguez-Morató, J., Xicota, L., Fitó, M., Farré, M., Dierssen, M., de la Torre, R. 2015. Potential role of olive oil phenolic compounds in the prevention of neuro degenerative diseases. *Molecules* 20:4655–4680.

Roychowdhury, D., Paul, M., Banerjee, S.K. 2014. Review on the effects of biofertilizers and biopesticides on rice and tea cultivation and productivity. *International Journal of Scientific Engineering and Technology* 2:96–108.

Singh, C., Tiwari, S., Boudh, S., Singh, J.S. 2017a. Biochar application in management of paddy crop production and methane mitigation. In: Singh, J.S., Seneviratne, G. (Eds.), *Agro-Environmental Sustainability: Managing Environmental Pollution*, 2nd ed. Switzerland: Springer, pp. 123–146.

Singh, C., Tiwari, S., Gupta, V.K., Singh, J.S. 2018. The effect of rice husk biochar on soil nutrient status, microbial biomass and paddy productivity of nutrient poor agriculture soils. *Catena* 171:485–493.

Singh, C., Tiwari, S., Singh, J.S. 2017b. Impact of rice husk biochar on nitrogen mineralization and methanotrophs community dynamics in paddy soil. *International Journal of Pure and Applied Bioscience* 5:428–435.

Singh, C., Tiwari, S., Singh, J.S. 2017c. Application of biochar in soil fertility and environmental management: A review. *Bulletin of Environment, Pharmacology and Life Sciences* 6:07–14.

Singh, C., Tiwari, S., Singh, J.S. 2019. Biochar: A sustainable tool in soil 2 pollutant bioremediation. In: Bharagava, R.N., Saxena, G. (Eds.), *Bioremediation of Industrial Waste for Environmental Safety*. Berlin: Springer, pp. 475–494.

Singh, J.S. 2019. *New and Future Developments in Microbial Biotechnology and Bioengineering: Microbes in Soil, Crop and Environmental Sustainability*. San Diego, CA: Elsevier.

Singh, J.S., Kumar, A., Singh, M. 2019. Cyanobacteria: A sustainable and commercial bioresource in production of bio-fertilizer and bio-fuel from waste waters. *Environmental and Sustainability Indicators* 3–4:100008.

Singh, J.S., Singh, D.P. 2019. *New and Future Developments in Microbial Biotechnology and Bioengineering: Microbial Biotechnology in Agro-Environmental Sustainability*. San Diego, CA: Elsevier.

Sinha, R.K., Valani, D., Chauhan, K. et al. 2014. Embarking on a second green revolution for sustainable agriculture by vermiculture biotechnology using earthworms. *International Journal of Agricultural Health Safety* 1:50–64.

Somasegaran, P., Hoben, H.J. 1994. *Handbook for Rhizobia: Methods in Legume Rhizobium Technology*. Berlin Heidelberg, New York, NY, USA: Springer, pp. 217–218.

Stephens, J.H., Rask, H.M. 2000. Inoculant production and formulation. *Field Crops Research* 65:249–258.

Taie, H.A., El-Mergawi, R., Radwan, S. 2008. Isofavonoids, favonoids, phenolic acids profiles and antioxidant activity of soybean seeds as affected by organic and bioorganic fertilization. *American–Eurasian Journal of Agricultural & Environmental Sciences* 4:207–213.

Tiwari, S., Singh, C., Boudh, S., Rai, P.K., Gupta, V.K., Singh, J.S. 2019a. Land use change: A key ecological disturbance declines soil microbial biomass in dry tropical uplands. *Journal of Environmental Management* 242:1–10.

Tiwari, S., Singh, C., Singh, J.S. 2018. Land use changes: A key ecological driver regulating methanotrophs abundance in upland soils. *Energy, Ecology, and the Environment* 3:355–371.

Tiwari, S., Singh, C., Singh, J.S. 2019b. Wetlands: A major natural source responsible for methane emission. In: Upadhyay, A.K. et al. (Eds.), *Restoration of Wetland Ecosystem: A Trajectory Towards a Sustainable Environment*. Berlin: Springer, pp. 59–74.

Trivedi, P., Singh, K., Pankaj, U., Verma, S.K., Verma, R.K., Patra, D.D. 2017. Effect of organic amendments and microbial application on sodic soil properties and growth of an aromatic crop. *Ecological Engineering* 102:127–136.

United States Environmental Protection Agency (USEPA). 2016. *Melathion Human Health Risk Assessment*. Washington, DC: USEPA, pp. 258.

van Veen, J.A., van Overbeek, L.S., van Elsas, J.D. 1997. Fate and activity of microorganisms introduced into soil. *Microbiology and Molecular Biology Reviews* 61:121–135.

Verma, S.K., Pankaj, U., Khan, K., Singh, R., Verma, R.K. 2016. Bio-inoculants and vermicompost improve *Ocimum basilicum* yield and soil health in a sustainable production system. *Clean – Soil, Air, Water* 44:1–8.

Vimal, S.R., Singh, J.S. 2019. Salt tolerant pgpr and fym application in saline soil paddy agriculture sustainability. *Climate Change and Environment Sustainability* 7:23–33.

5 Crop Residue Burning and Its Effects on the Environment and Microbial Communities

Indra Jeet Chaudhary and Sunil Soni

CONTENTS

5.1 INTRODUCTION

Air pollution is one of the key environmental issues affecting life on earth. Transportation, vehicular traffic, and continuous industrial activity have resulted in a further increase in the concentration of gaseous and particulate matter pollutants (Chen et al. 2017; Chaudhary and Rathore 2018a, b, 2019). Gases and particles are released by all types of combustion into the air, including sulfur and nitrogen oxides, CO and soot particles, as well toxic metals, organic molecules, and radioactive isotope (Agbaire and Esiefarienrhe 2009). In 2007, nearly 52% of cities were at critical PM_{10} levels (≥ 1.5 times limit). Higher PM_{10} levels were found in northern Indian cities, with continuous increases in Mumbai, Faridabad, Lucknow, Bangalore, and Delhi (Smith et al. 2004). There are numerous sources for the occurrence of air pollution, such as in industry > traffic > commercial areas > and residential areas (Chaudhary and Rathore 2018a, b, 2019).

Biomass burning is a global phenomenon that releases large quantities of gaseous and particulate pollutants into the atmosphere (Streets et al. 2003; Li et al. 2007; Yamaji et al. 2009; Shi and Yamaguchi 2014; Zhang et al. 2015a; Ding et al. 2016b). The tremendous annual amounts of combustion products from biomass burning in the atmosphere pose a threat to air quality. Literature in the field indicates that biomass burning has adverse environmental effects, both locally and at large distances downwind. It is necessary to prohibit the open burning of crop residues in order to protect public health and the environment (Zhang and Cao 2015b; Gustafsson et al. 2009).

5.2 CROP RESIDUE BURNING

China is among the major agricultural nations in the world. Agricultural crop production generates tremendous amounts of agricultural crop residues such as rice, wheat, and corn straw, etc., which account for 17.3% of the global crop residue production and rank the first in the world (Bi et al. 2010). During the summer/autumn harvest season, a large amount of agricultural straw is removed by burning in a short period in order to prepare the next crop planting. Open burning is the most convenient and less expensive way to eliminate agricultural straw. In China, studies on gaseous and particulate pollutant emissions from open burning of agricultural straw have been presented in previous publications (Li et al. 2007; Zhang and Smith 2007; Zhang et al. 2008a, 2011a; Huang et al. 2012; Tian et al. 2015). Efforts have been also made to characterize particle number emission factors and size distributions from agricultural straw burning in laboratory simulation experiments (Hays et al. 2005; Zhang et al. 2008a, 2011a). Particle size distribution from agricultural straw open burning is mainly dominated by an accumulation mode, with a count median diameter of 0.10–0.15 μm (Zhang et al. 2011a). In addition, trace gas emission inventories (CO_2, CO, NO_x, and BC, etc.) from agricultural straw open burning have been estimated in China (Zhang et al. 2008a; Huang et al. 2016; Sun et al. 2016; Singh et al. 2018, 2017a, b, c, 2019; Tiwari et al. 2018, 2019a, b, c; Kour et al. 2019).

In addition to emission characteristics, understanding of the impact of agricultural straw open burning on urban and regional air quality is essential. In China, especially during and shortly after the harvest seasons, open burning of agricultural straw has a significant impact on urban and regional air quality. In extreme cases, agricultural straw open burning may trigger the explosive growth of secondary $PM_{2.5}$ and accelerate heavy haze formation in the urban and regional atmosphere (Nie et al. 2015; Xie et al. 2015; Singh et al. 2016). However, detailed information on the effects of smoke from agricultural straw open burning on urban and regional air quality is still rare (Zhang et al. 2010a; Li et al. 2010a). It is well known that the impact of agricultural straw open burning on heavy haze formation during and shortly after the harvest seasons is complex, and not only contributes to primary $PM_{2.5}$ emissions but also includes a potential contribution to secondary $PM_{2.5}$ formation.

For instance, during the process of smoke plume transport, organic compounds (such as VOCs) in the presence of NOx can be oxidized to generate secondary organic aerosol (SOA) (Wang et al. 2009a, b; Li et al. 2014a). Similarly, atmospheric gases, such as SO_2 and NOx, can also be oxidized to form a secondary inorganic aerosol

(SIA, such as sulfate and nitrate) (Cheng et al. 2013; Zha 2013; Tao et al. 2013; Cheng et al. 2014a, b; Chen and Xie 2014; Zhang et al. 2016a, b). Heterogeneous reactions in BB plume also play important roles in the formation of HONO (Nie et al. 2015). A high concentration of NO_2 together with a high concentration of NH_3 in the BB plume has been found to enhance sulfate formation through aqueous phase reactions and to produce HONO as a by-product (Nie et al. 2015). Both SOA and SIA are the most important components of secondary $PM_{2.5}$. Therefore, when smoke plume is transported to the urban atmosphere, secondary $PM_{2.5}$ can rapidly increase in a short time under stagnant weather conditions, and can further aggravate haze pollution and/or result in the increase of the frequency of heavy haze pollution through the interactions between physical and chemical processes (Ding et al. 2013, 2016a; Huang et al. 2016). That is why heavy haze pollution often occurs in North, Central, and eastern China, especially during and shortly after the harvest season.

5.3 POLLUTANTS FROM AGRICULTURAL RESIDUE BURNING

Particulate matters refer to a mixture of solid particles and liquid droplets in the air (Hinds 2012), with varying physical and chemical properties. Particulate matter is a principal contributor to air pollution in China (Fang et al. 2009). In particular, Beijing-Tianjin-Hebei (BTH) Province, the Yangtze River Delta (YRD), and the Pearl River Delta (PRD) are subject to severe PM pollution. Daily average $PM_{2.5}$ concentrations during severe haze periods in 2013 were 159, 91, 69, and 345 $\mu g/m^3$ for Beijing, Shanghai, Guangzhou, and Xi'an, respectively (Huang et al. 2014). All $PM_{2.5}$ values are significantly higher than the mean concentration of 25 $\mu g/m^3$, as recommended by the World Health Organization (WHO). In order to address the serious PM pollution issue, the Chinese government announced a long-term plan to reduce the $PM_{2.5}$ concentration by 25%, 20%, 15%, and 10% compared to 2012 levels by 2017, in BTH, YRD, PRD, and other cities, respectively (http://www.gov.cn /zwgk/2013-09/12/content_2486773.htm). For instance, the annual value for $PM_{2.5}$ in Beijing is targeted as 60 $\mu g/m^3$, which would be still high compared to the values recommended by the WHO (an annual mean of 10 $\mu g/m^3$), but a big step in air pollution regulation in China. It will certainly be a challenge to achieve this level because of the heavy pollution prevailing in recent years; for example, it was observed in Beijing in 2014 that the $PM_{2.5}$ mean annual value was 86 $\mu g/m^3$ (Zhang et al. 2016a) (Table 5.1)

From the above references it may be safely concluded that crop residue/biomass residue burning not only emits poisonous gases such as SO_2, CH_4, CO_2, CO, N_2O, NOx, NO, NO_2, OC, BC, TC, NMHCs, SVOCs, VOCs, O_3, etc., but also influences the quality of the environment at large.

5.4 IMPACT ON THE ENVIRONMENT

5.4.1 AIR QUALITY

Biomass burning is a global phenomenon that releases large quantities of gaseous and particulate pollutants into the atmosphere, including CO_2, CO, VOCs, PM_{10},

TABLE 5.1
Pollutants from Rice Straw and other Agricultural Residue Burning

Name of Pollutant	Pollutants from Rice Straw Burning				Pollutants from Other Agricultural Residues Burning			
	EF (g/kgdm)	India (Gg)	Thailand (Gg)	Philippines (Gg)	EF (g/kgdm)	India (Gg)	Thailand (Gg)	Philippines (Gg)
CO_2	1,460	16,253	11,850	11,850	1,515	127,260	11,666	10,757
CH_4	1.20	13	10	10	2.70	227	21	19
N_2O	0.07	1	1	1	0.07	5.88	0.54	0.50
CO	34.70	386	290	282	92.00	7,728	708	653
NMHC	4.00	45	33	32	7.00	588	54	50
NOx	3.10	35	26	25	3.38	322	29	27
SO_2	2.00	22	17	16	0.40	34	3	3
Total particulate matter (TPM)	13.00	145	109	106	13.00	1092	100	92
Fine particulate matter ($PM_{2.5}$)	12.95	144	108	105	3.90	328	30	28

Gg = Giga gram, g/kgdm = gram per kg of dry matters (Tripathi et al. 2013)

$PM_{2.5}$, BC, OC, EC, and other compounds (Streets et al. 2003; Li et al. 2007; Yamaji et al. 2009; Shi and Yamaguchi 2014; Zhang et al. 2015a; Ding et al. 2016b). In China, it has been estimated that the total annual emissions due to crop residue burning were 120 Tg CO_2, 4.6 Tg CO, 0.88 Tg $PM_{2.5}$, 0.39 Tg OC, and 0.02 Tg EC in the year 2008 (Ni et al. 2015). The tremendous annual amounts of combustion products from BB into the atmosphere pose a threat to China's air quality. Literature indicates that BB has adverse environmental effects both locally and at large distances downwind. It is necessary to prohibit the open burning of crop residues in order to protect public health and the environment (Zhang and Cao 2015a, b; Gustafsson et al. 2009).

The annual and seasonal contributions of the biomass-burning source to ambient $PM_{2.5}$ have been reported for Beijing, Dongying, and Chengdu (Zhang et al. 2013; Tao et al. 2014; Yao et al. 2016a), three cities located in North, Central East, and Southwest China, respectively. Combining the PMF receptor model and BB markers (levoglucosan and K^+), it was possible to show that BB had higher contributions during intense farming (spring and autumn) and cold (winter) seasons, than in the hot season (summer), consistent with the timings of open burning of crop residues for the elimination of agricultural waste and domestic combustion of biomass fuels for heating and cooking. On an annual average basis, BB contributed 12%, 15.8%, and 11% of $PM_{2.5}$ mass in Beijing, Dongying, and Chengdu, respectively.

5.4.2 HUMAN HEALTH

As a renewable energy source, biomass materials include wood, animal waste, crops, and seaweed. Since the earliest stages of civilization, people have been using biomass energy for survival. Biomass-based energy was the most popular energy before the Industrial Revolution (Fernandes et al. 2007). The use of biomass energy is increasingly popular in the current century (Field et al. 2008). According to annual global energy consumption Gtoe (2010), the total biomass energy usage all over the world accounts for approximately 8–14% (fossil fuels 10.45, oil 4.03, coal 3.56, natural gas 2.86, renewable 0.94, hydro 0.78, commercial biomass 0.16, estimated biomass 1–2) (Berndes et al. 2003; Parikka 2004; Hoogwijk et al. 2005; Agarwal 2007; Bioenergy 2007; Demirbas 2007; Wit and Faaij 2009; Williams et al. 2012). Ethanol and biodiesel fuel, wood and agricultural products, gas and biogas, and solid waste are the most popular types of biomass in the current century. Thermal, chemical, electrochemical, and biochemical conversion processes are commonly used to convert biomass into other forms of energy. During the conversion process, biomass emits carbon dioxide, carbon monoxide, voltage organic compounds, nitrogen oxides, and other pollutants including particles (Andreae and Merlet 2001; Baxter 2005; Zhang et al. 2007). The open burning of biomass emits PAHs, which is a group of more than 100 different chemicals (Jenkins et al. 1996). PAHs are usually formed during the incomplete burning of the biomass and contribute to the production of particulate matter. The common source of PAHs is cigarette smoke, asphalt road, coal exhaust, and vehicle engine exhaust, wood burning for cooking, agricultural burning, hazardous waste management, and bushfire (Mumtaz et al. 1996; Finlayson-Pitts and Pitts 1997). The air contains a significant amount of suspended particles, liquids as well as solids, organic and inorganic substances, viruses, and bacteria (Seaton et al.

1995; Seinfeld et al. 2008; Tena and Clarà 2012). PAHs can be in the form of volatile, semi-volatile and particulate phase (Allen et al. 1996; Finlayson-Pitts and Pitts
1997; Thrane and Mikalsen 1981). PAHs are mainly attached to dust particles in air
and thus particle-bound PAHs are treated as a threat to human health (WHO 2010).
PAHs have a significant impact on health, especially on the respiratory system and
cancer incidence (Leiter et al. 1942; Kim et al. 2013). During inhalation, the particle
attached-PAHs can enter the human respiratory system.

5.4.3 Soil Health and Climate

The biomass burning smoke particles serve as a major source of BC and Br C, which
have adverse health and climate effects. Indoor smoke particles emitted for cooking
and heating are of particular significance in China. Therefore, there is an urgent
need to compare the biological effect of indoor fresh and aged smoke particles that
carry various carcinogenic substances, as ambient fine biomass burning particles are
found to contain an amount of PAHs, in addition to their nitrated, hydroxylated, and
oxygenated derivatives (Lin et al. 2015a, b). Given that a significant fraction of the
Earth's biomass burning originated in China, studies probing the climate effects of
BC and Br C should be given high priority by the international research community.

5.5   EFFECT ON MICROBIAL COMMUNITIES

Microorganisms, and especially those organisms which live in soil, are present in
greater diversity (Maron et al. 2011) and density in the uppermost layers of the
soil. Therefore there is a significant effect on the biogeochemical cycling of the
nutrients by degrading the organic material in the soil, including the crop residues,
maintaining the soil structures and thereby decreasing the crop residue. This causes
an increase in the nutrient contents of the soil which further results in increased
fertility of the soil (Guo et al. 2014; Hobara et al. 2014). The open burning of crop
residue may pollute the air (Yin et al. 2019), and affects the diversity of the microorganism present in the soil by reducing their numbers. Furthermore, crop residue
elimination results in a reduction of soil organic materials (SOM), which ultimately
decreases the fertility of soil (Hamza and Anderson 2005). In the past few decades,
researchers have recognized the necessity of biodiversity of the soil, and the function and long-lasting sustainability of natural and managed terrestrial ecosystems.
Soil microorganisms play a vital role in mediating changes in the soil TOC via
mineralization–immobilization of soil organic matter (Breulmann et al. 2014). In
the process of burning, a huge quantity of elements like carbon, nitrogen, and sulfur
undergo vaporization, and are hence lost from the plant residues, due to volatilization (Raison 1979).

   Conventional agricultural techniques include tillage and crop residue removal
from the field which increases soil erosion (Ehigiator and Anyata 2011), and thereby
induces biodiversity loss from the soil (Rey Benayas and Bullock 2012) and thus
contributes to climate change (Bajželj et al. 2014). Conventional agricultural practices, such as tillage (Raiesi and Beheshti 2015) residue management and extensive
inorganic fertilizer applications (Chaudhry et al. 2012) change the physicochemical

characteristics of soil, such as pH (Tripathi et al. 2012), electrical conductivity (EC) (Rezapour et al. 2013), and organic material concentration in soil (Sul et al. 2013; Li et al. 2014). Tillage and crop residue management changes the structure of the microbial community (Carbonetto et al. 2014). Navarro-Noya et al. (2013) in his study found that the relative abundance of Actinobacteria and Betaproteobacteria increased and Gammaproteobacteria decreased by tillage, while tillage-crop residue management (incorporated or retained on the soil surface) had a significant positive effect on the relative abundance of Bacteroidetes and Betaproteo bacteria, and a significant negative effect on Cyanobacteria and Gemmatimonadetes. Crop residue return also significantly affects soil microbial community composition (Gouaerts et al. 2007; Zhao et al. 2016). The effects of crop residue management techniques such as burning, bailing and removing, and incorporating residues in soil were investigated by many researchers and the chemical, biochemical, and microbiological properties of the soil have been analyzed (Bending et al. 2002; Pankhurst et al. 2002; Spedding et al. 2004; Graham and Haynes 2005). Other research has shown that residue and these management practices can affect the size, composition, and the activity of the soil microbial biomass (Gupta et al. 1994).

In most of the grain cultivating regions of the world, the post-harvest residues are considered a waste product requiring proper disposal before the production of the next crop. In such conditions, burning has historically been preferred over other methods of residue disposal by most farmers. However, in recent years, the burning of residue in crop fields has been reduced under scrutiny due to concern over air pollution, global warming caused due to higher carbon emissions, accelerated losses in organic matter of soil, and reductions in microbial-facilitated activities in the soil. (Biederbeck et al. 1980; Rasmussen et al. 1980). These changes in microbial activities, and the composition of soil MCs can in turn influence soil fertility and plant growth by increasing nutrient availability and turnover, disease incidence, or disease suppression (Pankhurst et al. 2005). Loss of soil organic matter and soil microbial activity under sugarcane subjected to pre-harvest burning is believed to be a major factor contributing to soil degradation in the South African sugar industry (Meyer et al. 1996). Hence, many workers have found an appreciable decrease in soil organic matter under long-term sugarcane production (Blair et al. 1998; Blair 2000), and this decrease is considered to be the most serious aspect of soil degradation that occurs under a sugarcane monoculture (Wood 1985).

Residue burning, however, causes considerable loss of organic C, N, and other nutrients by volatilization (Malhi and Kutcher 2007), which may detrimentally affect soil microorganisms (Wuest et al. 2005; Razafimbelo et al. 2006). In comparison to burning, residue retention increases soil carbon and nitrogen stocks (Govaerts et al. 2006; Ke and Scheu 2008), provides organic matter necessary for soil macro-aggregate formation (Six et al. 2000), and fosters cellulose-decomposing fungi (Franchini et al. 2002; Wardle et al. 2006) and thereby carbon cycling. It is thus believed that several million species of microorganisms are living in soils and most of them cannot be characterized or identified by the use of conventional culture techniques. Due to this, plate counts methods are not considered reliable for the measuring of soil microbial diversity and their activity (Roper and Opher-Keller 1997). This problem has led to new approaches which are based on the quantitative analysis of communities rather than the species; this community description can be examined by extracting and identifying the phospholipid fatty acids (Zelles 1999).

Field burning of agricultural crop residue is a traditional practice employed in South and East Asia for many years to clear and fertilize the lands, which emits a significant number of organic compounds in the atmosphere on a global scale (Andreae and Merlet 2001). Several studies have been carried out to ascertain its impact on organic aerosols in terms of various organic compounds (e.g. diacids, sugars, and lipids) (Wang et al. 2009a, b). However, studies focusing on the role of this anthropogenic source (i.e., biomass burning) in terms of soil microbes and higher plant waxes have been limited in literature (Yang et al. 2015). The airborne microorganisms (e.g., bacteria, viruses, and fungal spores) and their long-range atmospheric transport have received considerable attention owing to their potential impact on human health, animals, and plants. In addition, these bioaerosols can act as cloud condensation and ice nuclei (Bauer et al. 2003; Bowers et al. 2009; Després et al. 2012; Huffman et al. 2013). However, knowledge of microbial aerosols is particularly deficient due to the absence of reliable measurement techniques and the limitations in culturing and quantification of microorganisms. Alternatively, the determination of certain chemical markers (e.g., hydroxy fatty acids) has been proven as an efficient means to assess the contribution from airborne gram-negative bacteria (GNB) and higher plant metabolites (Hines et al. 2003; Lee et al. 2004, 2007).

5.6 MANAGEMENT OF AGRICULTURAL RESIDUE

Burning is a simple and economical option for the management of crop/biomass residues. Due to a lack of awareness or the non-availability of suitable technologies, it is a common practice everywhere. Burning of crop residues not only degrades the atmospheric quality but also affects the climate, and ultimately human health. Crop residue and biomass burning (forest fires) are considered to be a major source of carbon dioxide (CO_2), carbon monoxide (CO), methane (CH_4), volatile organic compounds (VOC), nitrogen oxides, and halogen compounds.

The production of "parali" (a term used in India to designate crop residue), has a direct link with the cultivation of rice. India is the world's second largest producer of white and brown rice, accounting for 20% of all world rice production. Rice is India's predominant crop, and is the staple food of the people of the eastern and southern parts of the country (https://en.wikipedia.org/wiki/Rice_production_in_India#cite _note-LOC-1). In the year 2017, the area of cultivation under rice in the country was estimated approximately at 43.2 million hectares (https://www.statista.com/stat istics/765691/india-area-of-cultivation-for-rice/). Cultivation of paddy is carried out in 21 states of India, of which there are 10 top rice-producing states. These states represent 83.5 percent of total rice production in India. Punjab and Haryana both together produce 15.3 percent of the total rice production. However, cases of parali burning are broadly reported from these two states. In other states, a traditional method of harvesting of rice is adopted and under this system, rice is harvested manually. Paddy straw is cut from the root and small bundles of the paddy straw are made. The bundles are removed from the field and stored outside the field in one place. But it is a hard task to undertake for the next crop cultivation, and also is very time-consuming. For better management of agricultural straw, various alternatives are used in a sustainable way.

5.6.1 Alternative Uses of Agricultural Residue

A large amount of agricultural residue is used as a by-product in the country. It is estimated that over 500 million tons of parali is produced annually in India. Rice straw is rich in polysaccharides and has a high lignin and silica content, limiting voluntary intake, and reducing degradability by microorganisms. Presently, various elements of agricultural residue have been used for the removal of toxic metals from the environment, due to the higher adsorption capacity of metals. (Singh et al. 2019). (Table 5.2).

5.7 SUGGESTION FOR MANAGEMENT OF CROP RESIDUE

Researchers have suggested that agricultural biomass burning has a negative impact on both soil and air environment, and also poses a human health risk. The utilization of crop residue in a sustainable way is the best practice for agricultural crop productivity and environmental health. Crop residue burning is not a suitable option for crop residue management. There are no appropriate technologies available for managing crop residue. If crop residue is collected for its utilization, there are different technologies available both at a national and international level for the valuable utilization of biomass, such as thermo-chemical process and biochemical.

One of the most important techniques is power generation from biomass, which has huge potential to provide electricity for rural energy with sustainable environmental benefits. Therefore, the main constraint is the collection of leftover crop residue. To collect this leftover crop residue, technologies are available, but these are not economical in Indian conditions. Crops such as cotton leaves and woody residue (stalks) need a different type of arrangement for collection which is not available (Pathak 2004). Crop residue management options include practices such as burning, incorporation, surface retention and mulching, baling and removing the straw for use as industrial/domestic fuel, and fodder, etc. The incorporation of biomass shows better results for soil fertility and a crop production point of view.

A solution to such a problem lies in the utilization of crop residue through productive alternatives. Some of the alternates recommended by a small number of researchers relate to the conversion of the crop residues into products such as fodder, biogas, biofuel, composts, and the generation of electricity. However, little progress has been made in this direction so far. With such a lack of development, there has hardly been any reduction in biomass burning. Amongst the suggested alternative uses of agricultural biomass, the most effective way is to use it for the generation of electricity. There are two alternatives to producing electricity from agricultural biomass. It can be either used in the exiting Thermal Power Plants, or through the establishment of small Gasifier Power Plants at a village level.

Biomass burning is a major cause of particulate matter and smoke particles that cause environmental stress and affect both plant and human health. Once pollution enters the environment, then only plants can remove it. Various technologies are used for the removal of environmental stress. In which selection of pollution

TABLE 5.2
Alternate Uses of Agricultural Residue

S. No.	Alternate Uses of Agricultural Residue	Picture
1.	**Fodder for Animals:** One of the prominent uses of paddy straw can be fodder for animals. This has highly nutritional values such as polysaccharides, lignin, and silica content.	
2.	**Bedding Material for Cattle:** Paddy straw can be used as bedding material and to build house for cattle during winter. This helps improving the quality and quantity of milk as it contributes to animals' comfort and health.	
3.	**Mushroom Cultivation:** Paddy Straw can be used as making ground for mushroom cultivation.	
4.	**Briquettes:** The alternate use of straw is for making briquettes. The parali can be converted into bio-fuel through compressing and used as fuel briquettes. The briquettes can be used in industrial boilers. The plant may also process grass; wood saw dust; peanut shells; coconut fiber.	

(Continued)

TABLE 5.2 (CONTINUED)
Alternate Uses of Agricultural Residue

S. No.	Alternate Uses of Agricultural Residue	Picture
5.	**Paper and Board Production:** Paddy straw is also used for making paper and board. It is estimated that around 0.1 million tons of paddy straw are being used by the paper industry for papermaking. There is not only a need to continue this but also to increase its utilization to 0.2 million tons by 2017.	 (https://ecoboard.en.made-in-china.com/)
6.	**Organic Manure:** Organic manuring is the best method for the management of agricultural wastes and also improving soil, plants and environmental health.	
7.	**Electricity Generation:** One of the effective procreative alternates to control burning of parali to generate electricity by utilizing paddy crop residue. Paddy stubble is bio-mass which is a renewable source of energy.	
8.	**Ploughing Field with a Rotavator:** Ploughing field by rotavator is the best example of mixing waste residue in soil and after then decomposed in soil as an organic fertilizer.	

tolerance cultivars and the application of various protectants is one of the most important technologies that enhanced plant growth and productivity (Rathore and Chaudhary 2019; Chaudhary and Singh 2020). Various agricultural waste is also used for the removal of heavy metals and preparation of various kinds of fertilizers (Chaudhary and Singh 2018; Singh et al. 2019; Shah et al. 2019a, b; Chaudhary et al. 2020). In this context, agricultural biomass is the best technique for the removal of heavy metals and the utilization of composting. It also fulfills the need to improve the utilization of agricultural waste in a sustainable way.

5.8 CONCLUSION

This chapter focuses on literature findings on biomass burning, and its impacts on soil, air quality, and microbial health. The main conclusion of this chapter is that biomass burning causes a negative impact on the atmosphere and soil microbe health. Therefore, the management of biomass burning is a global issue. The above technique described the management of biomass burning. Agricultural crises are major issues in many countries. For this reason, this technique is helpful for sustainable agriculture production, in addition to improving soil, water, and air quality and microbial activity.

REFERENCES

Agarwal, A.K. 2007. Biofuels (alcohols and biodiesel) applications as fuels for internal combustion engines. *Progress in Energy and Combustion Science* 33:233–271.
Agbaire, P.O., Esiefarienrhe, E. 2009. Air pollution tolerance indices (APTI) of some plants around Otorogun Gas Plant in Delta State, Nigeria. *Journal of Applied Science and Environment Management* 13:11–14.
Allen, J.O., Dookeran, N.M., Smith, K.A., Sarofim, A.F., Taghizadeh, K., Lafleur, A.L. 1996. Measurement of polycyclic aromatic hydrocarbons associated with size-segregated atmospheric aerosols in Massachusetts. *Environmental Science and Technology* 30:1023–1031.
Andreae, M.O., Merlet, P. 2001. Emission of trace gases and aerosols from biomass burning. *Global Biogeochemical Cycles* 15:955–966.
Bajželj, B., Richards, K.S., Allwood, J.M., Smith, P., Dennis, J.S., Curmi, E., Gilligan, C.A. 2014. Importance of food-demand management for climate mitigation. *Nature Climate Change* 4:924–929.
Bauer, H., Giebl, H., Hitzenberger, R., Kasper-Giebl, A., Reischl, G., Zibuschka, F., Puxbaum, H. 2003. Airborne bacteria as cloud condensation nuclei. *Journal of Geophysical Research* 108:4658. doi:10.1029/2003JD003545.
Baxter, L. 2005. Biomass-coal co-combustion: Opportunity for affordable renewable energy. *Fuel* 84:1295–1302.
Bending, G.D., Turner, M.K., Jones, J.E. 2002. Interactions between crop residue and soil organic matter quality and the functional diversity of soil microbial communities. *Soil Biology and Biochemistry* 34:1073–1082.
Berndes, G., Hoogwijk, M., Broek, R.V.D. 2003. The contribution of biomass in the future global energy supply: A review of 17 studies. *Biomass Bioenergy* 25:1–28.
Bi, Y.Y., Wang, Y.J., Gao, C.Y. 2010. Straw resource quantity and its regional distribution in China. *Journal of Agricultural Mechanization Research* 3:1–7.

Biederbeck, V.O., Campbell, C.A., Bowren, K.E., Schnitzer, M., McIver, R.N. 1980. Effect of burning cereal straw on soil properties and grain yields in Saskatchewan. *Soil Science Society of America Journal* 44:103–111.

Bioenergy, I. 2007. *Potential Contribution of Bioenergy to the World's Future Energy Demand*. IEA Bioenergy: Exco: 2007:02, Rotorua, New Zealand.

Blair, G.J., Chapman, L., Whitbread, A.M., Ball-Coelho, B., Larsen, P., Tiessen, H. 1998. Soil carbon changes resulting from sugarcane trash management at two locations in Queensland, Australia, and in north-east Brazil. *Australian Journal of Soil Research* 36:8773–8881.

Blair, N. 2000. Impact of cultivation and sugarcane green trash management on carbon fractions and aggregate stability for a Chromic Luvisol in Queensland, Australia. *Soil and Tillage Research* 55:183–191.

Bowers, R.M., Lauber, C.L., Wiedinmyer, C., Hamady, M., Hallar, A.G., Fall, R., Knight, R., Fierer, N. 2009. Characterization of airborne microbial communities at a high-elevation site and their potential to act as atmospheric ice nuclei. *Applied and Environmental Microbiology* 75:5121–5130.

Breulmann, M., Masyutenko, N.P., Kogut, B.M., Schroll, R., Dorfler, U., Buscot, F., Schulz, E. 2014. Short-term bioavailability of carbon in soil organic matter fractions of different particle sizes and densities in grassland ecosystems. *Science of the Total Environment* 497–498:29–37.

Carbonetto, B., Rascovan, N., Alvarez, R., Mentaberry, A., Vazquez, M.P. 2014. Structure, composition and metagenomic profile of soil microbiomes associated to agricultural land use and tillage systems in Argentine pampas. *PLoS One* 9:99949.

Chaudhary, I.J., Neeraj, A., Siddiqui, M.A., Singh, V. 2020. Nutrient management technologies and the role of organic matrix-based slow-release biofertilizers for agricultural sustainability: A review. *Agricultural Reviews* 41:1–13.

Chaudhary, I.J., Rathore, D. 2018a. Suspended particulate matter deposition and its impact on urban trees. *Atmospheric Pollution Research* 9:1072–1082.

Chaudhary, I.J., Rathore, D. 2018b. Phytomonitoring of dust load and its effect on foliar micro morphological characteristics of urban trees. *Journal of Plant Science* 2:170–179.

Chaudhary, I.J., Rathore, D. 2019. Dust pollution: Its removal and effect on foliage physiology of urban trees. *Sustainable Cities and Society* (51):101696. doi:10.1016/j.scs.2019.101696.

Chaudhary, I.J., Singh, R.P. 2018. Studies on growth, mobilization of nutrients and yield of wheat (*Triticum aestivum* L. PBW −343) applied with organic matrix based slow release bio fertilizers. *International Journal of Current Microbiology and Applied Sciences* 3(7):3221–3238. ISSN: 2319-7706.

Chaudhary, I.J., Singh, V. 2020. Titanium dioxide nanoparticles and its impact on growth, biomass and yield of agricultural crops under environmental stress: A review. *Research Journal of Nanoscience and Nanotechnology* 10:1–8.

Chaudhry, V., Rehman, A., Mishra, A., Chauhan, P., Nautiyal, C. 2012. Changes in bacterial community structure of agricultural land due to long-term organic and chemical amendments. *Microbial Ecology* 64:450–460.

Chen, J., Li, C., Ristovski, Z., Milic, A., Gu, Y., Islam, M.S., Wang, S., Hao, J., Zhang, H., He, C., Guo, H., Fu, H., Miljevic, B., Morawska, L., Thai, P., Lam, Y.F., Gavin Pereira, G., Ding, A., Huang, X., Dumka, U.C. 2017. A review of biomass burning: Emissions and impacts on air quality, health and climate in China. *Science of the Total Environment* 579:1000–1034.

Chen, Y., Xie, S.D. 2014. Characteristics and formation mechanism of a heavy air pollution episode caused by biomass burning in Chengdu, Southwest China. *Science of the Total Environment* 473–474:507–517.

Cheng, Y., Engling, G., He, K., Duan, F., Du, Z., Ma, Y., Liang, L., et al. 2014b. The characteristics of Beijing aerosol during two distinct episodes: Impacts of biomass burning and fireworks. *Environmental Pollution* 185:149–157.

Cheng, Y., Engling, G., He, K., Duan, F., Ma, Y., Du, Z., Liu, J., et al. 2013. Biomass burning contribution to Beijing aerosol. *Atmospheric Chemistry and Physics* 13:7765–7781.

Cheng, Z., Wang, S., Fu, X., Watson, J.G., Jiang, J., Fu, Q., Chen, C., et al. 2014a. Impact of biomass burning on haze pollution in the Yangtze River delta, China: A case study in summer 2011. *Atmospheric Chemistry and Physics* 14:4573–4585.

Demirbas, A. 2007. Progress and recent trends in biofuels. *Progress in Energy and Combustion Science* 33:1–18.

Després, V.R., Huffman, J.A., Burrows, S.M., Hoose, C., Safatov, A.S., Buryak, G., Fröhlich-Nowoisky, J., Elbert, W., Andreae, M.O., Pöschl, U., Jaenicke, R. 2012. Primary biological aerosol particles in the atmosphere: A review. *Tellus B* 64. doi:10.3402/tellusb.v64i0.15598.

Ding, A.J., Fu, C.B., Yang, X.Q., Sun, J.N., Petäjä, T., Kerminen, V., Wang, T., et al. 2013. Intense atmospheric pollution modifies weather: A case of mixed biomass burning with fossil fuel combustion pollution in eastern China. *Atmospheric Chemistry and Physics* 13:10545–10554.

Ding, A.J., Huang, X., Nie, W., Sun, J.N., Kerminen, V.-M., Petaja, T., Su, H., Cheng, Y.F., Yang, X.Q., Wang, M.H., Chi, X.G., Wang, J.P., Virkkula, A., Guo, W.D., Yuan, J., Wang, S.Y., Zhang, R.J., Wu, Y.F., Song, Y., Zhu, T., Zilitin Kevich, S., Kulmala, M., Fu, C.B. 2016a. Enhanced haze pollution by black carbon in megacities in China. *Geophysical Research Letters* 43:2873–2879. doi:10.1002/2016GL067745.

Ding, X., He, Q.F., Shen, R.Q., Yu, Q.Q., Zhang, Y.Q., Xin, J.Y., Wen, T.X., et al. 2016b. Spatial and seasonal variations of isoprene secondary organic aerosol in China: Significant impact of biomass burning during winter. *Scientific Reports (UK)*. doi:10.1038/srep20411.

Ehigiator, O.A., Anyata, B.U. 2011. Effects of land clearing techniques and tillage systems on runoff and soil erosion in a tropical rain forest in Nigeria. *Journal of Environmental Management* 92:2875–2880.

Fang, M., Chan, C.K., Yao, X. 2009. Managing air quality in a rapidly developing nation: China. *Atmospheric Environment* 43:79–86.

Fernandes, S.D., Trautmann, N.M., Streets, D.G., Roden, C.A., Bond, T.C. 2007. Global biofuel use. *Global Biogeochemical Cycles* 2:241–253.

Field, C.B., Campbell, J.E., Lobell, D.B. 2008. Biomass energy: The scale of the potential resource. *Trends in Ecology and Evolution* 2:65–72.

Finlayson-Pitts, B.J., Pitts, J.N. 1997. Tropospheric air pollution: Ozone, airborne toxics, polycyclic aromatic hydrocarbons, and particles. *Science* 276:1045–1052.

Franchini, J.C., Gonzalez-Vila, F.J., Rodriguez, J. 2002. Decomposition of plant residues used in no-tillage systems as revealed by flash pyrolysis. *Journal of Analytical and Applied Pyrolysis* 62:35–43.

Gouaerts, B., Mezzalama, M., Unno, Y., Sayre, K.D., Marco, L.G., Vanherck, K., Dendooven, L., Deckers, J. 2007. Influence of tillage, residue management, and crop rotation on soil microbial biomass and catabolic diversity. *Applied Soil Ecology* 37:18–30.

Govaerts, B., Mezzalama, M., Sayre, K.D., Crossa, J., Nicol, J.M., Deckers, J. 2006. Longterm consequences of tillage, residue management, and crop rotation on maize/wheat root rot and nematode populations in subtropical highlands. *Applied Soil Ecology* 32:305–315.

Graham, M.H., Haynes, R.J. 2005. Organic matter accumulation and fertilizer-induced acidification interact to affect soil microbial and enzyme activity on a long-term sugarcane management experiment. *Biology and Fertility of Soils* 41:249–256.

Guo, Y.H., Gong, H.L., Guo, X.Y. 2014. Rhizosphere bacterial community of *Typha angustifolia* L. and water quality in a river wetland supplied with reclaimed water. *Applied Microbiology and Biotechnology* 99:2883–2893.

Gupta, V.V.S.R., Roper, M.M., Kirkegaard, J.A., Angus, J.F. 1994. Changes in microbial biomass and organic matter levels during the first year of modified tillage and stubble management practices on a red earth. *Soil Research* 32:1339–1354.

Gustafsson, Ö., Kruså, M., Zencak, Z., Sheesley, R.J., Granat, L., Engström, E., Praveen, P.S., et al. 2009. Brown clouds over South Asia: Biomass or fossil fuel combustion. *Science* 323:495–498.

Hamza, M.A., Anderson, W.K. 2005. Soil compaction in cropping systems—A review of the nature: Causes and possible solutions. *Soil & Tillage Research* 82:121–145.

Hays, M.D., Fine, P.M., Geron, C.D., Kleeman, M.J., Gullett, B.K. 2005. Open burning of agricultural biomass: Physical and chemical properties of particle-phase emissions. *Atmospheric Environment* 39:6747–6764.

Hinds, W.C. 2012. *Aerosol Technology: Properties, Behavior, and Measurement of Airborne Particles*, 2nd Edition. John Wiley & Sons, Hoboken, NJ.

Hines, C.J., Waters, M.A., Larsson, L., Petersen, M.R., Saraf, A., Milton, D.K. 2003. Characterization of endotoxin and 3-hydroxy fatty acid levels in air and settled dust from commercial aircraft cabins. *Indoor Air* 13:166–173.

Hobara, S., Osono, T., Hirose, D., Noro, K., Hirota, M., Benner, R. 2014. The roles of microorganisms in litter decomposition and soil formation. *Biogeochemistry* 118:471–486.

Hoogwijk, M., Faaij, A., Eickhout, B., Vries, B.D., Turkenburg, W. 2005. Potential of biomass energy out to 2100, for four IPCC SRES land-use scenarios. *Biomass Bioenergy* 29:225–257.

Huang, K., Zhuang, G., Lin, Y., Fu, J.S., Wang, Q., Liu, T., Zhang, R., et al. 2012. Typical types and formation mechanisms of haze in an Eastern Asia megacity, Shanghai. *Atmospheric Chemistry and Physics* 12:105–124.

Huang, R., Zhang, Y., Bozzetti, C., Ho, K., Cao, J., Han, Y., Daellenbach, K.R., et al. 2014. High secondary aerosol contribution to particulate pollution during haze events in China. Nature 7521:218–222.

Huang, X., Ding, A., Liu, L., Liu, Q., Ding, K., Niu, X., Nie, W., Xu, Z., Chi, X., Wang, M., Sun, J., Guo, W., Fu, C. 2016. Effects of aerosol–radiation interaction on precipitation during biomass-burning season in East China. *Atmospheric Chemistry and Physics* 16:10063–10082.

Huffman, J.A., Prenni, A.J., DeMott, P.J., Pöhlker, C., Mason, R.H., Robinson, N.H., Fröhlich Nowoisky, J., Tobo, Y., Després, V.R., Garcia, E., Gochis, D.J., Harris, E., Müller Germann, I., Ruzene, C., Schmer, B., Sinha, B., Day, D.A., Andreae, M.O., Jimenez, J.L., Gallagher, M., Kreidenweis, S.M., Bertram, A.K., Pöschl, U. 2013. High concentrations of biological aerosol particles and ice nuclei during and after rain. *Atmospheric Chemistry and Physics* 13:6151–6164.

Jenkins, B.M., Jones, A.D., Turn, S.Q., Williams, R.B. 1996. Particle concentrations, gas particle partitioning, and species inter correlations for polycyclic aromatic hydrocarbons (PAH) emitted during biomass burning. *Atmospheric Environment* 30:3825–3835.

Ke, X., Scheu, S. 2008. Earthworms, Collembola and residue management change wheat (*Triticum aestivum*) and herbivore pest performance (Aphidina: Rhophalosiphum padi). *Oecologia* 157:603–617.

Kim, K., Jahan, S.A., Kabir, E., Brown, R.J. 2013. A review of air borne polycyclic aromatic hydrocarbons (PAHs) and their human health effects. *Environment International* 60:71–80.

Kour, D., Rana, K.L., Yadav, N., Yadav, A.N., Rastegari, A.A., Singh, C., Negi, P., Singh, K., Saxena, A.K. 2019. Technologies for biofuel production: Current development, challenges, and future prospects. In: Rastegari, A.A. et al. (Eds.), *Prospects of Renewable Bioprocessing in Future Energy Systems, Biofuel and Biorefinery Technologies*, Vol. 10. Springer, Berlin, pp. 1–50.

Lee, A.K.Y., Lau, A.P.S., Cheng, J.Y.W., Fang, M., Chan, C.K. 2007. Source identification analysis for the airborne bacteria and fungi using a biomarker approach. *Atmospheric Environment* 41:2831–2843.

Lee, K.Y.A., Chan, C.K., Fang, M., Lau, P.S.A. 2004. The 3-hydroxy fatty acids as biomarkers for quantification and characterisation of endotoxins and gram-negative bacteria in atmospheric aerosols in Hong Kong. *Atmospheric Environment* 38:6307–6317.

Leiter, J., Shimkin, M., Shear, M. 1942. Production of subcutaneous sarcomas in mice with tars extracted from atmospheric dusts. *Journal of the National Cancer Institute* 3:155–165.

Li, C., Yan, K., Tang, L., Jia, Z., Li, Y. 2014. Change in deep soil microbial communities due to long-term fertilization. *Soil Biology and Biochemistry* 75:264–272.

Li, H., Han, Z., Cheng, T., Du, H., Kong, L., Chen, J., Zhang, R., et al. 2010a. Agricultural fire impacts on the air quality of shanghai during summer harvest time. *Aerosol and Air Quality Research* 10:95–101.

Li, J., Song, Y., Mao, Y., Mao, Z., Wu, Y., Li, M., Huang, X., et al. 2014a. Chemical characteristics and source apportionment of $PM_{2.5}$ during the harvest season in eastern China's agricultural regions. *Atmospheric Environment* 92:442–448.

Li, X., Wang, S., Duan, L., Hao, J., Li, C., Chen, Y., Yang, L. 2007. Particulate and trace gas emissions from open burning of wheat straw and corn stover in China. *Environmental Science and Technology* 41:6052–6058.

Lin, Y., Ma, Y., Qiu, X., Li, R., Fang, Y., Wang, J., Zhu, Y., et al. 2015a. Sources, transformation, and health implications of PAHs and their nitrated, hydroxylated, and oxygenated derivatives in $PM_{2.5}$ in Beijing. *Journal of Geophysical Research: Atmospheres* 120:7219–7228.

Lin, Y., Qiu, X., Ma, Y., Ma, J., Zheng, M., Shao, M. 2015b. Concentrations and spatial distribution of polycyclic aromatic hydrocarbons (PAHs) and nitrated PAHs (NPAHs) in the atmosphere of North China, and the transformation from PAHs to NPAHs. *Environmental Pollution* 196:164–170.

Malhi, S.S., Kutcher, H.R. 2007. Small grains stubble burning and tillage effects on soil organic C and N, and aggregation in northeastern Saskatchewan. *Soil and Tillage Research* 94:353–361.

Maron, P.A., Mougel, C., Ranjard, L. 2011. Soil microbial diversity: Methodological strategy, spatial overview and functional interest. *Comptes Rendus Biologies* 334:403–411.

Meyer, J.H., van Antwerpen, R., Meyer, E. 1996. A review of soil degradation and management research under intensive sugarcane cropping. *Proceedings of the South African Sugar Technologists Association* 70:1–7.

Mumtaz, M.M., George, J.D., Gold, K.W., Cibulas, W., De Rosa, C.T. 1996. ATSDR evaluation of health effects of chemicals. IV. Polycyclic aromatic hydrocarbons (PAHs): Understanding a complex problem. *Toxicology and Industrial Health* 12:742–971.

Navarro-Noya, Y.E., Gómez-Acata, S., Montoya-Ciriaco, N., Rojas-Valdez, A., Suárez Arriaga, M.C., Valenzuela-Encinas, C., Jiménez-Bueno, N., Verhulst, N., Govaerts, B., Dendooven, L. 2013. Relative impacts of tillage: Residue management and crop-rotation on soil bacterial communities in a semi-arid agro ecosystem. *Soil Biology and Biochemistry* 65:86–95.

Ni, H., Han, Y., Cao, J., Chen, L.W.A., Tian, J., Wang, X., Chow, J.C., et al. 2015. Emission characteristics of carbonaceous particles and trace gases from open burning of crop residues in China. *Atmospheric Environment* 123:399–406.

Nie, W., Ding, A.J., Xie, Y.N., Xu, Z., Mao, H., Kerminen, V., Zheng, L.F., et al. 2015. Influence of biomass burning plumes on HONO chemistry in eastern China. *Atmospheric Chemistry and Physics* 15:1147–1159.

Pankhurst, C.E., Blair, B.L., Magarey, R.C., Stirling, G.R., Bell, M.J., Garside, A.L. 2005. Effect of rotation breaks and organic matter amendments on the capacity of soils to develop biological suppression towards soil organisms associated with yield decline of sugarcane. *Applied Soil Ecology* 28:271–282.

Pankhurst, C.E., Kirkby, C.A., Hawke, B.G., Harch, B.D. 2002. Impact of a change in tillage and crop residue management practice on soil chemical and microbiological properties in a cereal-producing red duplex soil in NSW, Australia. *Biology and Fertility of Soils* 35:189–196.

Parikka, M. 2004. Global biomass fuel resources. *Biomass Bioenergy* 27:613–620.

Pathak, P.S. 2004. The relevance of biomass management. *Proceedings of the National Seminar on Biomass Management for Energy Purposes-Issues and Strategies*, Anand, 11–12 December, 2004, pp. 1–9.

Raiesi, F., Beheshti, A. 2015. Microbiological indicators of soil quality and degradation following conversion of native forests to continuous croplands. *Ecological Indicators* 50:173–185.

Raison, R.J. 1979. Modification of the soil environment by vegetation fires, with particular reference to nitrogen transformation: A review. *Plant Soil* 51:73–108.

Rasmussen, P.E., Allmaras, R.R., Rohde, C.R., Roager Jr, N.C. 1980. Crop residue influences on soil carbon and nitrogen in a wheat-fallow system. *Soil Science Society of America Journal* 44:596–600.

Rathore, D., Chaudhary, I.J. 2019. Ozone risk assessment of castor (*Ricinus communis* L.) cultivars using open top chamber and ethylenediurea (EDU). *Environmental Pollution* 244:257–269.

Razafimbelo, T., Barthès, B., Larré-Larrouy, M.C., Luca, E.F.D., Laurent, J., Cerri, C.C., Feller, C. 2006. Effect of sugarcane residue management (mulching versus burning) on organic matter in a clayey Oxisol from southern Brazil. *Agriculture, Ecosystems and Environment* 115:285–289.

Rey Benayas, J.M., Bullock, J. 2012. Restoration of biodiversity and ecosystem services on agricultural land. *Ecosystem* 15:883–899.

Rezapour, S., Taghipour, A., Samadi, A. 2013. Modifications in selected soil attributes as influenced by long-term continuous cropping in a calcareous semiarid environment. *Natural Hazards* 69:1951–1966.

Roper, M.M., Opher-Keller, K.M. 1997. Soil microflora as indicators of soil health. In: Pankhurst, C.E., Doube, B.M., Gupta, V.V.S.R. (Eds.), *Biological Indicators of Soil Health*. CAB International, Wallingford, pp. 157–177.

Seaton, A., Godden, D., Macnee, W., Donaldson, K. 1995. Particulate air pollution and acute health effects. *The Lancet* 345:176–178.

Seinfeld, J.H., Pandis, S.N., Seinfeld, J.H., Pandis, S.N. 2008. *Atmospheric Chemistry and Physics: From Air Pollution to Climate Change* 51:212–214.

Shah, K.N., Chaudhary, I.J., Rana, D.K., Singh, V. 2019a. Growth, yield and quality of knol-khol (*Brassica oleracea* var. gongylodes) as affected by fertilizer management. *Fundamental and Applied Agriculture* 4:1–11.

Shah, K.N., Chaudhary, I.J., Rana, D.K., Singh, V. 2019b. Impact assessment of different organic manures on growth, morphology and yield of Onion (*Allium cepa* L.) cultivar. *Asian Journal of Agricultural Research* 13:1819–1894.

Shi, Y., Yamaguchi, Y. 2014. A high-resolution and multi-year emissions inventory for biomass burning in Southeast Asia during 2001–2010. *Atmospheric Environment* 98:8–16.

Singh, C., Chowdhary, P., Singh, J.S., Chandra, R. 2016. Pulp and paper mill wastewater and coliform as health hazards: A review *Microbiology Research International* 4:28–39.

Singh, C., Tiwari, S., Boudh, S., Singh, J.S. 2017a. Biochar application in management of paddy crop production and methane mitigation. In: Singh, J.S., Seneviratne, G. (Eds.), *Agro-Environmental Sustainability: Managing Environmental Pollution*, 2nd Edition. Springer, Switzerland, pp. 123–146.

Singh, C., Tiwari, S., Gupta, V.K., Singh, J.S. 2018. The effect of rice husk biochar on soil nutrient status, microbial biomass and paddy productivity of nutrient poor agriculture soils. *Catena* 171:485–493.

Singh, C., Tiwari, S., Singh, J.S. 2017b. Impact of rice husk biochar on nitrogen mineralization and methanotrophs community dynamics in paddy soil. *International Journal of Pure and Applied Bioscience* 5:428–435.

Singh, C., Tiwari, S., Singh, J.S. 2017c. Application of biochar in soil fertility and environmental management: A review. *Bulletin of Environment, Pharmacology and Life Sciences* 6:07–14.

Singh, C., Tiwari, S., Singh, J.S. 2019. Biochar: A sustainable tool in soil 2 pollutant bioremediation. In: Bharagava, R.N., Saxena, G. (Eds.), *Bioremediation of Industrial Waste for Environmental Safety*. Springer, Berlin, pp. 475–494.

Singh, S., Chaudhary, I.J., Pankaj Kumar, P. 2019. Utilization of low-cost agricultural waste for removal of toxic metals from environment: A review. *International Journal of Scientific Research in Biological Sciences* 6(4):56–61.

Six, J., Elliott, E.T., Paustian, K. 2000. Soil macroaggregate turnover and microaggregate formation: A mechanism for C sequestration under no-tillage agriculture. *Soil Biology and Biochemistry* 32:2099–2103.

Smith, K.R., Mehta, S., Maeusezahl-Feuz, M. 2004. Indoor air pollution from household use of solid fuels. In: *Comparative Quantification of Health Risks: Global and Regional Burden of Disease Attributable to Selected Major Risk Factors*. World Health Organization, Geneva, Switzerland, pp. 1435–1493.

Spedding, T.A., Hamel, C., Mehuys, G.R., Madramootoo, C.A. 2004. Soil microbial dynamics in maize-growing soil under different tillage and residue management systems. *Soil Biology and Biochemistry* 36:499–512.

Streets, D.G., Yarber, K.F., Woo, J.H., Carmichael, G.R. 2003. Biomass burning in Asia: Annual and seasonal estimates and atmospheric emissions. *Global Biogeochemical Cycles* 17:1759–1768.

Sul, W.J., Asuming-Brempong, S., Wang, Q., Tourlousse, D.M., Penton, C.R., Deng, Y., Rodrigues, J.L.M., Adiku, S.G.K., Jones, J.W., Zhou, J., Cole, J.R., Tiedje, J.M. 2013. Tropical agricultural land management influences on soil microbial communities through its effect on soil organic carbon. *Soil Biology and Biochemistry* 65:33–38.

Sun, J., Peng, H., Chen, J., Wang, X., Wei, M., Li, W., Yang, L., et al. 2016. An estimation of CO2 emission via agricultural crop residue open field burning in China from 1996 to 2013. *Journal of Cleaner Production* 112:2625–2631.

Tao, J., Gao, J., Zhang, L., Zhang, R., Che, H., Zhang, Z., Lin, Z., et al. 2014. PM2.5 pollution in a megacity of southwest China: Source apportionment and implication. *Atmospheric Chemistry & Physics* 14:8679–8699.

Tao, J., Zhang, L., Engling, G., Zhang, R., Yang, Y., Cao, J., Zhu, C., et al. 2013. Chemical composition of PM2.5 in an urban environment in Chengdu, China: Importance of springtime dust storms and biomass burning. *Atmospheric Research* 122:270–283.

Tena, A.F., Clarà, P.C. 2012. Deposition of inhaled particles in the lungs. *Archivos de Bronconeumología* 48:240–246.

Thrane, K.E., Mikalsen, A. 1981. High-volume sampling of airborne polycyclic aromatic hydrocarbons using glass fibre filters and polyurethane foam. *Atmospheric Environment* 15:909–918.

Tian, J., Chow, J.C., Cao, J., Han, Y., Ni, H., Chen, L.A., Wang, X., et al. 2015. A biomass combustion chamber: Design, evaluation, and a case study of wheat straw combustion emission tests. *Aerosol Air Quality Research* 15:2104–2114.

Tiwari, A.K., Chaudhary, I.J., Pandey, A.K. 2019a. Indian traditional trees and their scientific relevance. *Journal of Medicinal Plants* 7:29–32.

Tiwari, S., Singh, C., Boudh, S., Rai, P.K., Gupta, V.K., Singh, J.S. 2019b. Land use change: A key ecological disturbance declines soil microbial biomass in dry tropical uplands. *Journal of Environmental Management* 242:1–10.

Tiwari, S., Singh, C., Singh, J.S. 2018. Land use changes: A key ecological driver regulating methanotrophs abundance in upland soils. *Energy, Ecology, and the Environment* 3:355–371.

Tiwari, S., Singh, C., Singh, J.S. 2019c. Wetlands: A major natural source responsible for methane emission. In: Upadhyay, A.K. et al. (Eds.), *Restoration of Wetland Ecosystem: A Trajectory Towards a Sustainable Environment*. Springer, Berlin, pp. 59–74.

Tripathi, B.M., Kim, M., Singh, D., Lee-Cruz, L., Lai-Hoe, A., Ainuddin, A.N., Go, R., Rahim, R.A., Husni, M.H.A., Chun, J., Adams, J.M. 2012. Tropical soil bacterial communities in Malaysia: pH dominates in the Equatorial Tropics too. *Microbial Ecology* 64:474–484.

Wang, G., Kawamura, K., Xie, M., Hu, S., Cao, J., An, Z., Waston, J.G., et al. 2009a. Organic molecular compositions and size distributions of Chinese summer and autumn aerosols from Nanjing: Characteristic haze event caused by wheat straw burning. *Environmental Science and Technology* 43:6493–6499.

Wang, Q., Shao, M., Zhang, Y., Wei, Y., Hu, M., Guo, S. 2009b. Source apportionment of fine organic aerosols in Beijing. *Atmospheric Chemistry & Physics Discussions* 9:8573–8585.

Wardle, D.A., Yeates, G.W., Barker, G.M., Bonner, K.I. 2006. The influence of plant litter diversity on decomposer abundance and diversity. *Soil Biology and Biochemistry* 38:1052–1062.

Williams, A., Jones, J.M., Ma, L., Pourkashanian, M. 2012. Pollutants from the combustion of solid biomass fuels. Energy fuels. *Progress in Energy and Combustion Science* 38:113–137.

Wit, M.D., Faaij, A. 2009. European biomass resource potential and costs. *Biomass and Bioenergy* 34:188–202.

Wood, A.W. 1985. Soil degradation and management under intensive sugarcane cultivation in north Queensland. *Soil Use and Management* 1:120–123.

World Health Organization. 2010. *WHO Guidelines for Indoor Air Quality: Selected Pollutants*. World Health Organization, Geneva, Switzerland.

Wuest, S.B., Caesar-TonThat, T.C., Wright, S.F., Williams, J.D. 2005. Organic matter addition, N, and residue burning effects on infiltration, biological, and physical properties of an intensively tilled silt-loam soil. *Soil and Tillage Research* 84:154–167.

Xie, Y., Ding, A., Nie, W., Mao, H., Qi, X., Huang, X., Xu, Z., et al. 2015. Enhanced sulfate formation by nitrogen dioxide: Implications from in situ observations at the SORPES station. *Journal of Geophysical Research: Atmospheres* 24:12679–12694.

Yamaji, K., Li, J., Uno, I., Kanaya, Y., Irie, H., Takigawa, M., Komazaki, Y., et al. 2009. Impact of open crop residual burning on air quality over Central Eastern China during the Mount Tai Experiment 2006 (MTX2006). *Atmospheric Chemistry & Physics* 696:69–80.

Yang, H., Yang, B., Dai, Y., Xu, M., Koide, R.T., Wang, X., Liu, J., Bian, X. 2015. Soil nitrogen retention is increased by ditch-buried straw return in a rice–wheat rotation system. *European Journal of Agronomy* 69:52–58.

Yao, H., Song, Y., Liu, M., Xu, T., Li, J., Wu, Y., Hu, M., et al. 2016a. Direct radiative effect of carbonaceous aerosols from crop residue burning during the summer harvest season in East China. *Atmospheric Chemistry & Physics* 69:52–58.

Yin, S., Wang, X., Zhang, X., Zhang, Z., Xiao, Y., Tani, H., Sun, Z. 2019. Exploring the effects of crop residue burning on local haze pollution in Northeast China using ground and satellite data. *Atmospheric Environment* 199:189–201.

Zelles, L. 1999. Fatty acid pattern of phospholipids and lipopolysaccharides in the characterisation of microbial communities in soil: A review. *Biology and Fertility of Soils* 29:111–129.

Zha, S. 2013. Agricultural fires and their potential impacts on regional air quality over China. *Aerosol Air Quality Research* 3:992–1001.

Zhang, G., Li, J., Li, X., Xu, Y., Guo, L., Tang, J., Lee, C., et al. 2010a. Impact of anthropogenic emissions and open biomass burning on regional carbonaceous aerosols in South China. *Environmental Pollution* 158:3392–3400.

Zhang, H., Hu, D., Chen, J., Ye, X., Wang, S.X., Hao, J.M., Wang, L., et al. 2011a. Particle size distribution and polycyclic aromatic hydrocarbons emissions from agricultural crop residue burning. *Environmental Science and Technology* 45:5477–5482.

Zhang, H., Ye, X., Cheng, T., Chen, J., Yang, X., Wang, L., Zhang, R. 2008a. A laboratory study of agricultural crop residue combustion in China: Emission factors and emission inventory. *Atmospheric Environment* 42:8432–8441.

Zhang, J., Smith, K.R. 2007. Household air pollution from coal and biomass fuels in China: Measurements, health impacts, and interventions. *Environmental Health Perspectives* 115:848–855.

Zhang, J.K., Cheng, M.T., Ji, D.S., Liu, Z.R., Hu, B., Sun, Y., Wang, Y.S. 2016a. Characterization of submicron particles during biomass burning and coal combustion periods in Beijing, China. *Science of the Total Environment* 562:812–821.

Zhang, T., Wooster, M.J., Green, D.C., Main, B. 2015a. Newfield-based agricultural biomass burning trace gas, PM2.5, and black carbon emission ratios and factors measured in situ at crop residue fires in Eastern China. *Atmospheric Environment* 121:22–34.

Zhang, Y., Cao, F. 2015a. Fine particulate matter (PM2.5) in China at a city level. *Scientific Reports (UK)* 5:14884. doi:10.1038/srep14884.

Zhang, Y., Cao, F. 2015b. Is it time to tackle PM2.5 air pollutions in China from biomass burning emissions? *Environmental Pollution* 202:217–219.

Zhang, Y., Lin, Y., Cai, J., Liu, Y., Hong, L., Qin, M., Zhao, Y., et al. 2016b. Atmospheric PAHs in North China: Spatial distribution and sources. *Science of the Total Environment* 565:994–1000.

Zhang, Y., Min, S., Zhang, Y., Zeng, L., He, L., Bin, Z., Wei, Y., et al. 2007. Source profiles of particulate organic matters emitted from cereal straw burnings. *Journal of Environmental Sciences* 19:167–175.

Zhang, Y., Shao, M., Lin, Y., Luan, S., Mao, N., Chen, W., Wang, M. 2013. Emission inventory of carbonaceous pollutants from biomass burning in the Pearl River Delta Region, China. *Atmospheric Environment* 76:189–199.

Zhao, S., Li, K., Zhou, W., Qiu, S., Huang, S., He, P. 2016. Changes in soil microbial community, enzyme activities and organic matter fractions under long-term straw return in north-central China. *Agriculture, Ecosystems & Environment* 216:82–88.

6 Bioconversion of Solid Organic Wastes and Molecular Characterization of Bacterial Population During the Decomposition Process

Mohd Arshad Siddiqqui, R. Hiranmai Yadav, and B. Vijayakumari

CONTENTS

6.1 INTRODUCTION

Soil forms the resource of agriculture with its organic and inorganic content reflecting the fertility of soil and thereby productivity of crops (Nagarathinam 2004). Soil aggregates play a significant role in water retention and movement, a gas diffusion that helps in root growth that in turn increases nutrient-holding capacity (Sultan 2001). The adverse effects of fertilizers are usually manifested when the high analysis fertilizers are added continuously to soil without monitoring the soil fertility health. Various reports indicate that soil health continues to be deteriorated, particularly its physical, chemical, and biological properties by the continuous, indiscriminate use of inorganic fertilizers. The use of inorganic fertilizers, though essential for increasing crop production, can also prove to be hazardous to the environment (Kumar and Sharma 1998). Higher levels of fertilizer application might result in certain antagonistic interactions between and among the plant nutrients within the soil which adversely affect the mobility of nutrients in the soil solution and as far as the roots (Saikia and Pathak 1997).

The biodegradation of organic substances by vermicomposting results in a stable eco-friendly soil conditioner that facilitates plant growth. The application of vermicompost helps to reduce pollution and minimizes the use of chemical fertilizers, as it contains organic carbon and matter, macro and micronutrients in addition to microbes, enzymes, and growth regulators (Gupta 2003).

There are numerous microorganisms that are involved in different stages of composting. They play a significant role in the degradation of waste, the stabilization of decomposed material, and the quality of the final product. Identification of these organisms is crucial for utilizing them effectively in waste management. From the conventional morphological and biochemical methods of identification, now technological advances help to identify them by molecular methods, which are more accurate.

6.2 ROLE OF ORGANIC WASTES IN SOIL FERTILITY

The addition of organic waste has increased soil porosity and modified the pore size pattern (Pagligai et al. 1981). Organic matter perhaps associated with relatively high sugar content is known to be attractive to earthworms (Lee 1985). It is also known that application of town refuse compost increases soil pH and electrical conductivity. The sustainable agricultural practices help to avoid soil degradation and declining crop yields and compensate rising fertilizer costs. The use of chemical fertilizers to increase crop production is coupled with the adverse effects on the environment. Vermicomposting converts solid organic wastes to manure and is used to improve soil fertility and crop yields (Chaudhary et al. 2004). Hemalatha (2012) observed an increased nutrient content and reduced C/N ratio in vermicompost prepared from partially decomposed fruit waste, paper, and tannery sludge.

Soil moisture retention facilitates root penetration and crop growth. This could be achieved by the amendment of organic matter that improves the water holding capacity of soil (Son 1995). The application of locally available organic waste materials after vermicomposting could replenish soil physio-chemical properties. Nitrogen, phosphorus, potassium, and micronutrients like iron, zinc, magnesium, and calcium are found to be high in vermicompost (Mary and Sivagami 2014). Angelova et al. (2013) have reported that composts and vermicompost application have a positive influence on soil quality and reduce metal toxicity. The conversion of organic waste materials by vermicomposting and applying them to soil could be an effective method for soil waste management (Bajsa et al. 2003). Vermicomposting is a feasible process that reduces waste and adds the organic matter to the soil to maintain soil fertility (Garg et al. 2006; Van Gestel et al. 1992; Edwards 1998; Kaur et al. 2010; Suriyanarayanam et al. 2010; Sinha et al. 2008). Soil quality was improved by vermicompost application that promises a beneficial effect (Ansari 2008) in sustainable agriculture. Integrated organic and inorganic manure application is found to be suitable to improve per capita income (Sharma and Singh 2012).

6.3 VERMICOMPOSTING

This is a biotechnological process to produce nutrient-rich material from organic waste that is converted by earthworm activity. This quality product with available forms of macro and micronutrients has a positive effect on growth and yield of plant, fertility of soil, and microbial population (Tharmaraj et al. 2011). Moradi et al. (2014) have reported that vermicomposting of organic piles results in a manure with rich humus, microbial population, and high content of nutrients and plant growth hormones. The bioconversion process of organic waste is where earthworms feed on the materials to produce more earthworms, vermicomposts, and vermiwash as products. The process ranges with varying temperature, pH, and moisture content to produce a manure with total nitrogen, available phosphorus, and exchangeable potassium that can be applied to plants as biofertilizers (Manyuchi and Phiri 2013).

Earthworms have traditionally been used in domestic compost heaps for breaking down organic waste to produce better quality compost. Compost produced by earthworm activity will have a higher level of plant nutrients than the normal organic manures (Lee 1985). It is generally known that the epigeic species *Eudrilus eugeniae* has greater potentiality for degrading organic waste (Dash and Senapati 1986). Epigeic species like *Eisenia foetida* and others are also identified as potential organic waste decomposers (Kale and Bano 1986). *Eudrilus eugeniae* serves as the most suitable species for the degradation of organic waste and production of vermicompost (Bano and Kale 1988). Earthworm, *Eudrilus eugeniae*, was found to grow in abundance in farmyard manure (FYM) and the compost recovered was 60–70%.

Eudrilus eugeniae species converted organic matter at a rate of 4.5 kg per 100 g of worms, when cultured in a combination of feed consisting of cattle dung, coir waste and vegetable waste in 10:1:1 ratio with a worm biomass increase of more than 75% (Bano 1997). Thakur et al. (2000) reported that composted domestic organic waste materials with *Eisenia foetida* increased potassium and carbon content significantly. Earthworms play a crucial role in the various soil processes, particularly in the improvement of soil structure, fertility, and mineralization (Divya 2001) (Figure 6.1).

The bio-oxidative process of waste conversion by the combined activity of earthworms, microorganisms, and fauna of a decomposer community modifies the substrate properties. The substrate fragmentation increases the surface area for activity

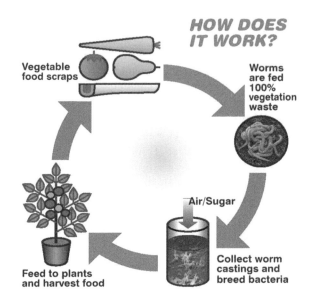

FIGURE 6.1   Schematic representation of various stages of vermicomposting.

by microbes and turnover and aeration. This leads to further degradation through enzymes released by microorganisms and the enrichment of organic and inorganic materials (Edwards et al. 1998; Aira et al. 2002; Loh et al. 2005; Molina et al. 2013). Vermicomposting using a suitable species has a significant influence on the degradation process. Vermicomposting is equally beneficial in the recycling of animal waste, crop residues, and industrial waste (Bansal and Kapoor 2000; Kaushik and Garg 2003; Yadav and Garg 2010; Garg et al. 2012).

6.4   PHYSICO-CHEMICAL CHARACTERS OF VERMICOMPOST

An important factor bearing influence is the stability of the final product of decomposition (Hu et al. 2008). Compost maturity is the level of decomposition that can be defined by the microbial biomass and decomposition of phytotoxic compounds in compost (Wu et al. 2000; Wu and Ma 2001). Vermicompost is prepared from cattle manure, food waste, and paper-contained plant growth regulating materials and humic acids that can increase the germination, growth, and yield of plants (Atiyeh et al. 2002; Arancon et al. 2006b).

The macronutrients were improved by the decomposition of Lantana with animal manures and earthworms (Singh and Angiras 2011). Pramod et al. (2009) observed a high rate of nutrient release during the decomposition of Lantana residue with oak pine leaf litter. Banta and Dev (2009) have also reported a low C/N ratio of the composts prepared from lantana. Ghadge and Jadhav (2013) also documented that the bioconversion of *Lantana camara* produced nutrient rich organic manure to improve soil fertility and increase the yield. Suthar and Sharma (2013) have vermicomposted

Lantana with cow dung using *Eisenia foetida* and recorded a decrease in pH, total organic carbon and C/N ratio; meanwhile, there was an increase in ash content, total nitrogen, available phosphorus, exchangeable potassium and calcium, and nitrate nitrogen. The microbial population was also increased and the population of actino-mycetes indicated the maturity of composts. Reddy et al. (2012) observed that the vermicompost extract preparation from weeds like Lantana can be a method for both weed control and litter management.

The application of vermicomposts changes the pH and electrical conductivity, improves the soil organic matter, organic carbon, available nitrogen carbonate ions, calcium ions, sodium ions, and exchangeable sodium percentage. These changes in the soil indicate the improvement of soil quality with the addition of vermicom-posts as an organic source (Ansari and Ismail 2008). Earlier reports suggest that vermicompost functions as an excellent soil conditioner by improving water holding capacity, thus releasing nitrogen (Harris et al. 1990; Christopher 1996), and resulting in total porosity and non-capillary porosity (Mathan 2000).

Vermicomposting is an eco-friendly practice for waste reduction (Hiranmai Yadav and Vijayakumari 2002). Organic manure prepared by vermicomposting Parthenium with cow dung using *Eudrilus eugeniae* exhibited changes in physical, chemical, and biological characters (Hiranmai Yadav et al. 2012) that are beneficial to plant growth. The pH, organic carbon, organic matter, total nitrogen, phosphorus, potassium, sodium, calcium, magnesium, copper, iron, zinc, and manganese con-tents were found to be improved in vermicomposted Parthenium with farm wastes and animal manures (cow dung, goat manure, poultry manure, and swine manure) using *Eisenia foetida,* i.e. the red worm. The results reveal the economic feasibility of vermicompost (the organic manure) production and certify the eco-friendly nature of its technology (Hiranmai Yadav and Argaw 2014).

Vermicomposting of cow dung and biogas plant slurry leads to the reduction of pH, total organic carbon, organic matter, and C/N ratio, and higher electrical con-ductivity, nitrogen, phosphorus, and potassium content than the substrate. The raw material used resulted in a homogeneous, odorless, and stabilized humus-like mate-rial (Yadav et al. 2013).

Prasad and Saini (2013) have reported that vermicomposting improves the nutri-ent content compared to substrates, and the product can serve as an ideal solution for solid waste management. As per the studies conducted by Arancon et al. (2008), paper wastes had 17.2% carbon, 1.0% nitrogen, 3.14 mg/g phosphorus, 0.10 mg/g of copper, 9.00 mg/g of iron, 0.24 mg/g of zinc, 0.30 mg/g of manganese after initial vermicomposting with *Eisenia foetida*. Animal manures have been effectively used as manure due to their high nitrogen content and they provide nutrients required by plants (Burton and Turner 2003). Inappropriate disposal of waste causes environ-mental pollution and the utilization of nutrients available in the animal manures can be utilized effectively for agriculture (Roeper et al. 2005; Williams et al. 1999; Mattson 1998). Co-composting is a preferred method to turn organic wastes into high nutrient content products that can be soil conditioners and amendments (Butler et al. 2001). It is an integrated waste management system which uses the biological stabilization of wastes that complement each other in decomposition, and the addi-tion of nutrients under a thermophilic condition with the elimination of pathogens and plant seeds (Ahring et al. 1992; Angelidaki and Ahring 1997; Gopinathan and Thirumurthy 2012). Co-composting also allows a humification process that miner-alizes the organic matter (Campitelli et al. 2006; Smidt et al. 2007). Organic sub-stances are the greatest carbon reservoirs on the earth that add organic matter to the soil (Pena-Mendez et al. 2005).

Earthworms and vermicomposts are helpful in maintaining the nutrient bal-ance in the soil cycling of organic waste. The African species *Eudrilus eugeniae* is used to decompose the waste paper and found to be effective with paper waste and cow dung in a 1:1 ratio (Basheer and Agarwal 2013). According to Arancon et al. (2008), vermicomposts produced from cattle manure, food wastes, and paper wastes were found to be beneficial for the growth of petunia plants. This might be due to the improvement of the physical structure of the potting medium, increased microbes, availability of plant growth-promoting substances like hormones and humates which probably contribute to the increased germination, growth, and flowering of petunia. The decomposition of *Parthenium hysterophorus*, *Eichornia crassipes*, and *Cannabis sativa* using *Eisenia foetida* was found to increase nitro-gen, phosphorus, and potassium and decrease the organic carbon, C/N ratio, and C/P ratio considerably (Chauhan and Joshi 2010). *Eisenia foetida* help to increase the microbial activity and release nitrogen, potassium, and calcium (Edwards 1995). Before vermicomposting, thermo-composting was beneficial for waste sta-bilization as well as for mass reduction, and was effective in reducing pathogens (Jaya et al. 2006).

The phosphorus and potassium content were high in cow dung added to vermi-compost which might be due to the mobilization of available P from the waste by the action of earthworm gut phosphates and phosphorus-solubilizing microbial activity. The enhanced microbial activity in worm casts results in a high potassium con-tent. The C/N ratios are less than 20. This indicates organic matter stabilization and reflects a satisfactory degree of decomposition and maturity of organic wastes. The loss of organic carbon as CO_2 due to microbial respiration and addition of earthworm N excrement results in good C/N ratios (Hait and Tare 2011). In a study by Anbalagan et al. (2012), vermicomposting of Parthenium with cow dung using *Eisenia fetida* resulted in a product which was high in nutrient contents with a low C/N ratio.

The sodium content was high in cow dung-added compost whereas a high con-tent of calcium was in poultry-added vermicompost. Yadav and Garg (2011) also reported that the Vermicomposting of Parthenium with *Eisenia fetida* is an effective technology for utilizing the weed biomass. They recorded a decrease in pH, total organic carbon, and C: N ratio, whereas the total nitrogen, available phosphorus, total calcium, and total potassium were high when Parthenium was vermicompos-ted with cow dung. The swine manure-added material exhibited high magnesium content compared to the other substrates. High micronutrients were recorded in cow dung-added vermicompost of Parthenium. This might be due to the nutrient content of cow dung. Vijaya and Seethalakshmi (2011) reported that application of 5t/ha Parthenium vermicompost resulted in enhanced growth, yield, and quality of okra plants, which denotes the beneficial effects of the manure. Similarly, it has been reported that Parthenium vermicomposting resulted in a high mineral nutrient con-tent product. In addition, it improved the growth of tomato.

According to Son (1995), the moisture content of the soil due to Parthenium compost application was 14.5–16.5% at 0–15 cm and 15–30 cm depths. This may be due to the buildup of organic carbon (C) status in soil. Earthworms modify soil pH and make it favorable for the growth of microbes, thus helping in the production of humus (Nayak and Rath 1996; Christopher 1996). Vermicomposted materials possess a good chemical and biological structure, porosity, and moisture-holding, coupled with reasonable quantities of plant nutrients (Christopher 1996).

Earlier studies by Srivastava et al. (1971) showed that manures having N above 1% are excellent. Studies by Gunathilagaraj and Ravignanam (1996) revealed that vermicomposting of mulberry (*Morus alba*) leaf litter resulted in increased levels of minerals like N, K, Mn, Zn, and Fe. Parthenium compost results in a higher availability of P content (Ramaswami 1997). Vermicomposting of a bio-digested slurry and weed combination resulted in a higher N content, followed by press mud and weeds. They contained appreciable quantities of micronutrients (Jeyabal and Kuppuswamy 1997). Talashilkar et al. (1997) reported that poultry manure contains total N 3.79%, total P_2O_5 3.09%, total K_2O 2.12%, total calcium (Ca) 1.71%, magnesium (Mg) 1.04%, total sulfur (S) 0.5% and total micronutrients namely Zn 198, Cu 306, Mn 447, and Fe 1978 ppm. The presence of micronutrients in appreciable amounts is of significance in the maintenance of soil fertility. The earthworm *Eisenia foetida* improves the N content of Parthenium compost (Ravankar et al. 1998).

The highest availability of micronutrients was due to the application of vermicompost followed by poultry manure which may be attributed to its higher organic C content. (Lee 1985). One ton of poultry manure deep litter contains N, P, K, Ca, and Mg in appreciable amounts besides the trace elements especially boron, Cu, Fe, S, and Zn (Channabasavanna and Biradar 2002; Channabasavanna 2003).

Poultry manure composted with varying amounts of sawdust had a C/N ratio ranging from 21:1 to 43:1 (Galler and Davey 1971). Higher C/N ratio leads to N-starvation in crops. Most of the Indian researchers have reported that chemical parameters like C/N ratio, weight loss, and ash content are indicators of compost maturity (Bharadwaj and Gaur 1985). Nagarajan et al. (1985) observed that Parthenium organic C content was reduced during degradation, which might be due to the microbial activity.

In the studies by Nagarajan et al. (1985), Rajendran (1991), and Thilagavathi (1992), in Parthenium compost the C/N ratio was markedly reduced. The reduction in C/N ratio was the result of a decrease in C content which was utilized as the energy source for the soil microflora and consequently converted into N content. Vermicomposting Parthenium using exotic earthworm, *Eisenia foetida* brought down the C/N ratio (Ravankar et al. 1998). Earthworm activities have important roles in maintaining C/N ratio and C/P ratio relationships for soil microorganisms (Deka et al. 2003). Earthworm activities bring down the C/N ratio, which is achieved by the combustion of C during respiration, and adding N with excretion in concentration that is readily assimilable to microbes. Reduction in ratios of C/N and C/P during vermicomposting is the main index to assess the rate of organic matter decomposition (Balamurugan and Vijayalakshmi 2004). Total C, C/N ratio, cellulose, hemicellulose, and the ratio of reducing sugar C to total C of city refuse compost decreased during a five-week maturation period; after that timeframe, their content did not change, while the total N content slightly increased and then maintained a constant value (Harada et al. 1981). The organic C content of wheat (*Triticum aestivum*) straw decreased continually with the age of the compost, while the dehydrogenase activity and N content increased (Abd et al. 1976).

The results of Sharma et al. (2004) have shown that during vermicomposting the pH of congress grass (*Parthenium hysterophorus*) heap was 8.9 and of Lantana (*Lantana camara*) 8.6. The increase in N content from 2.52 to 2.94% in Lantana and 2.04 to 2.38% in congress grass can be attributed to the enhanced microbial activity. This is because C is utilized by microorganisms for their energy requirements and causes the transformation of soluble N into microbial protein thereby preventing N loss. Vermicomposting also increased the P content of Lantana from 0.37 to 0.51% and 0.33 to 0.42% in congress grass. The K content of Lantana decreased from 1.92 to 0.80% and congress grass from 1.60 to 0.72%. Vermicomposting also narrowed down the C/N and C/P ratios substantially over the raw materials.

Nagarajan (1997) also documented that soils inhabited by earthworms have casts which in turn are richly inhabited by microorganisms. Nagarathinam et al. (2000) reported that vermicomposting pre-decomposed weeds using *Eudrilus eugeniae* showed more than a one-fold increase of colony-forming units of bacteria, fungi, and actinomycetes. Earthworm casts are rich in ammonia and partially digested organic matter and provide a good substrate for the growth of microorganisms.

As the decomposition progresses, the organic intermediates like various aliphatic acids, amino acids, and phenols will be converted to water-insoluble high molecular weight organic compounds. Therefore, the organic C to organic N ratio declines with the age of the composting process. During composting, high molecular weight compounds such as proteins, lipids, cellulose, lignin, and other compounds are enzymatically converted into low molecular weight water-soluble peptides, amino acids, fatty acids, phenols, and aliphatic acids. These are utilized by microorganisms; finally, the recalcitrant compounds are converted into a humus-like product (Kalaiselvi and Ramasamy 1996).

6.5   VERMICOMPOST AS MANURE

According to Kale et al. (1987), the material ingested by earthworms undergoes biochemical changes and ejected a cast which contained various plant nutrients in an assimilable form. Further vermicompost acts as a good medium for growth and the development of microorganisms which help in the enrichment of the micronutrients in soil and make them easily available for plant uptake. It is also claimed that vermicompost application to soil improves the soil fertility, increases yield, pest resistance, and improves the quality of produce (Bhawalkar and Bhawalkar 1991). Vermicompost as a "super manure" could be used to increase the air and water holding capacity of mine spoil (Lee 1985). Bhide (1988) reported an increase in the number and thickness of branches in grapevine (*Vitis vinifera*), when vermicompost was applied. The number of leaves also increased due to vermicompost application (Kasiviswanathan 1988).

The role of earthworm in improving the soil structure and thereby the bumper yields of conventional crops has been amply documented by Bhide (1988). When earthworm casts were used as a casing, there was an increase in the carpophore protein content of *Agaricus bisporus* (Galli et al. 1990). Lui et al. (1992) revealed that the application of vermicompost to the soil increased the available P and K content of soil. Vermicompost had a direct effect by supplying essential nutrients and an indirect effect by introducing earthworms in the field. An improvement in the enzymatic and microbial activities, and thus an improvement in the soil fertility due to the application of vermicompost, has been observed by Jambhekar and Bhinday (1992). Vermicompost has been reported to contain a large number of N-fixing, phosphate-solubilizing bacteria and other beneficial microbes, antibiotics, vitamins, hormones, enzymes, etc. which have a favorable effect on the growth and yield of plants (Bhawalkar 1992). Ravignanam and Gunathilagaraj (1996) observed that the application of earthworm, cow dung, and mulch significantly increased the plant height, number of leaves, leaf area, and leaf weight of mulberry (*Morusalba*).

Nainawat (1997) found that the application of vermicompost to two cultivars of wheat (*Triticum aestivum*), resulted in a higher total dry matter production and increased grain yield in comparison to organic manure and chemical fertilizers. According to Jeyabal and Kuppuswamy (1997), application of 50% N through vermicompost resulted in enhanced growth parameters, yield attributes, and yield of rice (*Oryza sativa*). The vermicompost derived from the combination of horse dung, mango litter, and pongamia litter was found to contain 2.17% N. The vermicompost application increased the seed production in sunflower (*Helianthus annuus*), fruit production in tomato (*Lycopersicon esculentum*), and increased flower production in ornamental plants (Bano 1997). Purekayastha and Bhatnagar (1997) observed an additional yield of 3.4 t ha^{-1} of lettuce (*Lactuca sativa*) and an increase in the total N in the experimental plot of rice (*Oryza sativa*) due to vermicompost application.

It was evident from the results of Thanunathan et al. (1997) that the incorporation of vermicomposted coir pith as an organic amendment increased the growth and yield of onion (*Allium cepa*) in mine spoil. Jadhav et al. (1997) recorded the maximum dry matter production of rice (*Oryza sativa*) in urea and vermicompost treatment. A considerable increase in the uptake of N, K, Ca, and Mg by rice was also observed in this treatment. The number of ears and total number of grains of rice were almost doubled in the vermicompost-applied plots. Thick, dark green leaves within fifteen days after application of vermicompost can be taken as a marker to indicate the effect of vermicompost (Sita and Hari 1997). Biddappa et al. (1997) noticed maximum bacterial population, actinomycete population, and fungal population on coffee husk vermicompost. Singh et al. (1997) opined that the application of 5 or 10 t ha^{-1} vermicompost was quite effective in increasing the growth and yield of wheat (*Triticum aestivum*). The results of Singh et al. (1998) indicate that application of 7.5–10 t ha^{-1} vermicompost saved about 20–50 kg N, 60 kg P_2O_5, and 25 kg $ZnSO_4$ ha^{-1}.

Tamak et al. (1998) found that the application of vermicompost @ 5 t ha^{-1} significantly improved the growth and yield of cotton (*Gossypium hirsutum*). According to Maheswarappa et al. (1998), vermicompost incorporation showed significantly superior growth attributes of arrow root (*Maranta arundinacea*). The vermicompost prepared from paper mill and dairy industries sludge were rich in N, P, and had good structure, low levels of heavy metals, low conductivity, high humic acid content, and good stability and maturity (Elvira et al. 1998). It is proposed that combined application of vermicompost, FYM and N, P, K significantly increases the number of branches, height of plant, number of leaves, and leaf yield in mulberry (*Morus alba*). The yield data of okra (*Abelmoschus esculentus*) revealed that vermicompost as an organic source along with the full recommended dose of inorganic fertilizers produced the highest yield (Ushakumari et al. 1999).

Vermicompost application increased the dehydrogenase activity of soil, which is the indicator of microbial activity (Maheswarappa et al. 1999). Vermicompost application is mainly attributed to an increase in organic C content, which in turn enhanced the microbial population. Karuna et al. (1999) stated that diluted vermi-wash spray was most effective in inducing a vegetative growth-like number of suckers, enhanced the length and breadth of leaves and length of the petiole, whilst also initiating early flowering in *Anthurium*.

Experiments conducted by Kumarasamy (2000) showed that the application of vermicompost at 5 t ha^{-1} along with recommended fertilizers significantly increased the grain and straw yields of rice (*Oryza sativa*), and improved the soil properties like bulk density, organic C content, cation exchange capacity, and soil fertility status. Sreenivas et al. (2000) documented a significant increase in the vine N content with increasing levels of vermicompost. They also observed a higher N uptake and highest yield in ridge gourd (*Luffa acutangula*), due to the application of vermicompost and N P K.

Senthilkumar et al. (2000) recorded an astonishing yield of 1800 t acre^{-1} paddy using vermicompost alone. According to Reddy et al. (2001), the leaf yield as well as the nutrient status (N, P and K) in mulberry (*Morusalba*) leaves were more pronounced in vermicompost. Ramalingam (2001) found that in the vermicomposted sugarcane trash and press mud mixture, the level of macro and micronutrients increased significantly with reduction in C/N and C/P ratios due to mineralization brought about by the combined action of earthworms and microbes.

Hangarge et al. (2002) observed a significant increase in yield of green chili (*Capsicum annuum*) and spinach (*Basella rubra*) due to vermicompost application. Vermicomposting of canteen waste gave value -products which help in improving soil texture and had a high bacterial count (Munnoli et al. 2002). Successful production of organic rice (*Oryza sativa*) using vermicompost as a major source of nutrients has been reviewed by Mohan et al. (2002).

Vegetable wastes from hostels when converted into compost using *Eudrilus eugeniae* showed a pH of 7.4, which is suitable for further bacteriological degradation. The organic matter constituents like nitrate, phosphate, K, and Ca were suitable for plant growth (Padma et al. 2002). Vermicomposting of 6% urea- treated pine needles recorded highest nutrient levels owing to a higher worm population, followed by higher palatability by earthworm and thus resulted in good quality vermicompost (Singh et al. 2003).

Press mud when vermicomposted had shown high percentage of N, P, K, Ca, Mg, Zn, and Fe. The C/N ratio was also narrowed down. Reddy et al. (2003) concluded that the use of vermicompost improved the leaf quality and quantity of mulberry (*Morusalba*). Thangavel et al. (2003) found that vermicast extract and vermiwash increased the growth and yield of paddy. According to Kikon and Sharma (2003), vermicomposting of paddy straw, vegetable waste, and weed resulted in a nutrient rich compost with high N, P, K (total and available), Ca, and Mg.

Umamaheswari and Vijayalakshmi (2003) reported that paper mill sludge vermicomposted with *Eudrilus eugeniae* resulted in increased macro and micronutrients as well as pH and pore space. Reduction in electrical conductivity and bulk density indicates the better degradation of organic waste. Deka et al. (2003) revealed that the vermicomposting of water hyacinth (*Eicchornia crassipes*) increased the N, P, K, Ca, and Mg contents and reduced the C/N ratio.

Vermicomposting of Parthenium improved the end product quality and when applied, improved the biometric, biochemical, and yield parameters of chili (Hiranmai Yadav and Vijayakumari 2003, 2004). Similarly, application of Parthenium compost and vermicompost was also beneficial in improving growth (Vijayakumari and Hiranmai Yadav 2008a) and yield of chickpea (Anto et al. 2008) and also the quality (Vijayakumari and Hiranmai Yadav 2013). The residual impact of chickpea and the vermicomposts, when assessed using radish as a test crop, was also found to be beneficial in improving the growth, quality, and yield of radish (Vijayakumari and Hiranmai Yadav 2008b, 2009a). The application of vermicomposted Parthenium and poultry manure has been beneficial to sesame plants (Vijayakumari et al. 2011; Vijayakumari and Hiranmai Yadav 2012) and radish (Vijayakumari and Hiranmai Yadav 2009b; Vijayakumari et al. 2009)

The growth and yield performance of *Vigna radiata* in vermicompost prepared in the ratio of 1:2:1 (cow dung : banyan leaf litter : goat dung) were significantly higher (Jayashree et al. 2004). According to Rachappaji et al. (2004), vermicompost application to mulberry (*Morusalba*) increased the growth attributes. *Lablab purpureus*, when supplemented with vermicompost gave increased leaf area index and yield (Karmegam and Thilagavathy 2004). Arunkumar (2004) documented that the best growth parameters of *Amaranthus dubius* were observed in vermicomposted sludge.

Neetha and Mira (1988) have also observed the best growth of maize (*Zea mays*) and wheat (*Triticum aestivum*) with applications of poultry waste alone and poultry waste with earthworms. Vermicomposted larval litter significantly increased the length and weight of shoot and root, shoot/root ratio, and NPK uptake of mulberry (*Morusalba*) (Gunathilagaraj and Ravignanam 1996). Studies by Govindan et al. (1995) and Manonmani and Anand (2002) showed that the application of vermicompost resulted in maximum vegetative growth of bhindi (*Abelmoschus esculentus*). A possible explanation for the beneficial effect of vermicompost may be the accumulation of mobile substances in earthworm casts. As per the results of Singh et al. (1997) application of 7.5–10 t ha^{-1} vermicompost to wheat (*Triticum aestivum*) field was able to save about 50 kg N, 60 kg P_2O_5 and 25 kg $ZnSo_4$ ha^{-1} and this may well be a step towards complete organic farming without the use of any chemical fertilizer or agrochemical.

Application of 10 t ha^{-1} vermicompost and the recommended dose of NPK recorded maximum plant height at harvest, days to initial flowering, and number of branches per plant. According to Reddy et al. (1998), the constant and optimal supply of nutrients through vermicompost and NPK has influenced better growth. Nethra et al. (1999) observed that the maximum plant height and number of leaves of China aster (*Callistephus chinensis*) were after application of 10 t ha^{-1} vermicompost. In situ vermiculture in a mulberry (*Morusalba*) field resulted in the improvement of growth parameters and yield of leaf (Jadhav et al. 1999). Compost prepared through worm-powered technology increases the growth and development of rice (*Oryza sativa*) plants (Nagarajan 1997).

A higher concentration of vermicomposted Parthenium application, dry weight of root, stem, and ear of wheat (*Triticum aestivum*) were reported to be maximum (Kavitha and Bhardwaj 2000). Vermicomposted market waste resulted in a significant increase in the morphological parameters of *Vigna mungo* (Balamurugan and Vijayalakshmi 2004). The presence of vermicompost enhances the macro and micronutrient uptake by plants, harbors rich amounts of microbes that degrade and mobilize the nutrients to available form, in addition to exudates of earthworm that support the microorganisms, and which secrete plant growth hormones.

Ravignanam and Gunathilagaraj (1996) reported that when mulberry (*Morus alba*) plants were treated with earthworm and cowdung and mulch exhibited maximum leaf weight and plant height. Plant height is an indicator of growth and an important contributor to leaf production in mulberry (*Morusalba*). Earthworm activity improves the availability of nutrients at the rhizosphere of the plants, leading to an increased shoot length of plants. Hashemimajd et al. (2004) observed greater biomass production of tomato (*Lycopersicon esculentum*) with vermicompost. Both lettuce (*Lactuca sativa*) and radish (*Raphanus sativus*) seedlings showed an increased capacity for protein synthesis when grown in the presence of earthworm casts, noted Tomati et al. (1990). The mechanism by which earthworms and their casts affect plant metabolism is still under discussion, although it is clear that microbes and microbial metabolites are involved (Tomati et al. 1988). In protein synthesis, microbial hormone-like substances could play an important role by regulating ionic absorption and inducing some enzymatic activities (Wareing 1982; Scott 1984).

Ravignanam and Gunathilagaraj (1996) have reported that the application of recommended doses of NPK fertilizers, earthworm, and cow dung and mulch significantly increased the contents of chlorophylls, protein, K, Fe, Mn, and Zn in mulberry (*Morus alba*) leaves. Total carbohydrate content and crude protein content were significantly higher in mulberry (*Morus alba*) leaves treated with vermicompost (Jadhav et al. 1998). The increase in carbohydrate content may be due to the presence of plant macronutrients, secondary elements, and micronutrients in vermicompost, as reported by Bano et al. (1987). Besides plant promoters (Gavrilov 1962), Jadhav et al. (1998) have reported increased crude protein content in the leaves of mulberry (*Morusalba*) by the application of vermicompost.

Manonmani and Anand (2002) found evidence that the overall increase in biochemical constituents like chlorophyll, protein, and carbohydrates from the leaf tissue of the lady's finger (*Abelmoschus esculentus*) were supplied with vermicompost. This may be attributed to the nutritive content of vermicompost (nitrate, chloride, P). The contents of chlorophylls, N, and protein were found to be high in the leaves of mulberry (*Morus alba*) treated with vermicompost (Reddy et al. 2003). This clearly indicates that vermicompost was rich in NPK, micronutrients, and enzymes. The presence of many beneficial microorganisms and enzymes influenced the N and protein synthesis which in turn improves the leaf quality. Govindan et al. (1995) showed that the application of 100% vermicompost resulted in a maximum yield of bhindi (*Abelmoschus esculentus*).

The application of 10 t ha^{-1} vermicompost and the recommended dose of NPK to pea (*Pisum sativum*) recorded a maximum number of pods, seeds per pod, and yield (Reddy et al. 1998). The constant and optimal supply of nutrients through vermicompost and recommended doses of NPK influenced better growth, and in turn manifested in improved yield parameters, and finally yield. The application of N through vermicompost and fertilizer gave a higher yield of maize (*Zea mays*), compared to application of N through fertilizer. This might be due to the presence of P, K, Ca, S, and micronutrients in organic source and bioactive compounds formed during the composting, which might have increased the yield of maize (*Zea mays*) and sustained soil health (Jeyabal et al. 1998). Recommended NPK, FYM with earthworms in sapota (*Achras sapota*) promoted the maximum number of fruits, a higher average weight of fruits per plant, also increasing the diameter of and length of fruit (Gawande et al. 1998).

Grain yield of soybean (*Glycine max*) was the highest in 100% vermicompost (Reddy and Reddy 1999). Krishnamoorthy and Vajranabaiah (1986) attributed that the positive effect of vermicompost in soybean (*Glycine max*) might be due to the presence of several plant growth hormones like cytokinins and auxins, which have positive a beneficial effect on all yield parameters. Nethra et al. (1999) observed maximum yield per ha of china aster (*Callistephus chinensis*) after application of 10 t ha^{-1} vermicompost and recommended NPK. Earlier studies by Kavitha and Bhardwaj (2000) also revealed that there was an increasing trend in grain weight per plant of wheat (*Triticum aestivum*) with a higher concentration of Parthenium vermicompost, as compared to other treatments.

The earthworms and cow dung had a significant impact on yield of the wheat (*Triticum aestivum*). Better yield is expected from healthy wheat (*Triticum aestivum*) plants with good vegetative growth because the green leaves assist in photosynthesis and produce carbohydrates which are later stored in the form of starch in the seed grains of cereal crops (Bhatia et al. 2000). Manonmani and Anand (2002) recorded a maximum yield of lady's finger (*Abelmoschus esculentus*) in vermicompost-treated plots. The minerals present in the vermicompost play a central role in the primary metabolism of the plants. They contribute to the overall yield and productivity. (Ghabbour 1966). Vermicomposted market waste resulted in a significant increase in yield parameters of *Vigna mungo* (Balamurugan and Vijayalakshmi 2004).

The practice of vermiculture is centuries old. Despite not being a new practice, it is currently receiving worldwide attention with diverse objectives like waste management, soil detoxification, regeneration, and sustainable agriculture. The use of earthworms for waste management results in the employment of vermicomposts, a potential organic input for sustainable agriculture. It contains beneficial microorganisms, major and micronutrients, enzymes, and hormones. The addition of vermicomposts to soil improves the chemical and biological properties of soil, and thereby soil fertility. Earthworms found in the soil act as soil aerators, grinders, crushers, chemical degraders, and biological stimulators (Probodini 1994; Purekayastha and Bhatnagar 1997; Sinha et al. 2002). Vermicomposting, the non-thermophilic decomposition of organic wastes by earthworms, is a popular waste management option in the Americas, Europe, and the Indian subcontinent.

6.6 BIOCONVERSION OF ORGANIC WASTES FROM FRUIT AND VEGETABLE MARKET

Vegetable and fruit waste is a biodegradable material generated in large quantities, much of which is dumped on land to rot in the open, which not only emits a foul odor, but also creates a big nuisance by attracting birds, rats, and pigs – vectors of various diseases. Apart from post-harvest losses due to lack of storage capacity, processing, and packaging of vegetables according to customers' specifications also plays a major role in waste generation. Vegetable and fruit wastes include the rotten, peels, shells, and scraped portions of vegetables and fruit or slurries. These wastes can be treated for biofuel production through fermentation under controlled conditions, or else used for composting. The natural decomposition of wastes by microbes generates products with high humus content. Research activities have confirmed that this carbohydrate-rich biomass can be a potent substrate for renewable energy generation (Singh et al. 2018, 2017a, b, c, 2019; Tiwari et al. 2018, 2019a, b; Kour et al. 2019). A huge amount of vegetable and fruit waste is generated on a daily basis from diverse sectors such as vegetable markets, agro-based companies, hostels, hotels, restaurants, canteens, hospitals, housing societies, institutions, waste dumping yards, etc.

6.6.1 CHARACTERIZATION OF VEGETABLE AND FRUIT WASTES

Vegetable and fruit wastes are a special group of biomass that needs to be characterized to understand its nature for application as raw material and to propose the best methodology for its proper utilization. Waste composition also influences the overall yield and kinetics of the biologic reaction during digestion. Characterization of waste can be done physically, chemically, or biologically. Physical characterization of solid wastes include estimation of weight, volume, moisture, ash, total solid, volatile solid (VS), color, odor, temperature, etc., while dissolved and suspended solids are estimated for liquid wastes. Turbidity is another important parameter for liquid wastes, which needs to be considered. Chemical studies include the measurement of cellulose, hemicellulose, starch, reducing sugars, protein, total organic carbon, phosphorus, nitrogen, BOD, COD, pH, halogens, toxic metals, etc. Besides these biochemical parameters, carbon, phosphorous, potassium, sulfur, calcium, magnesium, etc. can also be tested. All these chemical and biochemical parameters provide an insight into the applicability of waste for employment in fertilizer production. Biological characterization indicates the presence of pathogens and organisms which are indicators of pollution. A common feature of various forms of food wastes includes high COD, richness in protein, carbohydrate, and lipid biomolecules with noticeable pH variation. It has been reported that wastes from vegetable industries (including carrots, peas, and tomatoes) have a high BOD and are a rich source of several nutrients like vitamins, minerals, fibers, etc. So, a detailed study of waste characteristics is essential for deciding its application and determination of the economic feasibility of the process.

6.6.2 MANAGEMENT OF VEGETABLE AND FRUIT WASTE

There are seven commonly used methods of managing fruit and vegetable waste. The list of methods provided here will define the method of management. This list cannot be easily arranged in order of best management practice from an environmental standpoint due to the individual circumstances of the farmer and the packing house where culls originate. The management options are provided as a means to help better explain how each may be used. The seven management methods are:

1) Store the culled fruit and vegetables on-site in a pile or bermed area for a limited time
2) Return fruit and vegetable waste to the field on which it was grown
3) Feed fruit and vegetable waste to livestock
4) Give the fruit and vegetable culls to local food banks
5) Compost fruit and vegetable culls
6) Process fruit and vegetable culls to separate juice from pulp
7) Dispose of fruit and vegetable waste in a local Sub-Title D landfill

6.6.3 STORE THE CULLED FRUIT AND VEGETABLES ON-SITE

Storing culled fruit and vegetable waste on-site is a temporary solution to final disposal or the reuse of materials. To use this method, the culls may be hauled or transferred via mechanical methods to a location that has been prepared for holding the culls. At a minimum, the holding area should be bermed to capture and hold rainfall and any liquids that have formed from the decomposition of the culled fruit and vegetables. Other options for such a site include storage in tanks or bunkers with easy access for moving liquids or solids for later management. The culls stored in the bermed area should be crushed, if possible, to allow available liquid to better evaporate. Crushing the culled fruit and vegetables and placing them in a bermed area helps control the leachate, run-on and runoff, makes managing the material easier, allows extra liquids to evaporate, and reduces the volume that will need to be managed at a later time.

6.6.4 RETURN FRUIT AND VEGETABLE WASTE TO THE FIELD

From an agricultural nutrient management and organic building viewpoint, returning fruit and vegetable waste to the field may be one of the better options. This management method returns the culls back to the growing field where the nutrients can be recycled, allowing the fruit and vegetable pulp and juice to help build or maintain the soil organic matter content. The cost can be very low, based on the distance to the field and the amount of liquid removed. The protocol for transporting the culls to a growing field will consist of storing the culls at the packing house or at the field site until final harvesting of the crop. After the final harvest, the culls or remaining solids and liquids can be loaded into spreader trucks and applied evenly across the field. As a matter of practice, the material should be incorporated, which will reduce the potential for problematic odors and runoff.

6.6.5   FEED FRUIT AND VEGETABLE WASTE TO LIVESTOCK

Managing culls by feeding fruit and vegetable waste to livestock may be a good option based on the overall management system of the livestock operation. One of the major issues that must be addressed involves the nutritional benefits and effects of feeding culls to livestock. Farmers should consult with animal scientists or veterinarians to confirm the effects of feeding culls to livestock.

6.6.6   GIVE GOOD FRUIT AND VEGETABLE CULLS TO LOCAL FOODBANKS

Food banks may be an option to manage some of the culls resulting from the sorting of the fruit and vegetables. Giving culls to a food bank may be an option and the Good Samaritan Law will protect the donating company. However, since fruit and vegetables are perishable, not all of the culls can be utilized by this method. The farmer should stay in contact with the local food bank coordinator to inform them of harvest dates and what may be available, as well as to determine whether anyone would be allowed in the packing house and whether bins of culls would be available for further off-site culling and packaging for distribution to other food banks. The coordinator would need to provide a means to safely transport the culls to a location for further processing, if needed, and the remaining culls would have to be disposed of using another method listed in this document.

6.6.7   COMPOST FRUIT AND VEGETABLE CULLS

Composting culled fruits and vegetables is one option that can reduce the volume of culls as well as other "waste" materials in a community, if the land and equipment are available. Culls used in the compost process would either be transferred in a truck to the composting facility or mechanically transported if the compost facility is on-site. The culls would be mixed in proper ratios with other organic materials as recommended by composting professionals to produce compost suitable for incorporation into fields or for selling.

6.6.8   PROCESS FRUIT AND VEGETABLE CULLS TO SEPARATE JUICE FROM PULP

The method of separating the fruit and vegetable culls into juice and pulp is accomplished by using a press. Typical systems are screw presses that can effectively separate the juice from the pulp. After separation, each fraction has its benefits for different reasons and purposes. If the culls are of good food quality they can be used as juices in food applications based on available markets. The pulp can also potentially be used as a component of foods. For those culls that are not of human food quality, the separated pulp can be used as one component of compost or animal food. (If the pulp is used for animal feed, check with an animal scientist or veterinarian prior to feeding.) The pulp can also be used as a soil amendment or as one component of a composting process. The juice can also be used as a feedstock for ethanol production or anaerobic digestion processes. For either process, there should be a market for the final products: ethanol or methane.

6.6.9   Disposal of Fruit and Vegetable Waste in Landfill

Disposal of culled fruit and vegetable waste in landfill is a method that should be considered after all other options. From a sustainability standpoint, disposal of these culls in a landfill is probability not the best option based on fees. If landfilling is the chosen option, management of the culls should reduce leakage of liquids from the transport truck.

6.6.10   Management of Fruit and Vegetable Waste for a Specific Packing House

The above methods for the disposal or reuse of fruit and vegetable waste are provided for all fruit and vegetable packing houses. The disposal method specific for any given packing house may be different and must be a decision based upon the particular location and situation. There may be more disposal methods available to specific packing houses. Individual packing houses will have to identify additional pros and cons and take them into account. This document is provided only as a guide to aid the individual packing house in identifying different options that may be suitable for disposing of culled fruits and vegetables.

The collected organic materials can be shredded, weighed in specific ratios, layered in mud pots, and kept in anaerobic conditions for pre-digestion. After a predecided duration, earthworms are introduced for further decomposition in semi-aerobic conditions. Moisture content is maintained by the sprinkling of water and periodical mixing (Table 6.1).

The compost sample was analyzed for nitrogen (micro- Kjeldahl method, Humphries 1956), phosphorous (Colorimetric method, Jackson 1973), potassium (Piper 1966), and micronutrients, (zinc, iron, manganese, and copper) following the method of Jackson (1973) (Table 6.2).

6.7   MOLECULAR CHARACTERIZATION OF BACTERIA

The determination of microbial presence and diversity using molecular techniques have made possible the discovery of previously unknown microbes. Based on the molecular analysis, the bacterial species have been identified (Table 6.3) (Figure 6.2).

Siddiqqui (2017) also isolated and identified different bacterial species (*Halomonas sp* (KY952739), *Bordetella petrii* (KY952740), *Luteimonas marina* (KY952741), *Bordetella petrii* (KY952742), *Bordetella petrii* (KY952743), and *Bacillus megaterium* (KY952744) by molecular characterization in the samples

TABLE 6.1
Composition of Organic Wastes Used for Composting and Vermicomposting

Treatment	Description
T_1	Vegetable waste and Fruit waste + Farmyard manure + Food waste + Biodegradable municipal solid waste + Earthworms

TABLE 6.2
Physicochemical Characterization of Compost

S. No.	Parameters	Nutrient Content
1.	Total Nitrogen (%)	0.0523
2.	Phosphorous (%)	0.0033
3.	Potassium (%)	0.08
4.	**C/N ratio**	**333.6520**
5.	Zinc (ppm)	7.08
6.	Ferrous (ppm)	12.80
7.	Manganese (ppm)	0.88
8.	Copper (ppm)	8.16

collected from a landfill site. In a study, Nakasaki et al. (1994) have documented that a thermophilic bacterium, *Bacillus licheniformis*, enhances the rate of decomposition of organic waste by preventing the decrease of pH of the composting system. Zaved et al. (2008) have also reported that useful bacteria might be isolated from the surrounding environment for eco-friendly bioconversion of solid organic waste. Similarly, Eida et al. (2012) have also isolated and characterized bacteria from the compost prepared from sawdust and coffee residue which are responsible for the decomposition of cellulose. Efficiency of cellulolytic bacteria are isolated from the soils in the bioconversion of solid organic waste (Barman et al. 2011). They observed Moraxella to be most effective in bioconversion. Saha and Santra (2014) isolated bacteria from municipal solid waste that produce industrial enzymes during the biodegradation of industrial waste.

Reddy et al. (2017) have also documented that the compost prepared from municipal solid waste from Guntur and Vijayawada contains a large number of bacteria having the capability to biodegrade municipal solid waste. Bacteria isolated from the compost prepared from water hyacinth are metabolically active in degrading organic matter, surviving in the heavy metal loaded compost environment. So they can be used for micro-bioremediation of heavy metals (Vishan et al. 2017). Ishak et al. (2011). Isolated bacteria from activated sludge and compost are helpful in the treatment of municipal solid waste. Boulter et al. (2002) documented that the bacteria isolated from the compost have the potential to suppress the turfgrass pathogen which is responsible for causing the turfgrass disease dollar spot. Rao et al. (2016)

TABLE 6.3
16s r-RNA Sequencing of Different Bacteria

S. No.	Marking	Bacteria	Accession No.	Sequence
1.	C_1	Staphylococcus hominis	MH671870	AGCGTTATCCGGAATTATTGGGCGTAAAGCGCGCGTAGGCGGTTTTAAGTCTGATGTGAAAGCC CACGGCTCAACCGTGGAGGGTCATTGGAAACTGGAAAACTTGAGTGCAGAAGAGGAAAGT GGAATTCCATGTGTAGCGGTGAAATGCGCAGAGATATGGAGGAACACCAGTGGCGAAGGCGACTT TCTGGTCTGTAACTGACGCTGATGTGCGAAAGCGTGGGGATCAAACAGGATTAGATACCCTGGTA GTCCACGCCGTAAACGATGAGTGCTAAGTGTTAGGGGGTTTCCGCCCCTTAGTGCTGCAGCTAAC GCATTAAGCACTCCGCCTGGGGAGTACGACCGCAAGGTTGAAACTCAAAGGAATTGACGGGGACC CGCACAAGCGGTGGAGCATGTGGTTTAATTCGAAGCAACGCGAAGAACCTTACCAAATCTTGACA TCCTTTGACCCTTCTAGAGATAGAAGTTTCCCCTTCGGGGACAAAGTGACAGGTGGTGCATGGT TGTCGTCAGCTCGTGTCGTGAGATGTTGGGTTAAGTCCCGCAACGAGCGCAACCCTTAAGCTTAG TTGCCATCATTAAGTTGGGCACTCTAAGTTGACTGCCGGTGACAAACCGGAGGAAGGTGGGGATG ACGTCAAATCATCATGCCCCTTATGATTTGGGCTACACACGTGCTACAATGGACAATACAAAG
2.	C_4	Streptomyces werraensis	MH671871	GTCGAACGATGAACCTCCTTCGGGAGGGGATTAGTGGCGAACGGGTGAGTAACACGTGGGCAATC TGCCCTGCACTCTGGGACAAGCCCTGGAAACGGGGTCTAATACCGGATACTGATCATCTTGGCA TCCTTGGTGATCGAAAGCTCCGGCGGTGCAGGATGAGCCCGCGGCCTATCAGCTTGTTGGTGAGG TAATGGCTCACCAAGGCGACGACGGGTAGCCGGCCTGAGAGGGCGACCGGCCACACTGGGACTGA GACACGGCCCAGACTCCTACGGGAGGCAGCAGTGGGGAATATTGCACAATGGGCGAAAGCCTGAT GCAGCGACGCCGCGTGAGGGATGACGGCCTTCGGGTTGTAAACCTCTTTCAGCAGGGAAGAAGCG AAAGTGACGGTACCTGCAGAAGAAGCGCCGGCTAACTACGTGCCAGCAGCCGCGGTAATACGTAG GGCGCGAGCGTTGTCCGGAATTATTGGGCGTAAAGAGCTCGTAGGCGGCTTGTCGCGTCGGTTGT GAAAGCCCGGGGCTTAACTCCGGGTCTGCAGTCGATACGGGCAGGCTAGAGTTCGGTAGGGGAGA TCGGAATTCCTGGTGTAGCGGTGAAATGCGCAGATATCAGGAGGAACACCGGTGGCGAAGGCGGA TCTCTGGGCCGATACTGACGCTGAGGAGCGAAAGCGTGGGGAGCGAACAG

(Continued)

TABLE 6.3 (CONTINUED)
16s r-RNA Sequencing of Different Bacteria

3.	C_6	*Streptomyces matensis*	MH671873

CGTCGTCCGTGCCGCAGCTAACGCCATTAAGTGCCCCGCCTGGGGAGTACGGCCGCAAGGCTAAAA
CTCAAAGGAATTGACGGGGGCCCGCACAAGCGGCGGAGCATGTGGCTTAATTCGACGCAACGCGA
AGAACCTTACCAAGGCTTGACATACACCGGAAAGCATCAGAGATGGTGCCCCCTTGTGGTCGGT
GTACAGGTGGTGCATGGCTGTCGTCAGCTCGTGTCGTGAGATGTTGGGTTAAGTCCCGCAACGAG
CGCAACCCTTGTCCCGTGTTGCCAGCAGGCCCTTGTGGTGCTGGGGACTCACTGGTTTACCGCCG
GGGTCAACTCGGAGGAAGGTGGGGACGACGTCAAGTCATCATGCCCCTTATGTCTTGGGCTGCAC
ACGTGCTACAATGCCGGATACAATGAGCTGCGATACCGCAGGTGGAGCGAATCTCAAAAAGCCG
GTCTCAGTTCGGATTGGGGTCTGCAACTCGACCCCATGAAGTCGGAGTCGTAGTAATCGCAGAT
CAGCATTGCTGCGGTGAATACGTTCCCGGGCCTTGTACACACCGCCGTCACGTCACGAAAGTCG
GTAACACCCGAAGCCGGTGGCCCAACCCCTTGTGGGGAGGGAGCTG

4.	C_7	*Bacillus licheniformis*	MH671869

GAGGGTGATCGGCCACACTGGGACTGGACACGGCCCAGACTCCTACGGGAGGCAGCAGTAGGGA
ATCTTCCGCAATGGACGAAAGTCTGACGGAGCAACGCCGCGTGAGTGATGAAGGGTTTTCGGATCG
TAAAACTCTGTTGTTAGGGAAGAACAAGTACCGTTCGAATAGGGCGGTACCTTGACGGTACCTAA
CCAGAAAGCCACGGCTAACTACGTGCCAGCAGCCGCGGTAATACGTAGGTGGCAAGCGTTGTCCG
GAATTATTGGGCGTAAAGCGCGCGCAGGCGGTTTCTTAAGTCTGATGTGAAAGCCCCGGCTCAA
CCGGGGAGGGTCATTGGAAACTGGGAAACTTGAGTGCAGAAGAGGAGAGTGGAATTCCACGTGTA
GCGGTGAAATGCGTAGAGATGTGGAGGAACACCAGTGGCGAAGGCGACTCTCTGGTCTGTAACTG
ACGCTGAGGCGCGAAAGCGTGGGGAGCGAACAGGATTAGATACCCTGGTAGTCCACGCCGTAAAC
GATGAGTGCTAAGTGTTAGAGGGTTTCCGCCCTTTAGTGCTGCAGCAAACGCATTAAGCACTCCG
CCTGGGGAGTACGGTCGCAAGACTGAAACTCAAAGGAATTGACGGGGGCCCGCACAAGCG
GTGGAGCATGTGGTTTAATTCGAAGCAACGCGAAGAACCTTACCAGGTCTTGACATCCTCTGACA
ACCCTAGAGATAGGGCTTCCCCTTCGGGGGCAGAGTGACAGGTGGTGCATGGTTGTCGTCAGCTC
GTGTCGTGAGATGTTGGGTTAAGTCCCGCAACGAGCGCAACCCTTGATCTTAGTTGCCAGCATTC
AGTTGGGCACTCTAAGGTGACTGCCGGTGACAAACCGGAGGAAGGTGGGGATGACGTCAAATCAT
CATGCCCCTTATGACCTGGGCTACACACGTGCTACAATGGACAGAACAAAGGGCAGCGAAGCCGC
GAGGCTAAGCCAATCCACAAAATCTGTTCTCAGTTCGGATCGCAGTCTGCAACTCGACTGCGTGA
AGCTGGAATCGCTAGTAATCGCGGATCAGCATGCCGCGGTGAATACGTTCCCGGGCCTTGTACAC
ACCGCCCGTCACACCACGAG

(Continued)

TABLE 6.3 (CONTINUED)

16s r-RNA Sequencing of Different Bacteria

5.	C$_8$	Bacillus cereus	MH671874

GTTGGTGAGGTAACGGCTCACCAAGGCAACGATGCGTAGCCGACCTGAGAGGGTGATCGGCCACA
CTGGGACTGAGACACGGCCCAGACTCCTACGGGAGGCAGCAGTAGGGAATCTTCCGCAATGGACG
AAAGTCTGACGGAGCAACGCCGCGTGAGTGATGAAGGCTTTCGGGTCGTAAAACTCTGTTGTTAG
GGAAGAACAAGTGCTAGTTGAATAAGCTGGCACCTTGACGGTACCTAACCAGAAAGCCACGGCTA
ACTACGTGCCAGCAGCCGCGGTAATACGTAGGTGGCAAGCGTTATCCGGAATTATTGGGCGTAAA
GCGCGCGCAGGTGGTTTCTTAAGTCTGATGTGAAAGCCCACGGCTCAACCGTGGAGGGTCATTGG
AAACTGGGAGACTTGAGTGCAGAAGAGGAAAGTGGAATTCCATGTGTAGCGGTGAAATGCGTAGA
GATATGGAGGAACACCAGTGGCGAAGGCGACTTTCTGGTCTGTAACTGACACTGAGGCGCGAAAG
CGTGGGGAGCAAACAGGATTAGATACCCTGGTAGTCCACGCCGTAAACGATGAGTGCTAAGTGTT
AGAGGGGTTTCCGCCCTTTAGTGCTGCAGTTAACGCATTAAGCACTCCGCCTGGGAGTACGGCCG
CAAGGCTGAAACTCAAAGGAATTGACGGGGGCCCGCACAAGCGGTGGAGCATGTGGTTTAATTCG
AAGCAACGCGAAGAACCTTACCAGGTCTTGACATCCTCTGAAAACCCTAGAGATAGGCTTCTCC
TTCGGGAGCAGAGTGACAGGTGGTGCATGGTTGTCGTCAGCTCGTGTCGTGAGATGTTGGGTTAA
GTCCCGCAACGAGCGCAACCCTTGATCTTAGTTGCCATCATTAAGTTGGGCACTCTAAGGTGACT
GCCGGTGACAAACCGGAGGAAGGTGGGGATGACGTCAAATCATCATGCCCCTTATGACCTGGGCT
ACACACGTGCTACAATGGACGGTACAAAGAGCTGCAAAGACCGCGAGGTGGAGCTAATCTCATAAA
ACCGTTCTCAGTTCGGATTGTAGGCTGCAACTCGCCTACATGAAGCTGGAATCGCTAGTAATCGC
GGATCAGCATGCCGC

6.	C$_{10}$	Bacillus ionarensis	MH671872

GGAATTATTGGGCGTAAAGCGCGCGCAGGCGGCTTCTTAAGTCTGATGTGAAATCTTGCGGCTCA
ACCGCAAGCGGCCATTGGAAACTGGGAAGCTTGAGTACAGAAGAGGAGAGTGGAATTCCACGTGT
AGCGGTGAAATGCGTAGATATGTGGAGGAACACCAGTGGCGAAGGCGACTCTCTGGTCTGTAACT
GACGCTGAGGCGCGAAAGCGTGGGGAGCAAACAGGATTAGATACCCTGGTAGTCCACGCCGTAAA
CGATGAGTGCTAGGTGTTAGGGGTTTCCGCCCTTTAGTGCTGCAGCTAACGCATTAAGCACTCC
GCCTGGGGAGTACGGTCGCAAGGCTGAAACTCAAAGGAATTGACGGGGACCCGCACAAGC
AGTGGAGCATGTGGTTTAATTCGAAGCAACGCGAAGAACCTTACCAGGTCTTGACATCCTTTGAC
CACCCTAGAGATAGGGCTTTCCCTTCGGGGACAAAGTGACAGGTGGTGCATGGTTGTCGTCAGCT
CGTGTCGTGAGATGTTGGGTTAAGTCCCGCAACGAGCGCAACCCTTGATCTTAGTTGCCAGCATT
CAGTTGGGCACTCTAAGGTGACTGCCGGTGACAAACCGGAGGAAGGTGGGGACGACGTCAAATCA
TCATGCCCCTTATGACCTGGGCTACACACGTGCTACAATGGATGGTACAAAGGGCAGCGAAACCG
CGAGGTGGAGCCAATCCCATAAAGCCATTCTCAGTTCGGATTGTAGGCTGCAACTCGCCTACATG
AAGCCGGAATTGCTAGTAATCGCGGATCATCGCGGTGAATACGTTCCCGGGTCTTGTACA

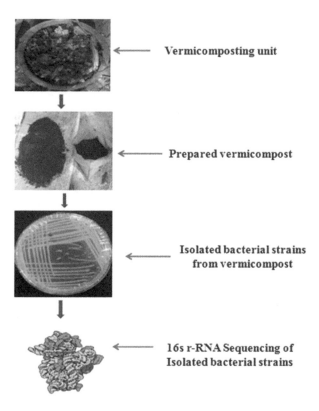

Vermicomposting unit

Prepared vermicompost

Isolated bacterial strains
from vermicompost

16s r-RNA Sequencing of
Isolated bacterial strains

FIGURE 6.2 Schematic representation of isolation and identification of bacterial strains from prepared vermicompost.

have also reported that the microorganisms isolated from the agro-waste compost could be used for biofertilizer production, for use as industrial microorganisms, and for pollution mitigation.

6.8   CONCLUSION

The process of decomposition of wastes is a natural process that aids in the reduction of waste. In nature, the wastes that are generated are reduced by activity of a variety of living organisms and biogeochemical processes. The major role of an annelid, earthworm in conversion of organic matter into nutrient rich organic matter takes place in nature. Solid organic wastes that pose a large threat to nature due to their accumulation need to be efficiently removed. The rate of generation of wastes is faster than nature's ability to recycle them. So, biodegradation through composting and vermicomposting could be a simple and efficient technique, a practice that has been in use for a long time. Microbes play a significant role in the decomposition of organic wastes. Molecular characterization of the bacterial species helps to identify the organisms and use them for further degradation studies.

The utilization of vermicompost in agriculture is a traditional method in reducing waste and improving soil fertility and crop production. The role of microbes in this process has a significant part to play, by secreting enzymes and degrading different components. These organisms can be isolated and identified for specific degradation process. Molecular biology for specific microbial identification and the further growth and preparation of degrading organisms can be an eco-friendly method for waste management.

ACKNOWLEDGMENTS

The School of Environment and Sustainable Development, Central University of Gujarat, Gandhinagar, 382030 Gujarat, and Department of Botany, Avinashilingam Institute for Home Science and Higher Education for Women, Coimbatore-641043, Tamil Nadu, India, and GSBTM, Gandhinagar, Gujarat, (for sequencing) are highly acknowledged.

REFERENCES

Abd, E.M.Y., Monif, M., Rizk, S.G., Shehata, S.M. 1976. Biological activities during ripening of composts. *Infektionskrankheiten und Hygiene* 131:744–750.

Ahring, B., Angelidaki, I., Johansen, K. 1992. Anaerobic treatment of manure together with industrial waste. *Water Science Technology* 25:311–318.

Aira, M., Monroy, F., Dominguez, J., Mato, S. 2002. How earthworm density affects microbial biomass and activity in pig manure. *European Journal of Soil Biology* 38:7–10.

Anbalagan, M., Manivannan, S., Arulprakasm, B. 2012. Biomanagement of *Parthenium hysterophorus* (Asteraceae) using an earthworm (Savigny) for recycling nutrients. *Advances in Applied Science Research* 3:3025–3031.

Angelidaki, I., Ahring, B.K. 1997. Co digestion of olive mill waste water with manure, house hold waste or sewage sludge. *Biodegradation* 8:221–226.

Angelova, V.R., Akova, V.I., Artinova, N.S., Ivanov, K.I. 2013. The effect of organic amendments on soil chemical characteristics. *Bulgarian Journal of Agriculture* 19:958–971.

Ansari, A., Ismail, S.A. 2008. Paddy cultivation in sodic soil through vermitech. *International Journal of Sustainable Crop Production* 3:1–4.

Ansari, A.A. 2008. Soil profile studies during bioremediation of sodic soils through the application of organic amendments (vermiwash, tillage, green manure, mulch, earthworms and vermicompost). *World Journal of Agricultural Sciences* 4:550–553.

Anto Paul, P., Jithesh, K.P., Hiranmai Yadav, R., Vijayakumari, B. 2008. Efficacy of fresh the decomposed parthenium and poultry droppings in improving the yield component and quantity of chickpea (*Cicer arietinum* var. CO 1). National Seminar *on* Recent Advances *in* Plant Biology *and* Biotechnology, held at Department Botany, Government Arts College, Coimbatore on 29th February, 2008, Abstract, p. 16.

Arancon, N.Q., Edwards, C.A., Babenko, A., Cannon, J., Galvis, P., Metzger, J.D. 2008. Influences of vermicomposts, produced by earthworms and microorganisms from cattle manure, food waste and paper waste, on the germination, growth and flowering of Petunias in the green house. *Applied Soil Ecology* 39:91–99.

Arancon, N.Q., Edwards, C.A., Lee, S., Byrne, R. 2006b. Effects of humic acids from vermicomposts on plant growth. *Eurasian Journal of Soil Biology* 46:65–69.

Arunkumar, J. 2004. Effect of vermicomposted sludge on growth of *Amaranthus dubius*. *Journal of Ecotoxicology Environment Monitoring* 14:157–160.

Atiyeh, R.M., Lee, S., Edwards, C.A., Arancon, N.Q., Metzger, J.D. 2002. The influence of humic acids derived from earthworms – Processed organic wastes on plant growth. *Bioresource Technology* 84:7–14.

Bajsa, O., Nair, J., Mathew, K., Ho, G.E. 2003. Vermiculture as a tool for domestic waste water management. *Water Science and Technology* 48:125–132.

Balamurugan, V., Vijayalakshmi, G.S. 2004. Recycling of market waste into vermicompost and its impact on the growth of *Vigna mungo*. National Seminar on Rural Biotechnology for Sustainable Development, The Gandhigram Rural Institute, Gandhigram, 19th and 20th February, 2004, Abstract, p. 28.

Bano, K. 1997. Vermicomposting in changing agricultural scenario. 3rd IFOAM – ASIA Scientific Conference and General Assembly on Food Security in Harmony with Nature, University of Agric. Sci., Hebbal Campus, Bangalore, 1st–4th December, 1997, Abstract, pp. 63–64.

Bano, K., Kale, R.D. 1988. Reproductive potential and existence of endogenous rhythm in reproduction in earthworm *Eudrillus eugeniae*. *Proceedings of Zoological Society, Calcutta* 38:9–14.

Bano, K., Kale, R.D., Gajanan, G.N. 1987. Culturing of earthworms *Eudrillus eugeniae* for cast production and assessment of worm cast as biofertilizer. *Journal of Soil Biology* 7:98–105.

Bansal, S., Kapoor, K.K. 2000. Vermicomposting of crop residues and cattle dung with *Eisenia foetida*. *Bioresource Technology* 73:95–98.

Banta, G., Dev, S.P. 2009. Field evaluation of nitrogen enriched with phosphocompost prepared from green biomass of *Lantana camara* in wheat. *Indian Journal of Ecology* 36:39–44.

Barman, D., Saud, Z.A., Habib, M.R., Islam, M.F., Hossain, K., Yeasmin, T. 2011. Isolation of cellulytic bacterial strains from soil for effective and efficient bioconversion of solid waste. *Life Sciences and Medicine Research* 2011:LSMR-25.

Basheer, M., Agarwal, O.P. 2013. Management of paper waste by vermicomposting using epigeic earthworm *Eudrillus eugeniae* in Gwalior, India. *International Journal of Current Microbiology and Applied Sciences* 2:42–47.

Bharadwaj, K.K.R., Gaur, A.C. 1985. *Recycling of Organic Wastes*. ICAR, New Delhi, p. 104.

Bhatia, S., Sinha, R.K., Sharma, R. 2000. Seeking alternatives to chemical fertilizers for sustainable agriculture: A study of the impact of vermiculture on the growth and yield of potted wheat crops (*Triticum aestivum* Linn.). *Environmental Education and Information* 19:295–304.

Bhawalkar, U.S. 1992. Chemical farming to natural farming – A quick change over without loss of yield. In: *Proc.* Nat. Sem. Natural Fmg. L.L. Somani, K.L. Tatawat, B.L. Baser (eds.). Rajasthan College Agric., Udaipur, pp. 79–88.

Bhawalkar, V., Bhawalkar, V.S. 1991. *Vermiculture Biotechnology*. Bhawalkar Earthworm Research Institute, Pune, MS.

Bhide, M.R. 1988. In: *Naisargik Khat Nirmiti Ek Navi Disha*. N.R. Bhide (ed.). Moreshwar Co-op., Housing Society, Baner Road, Pune, pp. 30–32.

Biddappa, C.C., Palaniswami, C., Upadhyay, A.K., Ramanujam, B., Bopaiah, M.G. 1997. Preparation of composts by various methods and its impact on the nutrition and productivity of coconut and arecanut. Annual Report 1996–97. Central Plantation Crop Research Institute (Indian Council of Agric. Res.), Kasargod, Kerala, pp. 75–76.

Boulter, J.I., Trevors, J.T., Boland, G.J. 2002. Microbial studies of compost: Bacterial identification, and their potential for turfgrass pathogen suppression. *World Journal of Microbiology & Biotechnology* 18:661–671.

Burton, C.H., Turner, C. 2003. Manure management – Treatment strategies for sustainable agriculture. *Proceedings of the MATRESA, EU Accompanying Measure Project*, 2nd Edn, Silsoe Research Institute, Wrest Park, Silsoe, Bedford, UK.

Butler, T.A., Sikora, L.J., Steinhilber, P.M., Douglass, L.W. 2001. Compost age and sample storage effects on maturity indicators of biosolids compost. *Journal of Environmental Quality* 30:2141–2148.

Campitelli, P.A., Velasco, M.I., Ceppi, S.B. 2006. Chemical and physico chemical characteristics of humic acids extracted from the compost, soil and amended soil. *Talanta* 69:1234–1239.

Channabasavanna, A.S. 2003. Organic farming to meet the demands of world trade organisation. *Kisan World* 30:37–38.

Channabasavanna, A.S., Biradar, D.P. 2002. Poultry by-product to avoid pollution. *Kisan World* 29:52–53.

Chaudhary, D.R., Bhandari, S.C., Shukla, M. 2004. Role of vermicompost in sustainable agriculture – A review. *Agricultural Review* 25:29–39.

Chauhan, A., Joshi, P.C. 2010. Composting of some dangerous and toxic weeds using *Eisenia foetida*. *American Science* 6:1–6.

Christopher, M.S.M. 1996. Recycling of plantation agro-wastes. *The Planters' Chronicle* 91:53–61.

Dash, M.C., Senapati, B.K. 1986. Vermitechnology, an option for organic waste management in India. In: *Proceedings of National Seminar on Organic Waste Utilization and Vermicomposting*. M.C. Dash, B.K. Senapati, P.C. Mishra (eds.). Sambalpur University, Orissa, India, pp. 157–172.

Deka, P.K., Paul, S.K., Borah, A. 2003. Vermitechnology for economic use and control of water hyacinth. *Pollution Research* 22:385–387.

Divya, U.K. 2001. Relevance of vermiculture in sustainable agriculture. *Kisan World* 28:9–12.

Edwards, C.A. 1995. Historical overview of vermicomposting. *Biocycle* 36:56–58.

Edwards, C.A. 1998. The use of earthworms in the breakdown and management of organic wastes. In: *Earthworm Ecology*. C.A. Edwards (ed.). Lewis, Boca Raton, FL, pp. 327–354.

Edwards, C.A., Dominguez, J., Neuhauser, E.F. 1998. Growth and reproduction of *Perionyx excavates* (Perr.) (Megascolecidae) as factors in organic waste management. *Biological Fertile Soils* 27:155–161.

Eida, M.F., Nagaoka, T., Wasaki, J., Kouno, K. 2012. Isolation and characterization of cellulose-decomposing bacteria inhabiting sawdust and coffee residue composts. *Microbes in Environment* 27:226–233.

Elvira, C., Sampedro, L., Benitez, E., Nogales, R. 1998. Vermicomposting of sludges from paper mill and dairy industries with *Eisenia andrei*: A pilot scale study. *Bioresource Technology* 63:205–211.

Galler, W.S., Davey, C.B. 1971. High rate poultry manure composting with saw dust. *Livestock Waste Management and Pollution Abatement*, ASAE Publication, PROC-271, pp. 159–162.

Galli, E., Tomati, U., Grappelli, A., Di Lena, A. 1990. Effect of earthworm casts on protein synthesis in *Agaricus bisporus*. *Biological Fertile Soils* 9:290–291.

Garg, P., Gupta, A., Satya, S. 2006. Vermicomposting of different types of waste using *Eisenia foetida*: A comparative study. *Bioresource Technology* 97:391–395.

Garg, V.K., Suthar, S., Yadav, A. 2012. Management of food industry waste employing vermicomposting technology. *Bioresource Technology* 126:437–444.

Gavrilov, K. 1962. Role of earthworm in enrichment of soil by biologically active substances. *Voprosy Ekologie Vyrshaya, Moscow* 7:34.

Gawande, S.S., Jiotode, D.J., Turkhede, A.B. 1998. Effect of organic and inorganic fertilizers on yield and quality of sapota. *Journal of Soils and Crops* 8:58–60.

Ghabbour, S.I. 1966. Earthworms in agriculture-a modern evaluation. *Review of Ecological and Biological Society* 3:259–271.

Ghadge, S., Jadhav, B. 2013. Effect of Lantana manures on nutrient content of fenugreek (*Trigonella foenum graecum* L.). *Bioscience Discovery* 4:189–193.

Gopinathan, M., Thirumurthy, M. 2012. Evaluation of phytotoxicity for compost from organic fraction of municipal solid waste and paper and pulp mill sludge. *Environmental Resources and Engineering Management* 1:47–51.

Govindan, M., Muralidharan, P., Sasikumaran, S. 1995. Influence of vermicompost in the field performance of bhindi (*Abelmoschus esculentus* L.) in a laterite soil. *Journal of Tropical Agriculture* 33:173–174.

Gunathilagaraj, K., Ravignanam, T. 1996. Effect of vermicompost on mulberry sapling establishment. *Madras Agricultural Journal* 83:476–477.

Gunathilagaraj, K., Ravignanam, T. 1996. Vermicomposting of sericultural wastes. *Madras Agricultural Journal* 83:455–457.

Gupta, P.K. 2003. *Vermicomposting for Sustainable Agriculture*. Agrobio, Jodhpur, India.

Hait, S., Tare, V. 2011. Optimizing vermistabilisation of waste activated sludge using vermicompost as a bulking material. *Waste Management* 31:502–511.

Hangarge, D.S., Raut, R.S., More, S.D., Birajdar, R.R., Pholane, L.P. 2002. Effect of vermicompost and soil conditioner on some physical properties of soil under chilli–spinach system. *Journal of Soils and Crops* 12:41–45.

Harada, Y., Inoko, A., Iwata, T. 1981. Maturing process of city refuse during piling. *Soil Science and Plant Nutrition* 27:357–364.

Harris, G.D., Platt, W.L., Price, B.C. 1990. Vermicomposting in a rural community. *Biocycle* 31:48–51.

Hashemimajd, K., Kalbasi, M., Golchin, A., Shariatmadari, H. 2004. Comparison of vermicompost and composts as potting media for growth of tomatoes. *Journal of Plant Nutrition* 27:1107–1123.

Hemalatha, B. 2012. Vermicomposting of fruit waste and industrial sludge. *International Journal of Advanced Engineering Technology* III:60–63.

Hiranmai Yadav, R., Argaw, A. 2014. Bioconversion of *Parthenium hysterophorus* into organic manure and its value addition using different animal manures. *A National Symposium on Science, Technology and Innovations for National Development*, Wollega University, Ethiopia, pp. 21–22.

Hiranmai Yadav, R., Mekonnen, E., Vijayakumari, B. 2012. Bioconversion of *Parthenium hysterophorus* into valuable organic manure. *Proceedings of the National Conference on Science, Technology, and Innovations for Prosperity of Ethiopia*, 16th–18th May, 2012, Bahir Dar, Ethiopia, East Africa, pp. 54–58.

Hiranmai Yadav, R., Vijayakumari, B. 2002. Vermicomposting of vegetable wastes. *XXV Indian Social Science Congress*, held at University of Kerala, Thiruvananthapuram, from 28th January to 1st February, 2002, Abstract, p. 39.

Hiranmai Yadav, R., Vijayakumari, B. 2003. Influence of vermicompost with organic and inorganic manures on biometrics and yield parameters of chilli (*Capsicum annuum* L. var. PLRI). *Crop Research* 25:236–243.

Hiranmai Yadav, R., Vijayakumari, B. 2004. Impact of vermicompost on biochemical characters of chilli (*Capsicum annuum*). *Journal of Ecotoxicology and Environment Monitoring* 14:51–56.

Hu, Z., Lane, R., Wen, Z. 2008. Composting clam processing wastes in a laboratory and pilot-scale in – Vessel system. *Waste Management* 29:180–185.

Ishak, W.M.F., Wan, S., Jamek, N.A., Jalanni, N.F., Jamaludin, M. 2011. Isolation and identification of bacteria from activated sludge and compost for municipal solid waste treatment system. International Conference on Biology, Environment and Chemistry IPCBEE, Vol. 24, ©IACSIT Press, Singapore.

Jackson, M.L. 1973. *Soil Chemical Analysis*. Prentice Hall India Pvt. Ltd., New Delhi, India.

Jadhav, A.D., Talashilkar, S.C., Powar, A.G. 1997. Influence of the conjunctive use of FYM, vermicompost and urea on growth and nutrient uptake in rice. *Journal of Maharashtra Agricultural University* 22:249–250.

Jadhav, B.J., Damke, M.M., Wagh, A.P., Joshi P.S., Kulkarni, P.M. 1998. Effect of N P K on growth and seed yield of radish (Cv. Pusa *chetki*). *PKV Research Journal* 22:178–179.

Jadhav, S.N., Geeta, L., Konda, C.R. 1999. In situ vermiculture in mulberry. *Kisan World* 26:53.

Jadhav, S.N., Patil, G.M., Kulkarni, B.S., Shivaprasad, M. 1998. Influence of vermicompost, inorganic fertilizer and their combinations on the nutritive value of M-5 mulberry. *Advanced Agricultural Research in India* 10:133–139.

Jambhekar, H.A., Bhinday, M.R. 1992. Effect of vermicompost on sugarcane planting in saline land. In: *Proc. Nat. Sem. Natural Fmg. L.* L. Somani, K.L. Tatawat, B.L. Baser (eds.). Rajasthan College Agric., Udaipur, pp. 275–285.

Jaya, N., Sekiozoic, V., Anda, M. 2006. Effect of precomposting on vermicomposting of kitchen waste. *Bioresource Technology* 97:2091–2095.

Jayashree, S., Rathinamala, J., Velmurugan, P., Lakshmanaperumalsamy, P. 2004. Effect of vermicompost prepared from organic wastes using *Eudrillus eugeniae* and its efficiency on the growth of *Vigna radiata*. National Seminar on Rural Biotechnology for Sustainable Development, The Gandhigram Rural Institute, Gandhigram, 19th and 20th February, 2004, Abstract, p. 9.

Jeyabal, A., Kuppuswamy, G. 1997. Recycling of agricultural and agro-industrial wastes for production of vermicompost and its effect on soil fertility management. 3rd IFOAM – ASIA Scientific Conference and General Assembly on Food Security in Harmony with Nature, University of Agric. Sci., Hebbal Campus, Bangalore, 1st–4th December, 1997, Abstract, pp. 61–62.

Jeyabal, A., Palaniappan, S.P., Chelliah, S. 1998. Evaluation of pressmud based bio-compost in maize. *Madras Agricultural Journal* 85:148–149.

Kalaiselvi, T., Ramasamy, K. 1996. Compost maturity: Can it be evaluated. *Madras Agricultural Journal* 83:609–618.

Kale, R.D., Bano, K. 1986. *Applied Biology and Ecology*, 2nd Edn. Oxford and IBH Publication Co., New Delhi, p. 103.

Kale, R.D., Bano, K., Sreenivas, M.N., Bhagyaraj, D.J. 1987. Influence of worm cast on the growth and mycorrhizal colonization of two ornamental plants. *South Indian Horticulture* 35:433–437.

Karmegam, N., Thilagavathy, D. 2004. Vermiculture biotechnology for agricultural waste recycling and crop production. National Seminar on Rural Biotechnology for Sustainable Development, The Gandhigram Rural Institute, Gandhigram, 19th and 20th February, 2004, Abstract, pp. 51–52.

Karuna, K., Patil, C.R., Narayanaswamy, P., Kale, R.D. 1999. Stimulatory effect of earthworm body fluid (vermiwash) on Crinkle Red variety of *Anthurium andreanum* Lind. *Crop Research* 17:253–257.

Kasiviswanathan, K. 1988. Green manuring. *Indian Silk* 26:25.

Kaur, A., Singh, J., Vig, A.P., Dhaliwal, S.S., Rup, P.J. 2010. Composting with and without *Eisenia foetida* for conversion of toxic paper mill sludge into soil conditioner. *Bioresource Technology* 101:8192–8198.

Kaushik, P., Garg, V.K. 2003. Vermicomposting of mixed solid textile mill sludge and cow dung with epigeic earthworm *Eisenia foetida. Bioresource Technology* 90:311–316.

Kavitha, G., Bhardwaj, N. 2000. Effect of vermicompost of parthenium on two cultivars of wheat. *Indian Journal of Ecology* 27:177–180.

Kikon, J.Z., Sharma, S.K. 2003. Relative efficiency of local and exotic earthworm species in vermicomposting. *Advanced in Plant Science* 16:421–427.

Kour, D., Rana, K.L., Yadav, N., Yadav, A.N., Rastegari, A.A., Singh, C., Negi, P., Singh, K., Saxena, A.K. 2019. Technologies for biofuel production: Current development, challenges, and future prospects. In: *Prospects of Renewable Bioprocessing in Future Energy Systems, Biofuel and Biorefinery Technologies*. A.A. Rastegari et al. (eds.), Vol. 10. Springer, Berlin, pp. 1–50.

Krishnamoorthy, R.V., Vajranabaiah, N. 1986. *Biological Activity of Earthworm Casts*. Indian Academic Science Press, pp. 341–351.

Kumar, J., Sharma, K.L. 1998. Potential sources of pollution. *Indian Farmers Digest* 31:29–32.

Kumarasamy, K. 2000. Environment-friendly recycling of organic wastes. *Kisan World* 27:37–39.

Lee, K.E. 1985. *Earthworms: Their Ecology and Relationship with Soil and Land Use*. Academic Press, Sydney, pp. 188–194.

Loh, T.C., Lee, Y.C., Liang, J.B., Tan, D. 2005. Vermicomposting of cattle and goat manures by *Eisenia foetida* and their growth and reproduction performance. *Bioresource Technology* 96:111–114.

Lui, S., Xiang, D., Wu, D. 1992. Studies on the effect of earthworm on the fertility of red arid soil. In: *Advances in Management and Conservation of Soil Fauna*. G.K. Veeresh, D. Rajagopal, C.A. Viraktamath (eds.). Oxford and IBH, New Delhi, pp. 543–546.

Maheswarappa, H.P., Nanjappa, H.V., Hegde, M.R. 1998. Effect of sett size, plant population and organic manures on growth components of arrow root (*Maranta arundinacea* L.) grown as intercrop in coconut garden. *Mysore Journal of Agricultural Science* 32:257–263.

Maheswarappa, H.P., Nanjappa, H.V., Hegde, M.R. 1999. Influence of organic manures on yield of arrow root, soil physico-chemical and biological properties when grown as intercrop in coconut garden. *Annals of Agricultural Research* 20:318–323.

Manonmani, M., Anand, R. 2002. Vermicompost – An uprising fertilizer for lady's finger (*Hibiscus esculentus*). *Kisan World* 29:40.

Manyuchi, M.M., Phiri, A. 2013. Vermicomposting in solid waste management: A review. *International Journal of Scientific Engineering and Technology* 2:1234–1242.

Mary, D.D., Sivagami, S. 2014. Effect of individual and combined application of biofertilisers, vermicomposts and inorganic fertilizers on soil enzymes and minerals during the post harvesting stage of chilli (NS 1701). *Research Journal of Agriculture and Environmental Management* 3:434–441.

Mathan, K.K. 2000. Impact of biological wastes on soil physical properties and yield of maize and finger millet. *Madras Agricultural Journal* 87:618–620.

Mattson, M. 1998. Influence of nitrogen nutrition and metabolism on ammonia volatilization in plants. *Nutrient Cycle Agroecosystems* 51:35–40.

Mohan, M., Subha, R., Ganesh, P., Vijayalakshmi, G.S. 2002. Vermicompost for organic rice production. *Kisan World* 29:33–34.

Molina, M.J., Soriano, M.D., Ingelmo, F., Llinares, J. 2013. Stabilisation of sewage sludge and vinasse bio wastes by vermicomposting with rabbit manure using *Eisenia foetida*. *Bioresource Technology* 137:88–97.

Moradi, H., Fahramand, M., Sobhkhizi, A., Adibian, M., Noori, M., Abdollahi, S., Rigi, K. 2014. Effect of vermicompost on plant growth and its relationship with soil properties. *International Journal of Farming and Allied Sciences* 3:333–338.

Munnoli, P.M., Arora, J.K., Sharma, S.K. 2002. Impact of vermiprocessing on soil characteristics. *Journal of Industrial Pollution Control* 18:87–92.

Nagarajan, R., Manickam, T.S., Kothandaraman, G.V., Ramaswamy, K., Palaniswamy, G.V. 1985. Coir pith as manure for groundnut. *TNAU Newsletter* 15:18–25.

Nagarajan, S.S. 1997. Vermitech in rice cultivation. *Kisan World* 24:49–50.

Nagarathinam, B., Karmegam, N., Thilagavathy, D. 2000. Microbial changes in some organic materials subjected to earthworm action. *Journal of Eco-Biology* 12:45–48.

Nagarathinam, P. 2004. Soil phenolics and their effect on plant growth. *Journal of Ecotoxicological Environment Monitoring* 14:89–92.

Nainawat, R. 1997. *Vermitechnological Studies on Organic Solid Waste Management*. Ph.D. Thesis, Rajasthan University, India.

Nakasaki, K., Fujiwara, S., Kubota, H. 1994. A newly isolated thermophilic bacterium, *Bacillus Licheniformis* HA1 to accelerate the organic matter decomposition in high rate composting. *Journal of Compost Science and Utilization* 2:2.

Nayak, A.K., Rath, L.K. 1996. Vermiculture and its application. *Kisan World* 23:61–62.

Neetha, S., Mira, M. 1988. Effects of various organic wastes alone and with earthworms on the total dry matter yield of wheat and maize. *Biological Wastes* 25:33–40.

Nethra, N.N., Jayaprasad, K.V., Kale, R.D. 1999. China aster (*Callistephus chinensis* (L.) Ness) cultivation using vermicompost as organic amendment. *Crop Resources* 17:209–215.

Padma, U., Rao, S.R., Srinivas, N. 2002. Eco-friendly disposal of vegetable wastes through vermitechnology. *Journal of Eco-Biology* 14:155–159.

Pagligai, M., Guindi, G., Marca, M.L., Giachetti, M., Lucamante, G. 1981. Effects of sewage sludge and composts on soil porosity and aggregates. *Journal of Environmental Quality* 10:556–561.

Pena-Mendez, E.M., Havel, J., Patocka, J. 2005. Humic substances-compounds of still unknown structure: Applications in agriculture, industry, environment and biomedicine. *Journal of Applied Biomedicine* 3:13–24.

Piper, C.S. 1966. *Chemical Analysis Saline Soil. Soil and Plants Analysis*. Hans Publication, Bombay, India.

Pramod, K., Madhuri, P., Negi, G.C.S. 2009. Soil physico chemical properties and crop yield improvement following Lantana mulching and reduced tillage in rainfed croplands in the Indian Himalayan Mountains. *Journal of Sustainable Agriculture* 33:636–657.

Prasad, A., Saini, S.K. 2013. Vermicomposting of city garbage with earthworm species *Eisenia foetida*: A better way of solid waste management with nutritional enhancement. *Indian Journal of Applied Research* III(V):Online.

Probodhini, J. 1994. Recycle kitchen waste into vermicompost. *India Farming* 43:34.

Purekayastha, T.J., Bhatnagar, R.K. 1997. Vermicompost: A promising source of plant nutrients. *India Farming* 46:35–37.

Rachappaji, K.S., Siddapaji, C., Chinnaswamy, K.P., Kale, R.D., Sundar, S.R. 2004. Influence of vermicompost supplemented with different levels of NPK on growth parameters of mulberry. National Seminar on Rural Biotechnology for Sustainable Development, The Gandhigram Rural Institute, Gandhigram, 19th and 20th February, Abstract, p. 10.

Rajendran, P. 1991. *Evaluation of Coir Waste and Farm Yard Manure in Different Forms and Methods of Application in Rainfed Cotton Based Intercropping System and Their Residual Effects on Sorghum*. Ph.D. Thesis, TNAU, Coimbatore.

Ramalingam, R. 2001. Vermicomposting of crop residue sugarcane trash using an Indian epigeic earthworm, *Perionyx excavatus*. *Asian Journal of Microbiology, Biotechnology and Environmental Sciences*, 3:269–273.

Ramaswami, P.P. 1997. Potential use of parthenium. In: *Proc. First International Conference on Parthenium Management*, Vol. 1, pp. 77–80.

Rao, M., Krishna, R., Sivagnanam, S.K., Syed, F.B. 2016. Isolation, characterization and identification of predominant microorganisms from agro-waste. *Scholars Research Library Der Pharmacia Lettre* 8:79–86.

Ravankar, H.N., Puranik, R.B., Kumar, G.P. 1998. Role of earthworm culture in rapid decomposition of farm waste. *PKV Research Journal* 22:19–22.

Ravignanam, T., Gunathilagaraj, K. 1996. Effect of earthworm on mulberry biochemical characters. *Madras Agricultural Journal* 83:451–454.

Ravignanam, T., Gunathilagaraj, K. 1996. Effect of earthworm on mulberry plant characters. *Madras Agricultural Journal* 83:381–384.

Reddy, G.S., Pranavi, S., Srimoukthika, B., Reddy, V.V. 2017. Isolation and characterization of bacteria from compost for municipal solid waste from Guntur and Vijayawada. *Journal of Pharmaceutical Science & Resources* 9:1490–1497.

Reddy, K.B., Rao, D.M.R., Reddy, D.C., Suryanarayana, N. 2003. Studies on the effect of farmyard manure and vermicompost on quantitative and qualitative characters of mulberry (*Morus* spp.) under semi-arid conditions of Andhra Pradesh. *Advanced Plant Science* 16:177–182.

Reddy, P.S., Rao, T.V.S.S., Venkatramana, P., Suryanarayana, S. 2001. Vermicompost in management of nutrients and leaf yield in V-1 mulberry variety. *Journal of Environmental Resources* 11:137–140.

Reddy, R.M., Reddy, A.N., Reddy, Y.T.N. 1998. Effect of organic and inorganic sources of NPK on growth and yield of pea (*Pisum sativum*). *Legume Research* 21:57–60.

Reddy, S.A., Bagyaraj, D.J., Kale, R.D. 2012. Vermicompost as a biocontrol agent in suppression of two soil-borne plant pathogens in the field. *Acta Biologica Indica* 1:137–142.

Roeper, H., Khan, S., Koerner, I., Stegman, R. 2005. Low tech options for chicken manure treatment and application possibilities in agriculture. *Proceedings Sardinia. Tenth International* Waste Management *and* Landfill Symposium, S. Margherita di Pula, Cagliari, Italy.

Saha, A., Santra, S.C. 2014. Isolation and characterization of bacteria isolated from municipal solid waste for production of industrial enzymes and waste degradation. *Journal of Microbiology & Experimentation* 1:1.

Saikia, J.K., Pathak, A.K. 1997. Effect of levels of NPK with or without soil test on grain yield of low land kharif rice and its residual effect on succeeding summer rice. *Annals of Agricultural Resources* 18:530–532.

Scott, T.K. 1984. Hormonal regulation of development II. In: *Encyclopedia of Plant Physiology*. T.K. Scott (ed.), New Series, Vol. 10. Springer, Berlin, Heidelberg, New York, NY, Tokyo.

Senthilkumar, M., Sivalingam, P.N., Suganthi, A. 2000. Vermitech. *Kisan World* 27:57–58.

Sharma, P.K., Singh, Y.V. 2012. Litter bag study of Lantana shrub decomposition in Alluvial soils of North India. *Libyan Agriculture Research Center Journal International* 2:26–28.

Sharma, V., Kamla, K., Dev, S.P. 2004. Efficient recycling of obnoxious weed plants (*Lantana camara* L.) and congress grass (*Parthenium hysterophorus* L.) as organic manure through vermicomposting. *Journal Indian Society of Soil Science* 52:112–114.

Siddiqqui, M.A. 2017. Molecular characterisation of bacterial species isolated from landfill site. *Research Journal of Environmental Sciences* 11(2). doi:10.3923/rjes.2017.65.70.

Singh, C., Tiwari, S., Boudh, S., Singh, J.S. 2017a. Biochar application in management of paddy crop production and methane mitigation. In: *Agro-Environmental Sustainability: Managing Environmental Pollution*. J.S. Singh, G. Seneviratne (eds.), 2nd Edn. Springer, Switzerland, pp. 123–146.

Singh, C., Tiwari, S., Gupta, V.K., Singh, J.S. 2018. The effect of rice husk biochar on soil nutrient status, microbial biomass and paddy productivity of nutrient poor agriculture soils. *Catena* 171:485–493.

Singh, C., Tiwari, S., Singh, J.S. 2017b. Impact of rice husk biochar on nitrogen mineralization and methanotrophs community dynamics in paddy soil. *International Journal of Pure and Applied Bioscience* 5:428–435.

Singh, C., Tiwari, S., Singh, J.S. 2017c. Application of biochar in soil fertility and environmental management: A review. *Bulletin of Environment, Pharmacology and Life Sciences* 6:07–14.

Singh, C., Tiwari, S., Singh, J.S. 2019. Biochar: A sustainable tool in soil 2 pollutant bio-remediation. In: *Bioremediation of Industrial Waste for Environmental Safety*. R.N. Bharagava, G. Saxena (eds.). Springer, Berlin, pp. 475–494.

Singh, K.P., Angiras, N.N. 2011. Allelopathic effect of wild sage (*Lantana camara* L.) compost on wheat and associated weeds under north western Himalayas. *Journal of Environment and Bio-Sciences* 25:91–92.

Singh, K.P., Rinwa, R.S., Singh, H., Kathuria, M.K. 1997. Substitution of chemical fertilizers with vermicompost in cereal based cropping systems. 3rd IFOAM – ASIA Scientific Conference and General Assembly on Food Security in Harmony with Nature, 1st–4th December, 1997, University of Agric. Sci., Hebbal Campus, Bangalore, Abstract, p. 59.

Singh, K.P., Singh, H., Ainwa, R.S., Kathuria, M.K., Singh, S.M. 1998. Relative efficacy of vermicompost and some other organic manures integrated with chemical fertilizers in cereal based cropping systems. *Haryana Journal of Agronomics* 14:34–40.

Singh, R., Bhakuni, D.S., Kumar, N. 2003. Pine needles useful for vermicompost in central Himalayas. *Indian Farming* 52:13–14.

Sinha, R.K., Heart, S., Agarwal, S., Asadi, R., Carretero, E. 2002. Vermiculture and waste management: Study of action of earthworms *Eisenia foetida, Eudrillus eugeniae* and *Perionyx excavatus* on biodegradation of some community wastes in India and Australia. *The Environmentalist* 22:261–268.

Sinha, R.K., Nair, J., Bharambe, G., Patil, S., Bapat, P.S. 2008. Vermiculture revolution: A low cost and sustainable technology for management of municipal and industrial organic wastes (solid and liquid) by earthworms with significantly low greenhouse gas emissions. In: *Progress in Waste Management Research*. J.I. Daven, R.N. Klein (eds.). NOVA Science Publishers, Hauppauge, pp. 159–227.

Sita, J.P., Hari, N.S. 1997. Vermicompost increases ears and grains of rice. *Indian Farming* 47:29.

Smidt, E., Meissl, K., Schmutzer, M., Hinterstoisse, B. 2007. Co composting of lignin to build of humic substances – Strategies in waste management to improve compost quality. *Indian Crop Production* 27:196–201.

Son, T.T.N. 1995. *Bioconversion of Organic Wastes for Sustainable Agriculture*. Ph.D. Thesis, TNAU, Coimbatore.

Sreenivas, C., Muralidhar, S., Singa Rao, M. 2000. Yield and quality of ridge guard fruits as influenced by different levels of inorganic fertilizers and vermicompost. *Annals of Agriculture Research* 21(1):262–266.

Srivastava, O.P., Mann, G.S., Bhatia, I.S. 1971. The evaluation of rural and urban compost. *Journal of Research (PAU)* 8:451–455.

Sultan, I. 2001. Soil a living organism – A "worm's eye" view. *Kisan World* 28:24–26.

Suriyanarayanan, S., Mailappa, A.S., Jayakumar, D., Nanthakumar, K., Karthikeyan, K., Balasubramania, S. 2010. Studies on characterization and possibilities of reutiliza-tion of solid wastes from a waste paper based paper industry. *Global Journal of Environmental Resources* 4:18–22.

Suthar, S., Sharma, P. 2013. Vermicomposting of toxic weed – *Lantana camara* biomass: Chemical and microbial properties changes and assessment of toxicity of end product using seed bioassay. *Ecotoxicology and Environmental Safety* 95:179–187.

Talashilkar, S.C., Dosani, A.A.K., Mehta, V.B., Powar, A.G. 1997. Integrated use of fertil-izers and poultry manure to groundnut crop. *Journal of Maharashtra Agricultural University* 22:205–207.

Tamak, S.K., Kadian, V.S., Singh, K.P. 1998. Effect of different levels of organic manures, agrispon and nitrogen on growth and yield of cotton (*Gossypium hirsutum* L.). *Haryana Journal of Agronomics* 14:158–163.

Thakur, C., Kaur, A., Jain, N., Banger, S.K. 2000. Utilization of organic waste through ver-micompost technique. *Journal of Agricultural Issues* 5:27–29.

Thangavel, P., Balagurunathan, R., Prabakaran, J. 2003. Nutrient dynamics of the tamarind leaf litter composted with various inoculants. *Indian Journal of Ecology* 30:117–120.

Thanunathan, K., Natarajan, S., Senthilkumar, R., Arulmurugan, K. 1997. Effect of different sources of organic amendment son growth and yield of onion in mine spoil. *Madras Agricultural Journal* 84:382–384.

Tharmaraj, K., Ganesh, P., Kumar, S., Anandan, A.R., Kolanjinathan, K. 2011. Vermicompost – A soil conditioner cum nutrient supplier. *International Journal of Pharmaceutical and Biological Archives* 2:1615–1620.

Thilagavathi, T. 1992. *Evaluation of Modified Method of Coir Pith Degradation and the Influence of Periodically Decomposed Coir Pith on Soil Properties and Yield of Rice.* M.Sc. (Agri.) Thesis, TNAU, Madurai.

Tiwari, S., Singh, C., Boudh, S., Rai, P.K., Gupta, V.K., Singh, J.S. 2019a. Land use change: A key ecological disturbance declines soil microbial biomass in dry tropical uplands. *Journal of Environmental Management* 242:1–10.

Tiwari, S., Singh, C., Singh, J.S. 2018. Land use changes: A key ecological driver regulating methanotrophs abundance in upland soils. *Energy, Ecology, and the Environment* 3:355–371.

Tiwari, S., Singh, C., Singh, J.S. 2019b. Wetlands: A major natural source responsible for methane emission. In: *Restoration of Wetland Ecosystem: A Trajectory Towards a Sustainable Environment.* A.K. Upadhyay et al. (eds.). Springer, Berlin, pp. 59–74.

Tomati, U., Galli, E., Grapelli, A.T., Lena, G.D. 1990. Effect of earthworm casts in protein synthesis in radish (*Raphanus sativus*) and lettuce (*Laectuca sativa*) seedlings. *Biological Fertile Soils* 9:1–2.

Tomati, U., Galli, E., Grapelli, A., Lena, G.D. 1990. Effect of earthworm casts on protein synthesis in radish (*Raphanus sativum*) and lettuce (*Lectuca sativa*) seedlings. *Biological Fertile Soils* 9:288–289.

Tomati, U., Grappelli, A., Galli, E. 1988. The hormone like effect of earthworm casts. *Biological Fertile Soils* 5:288–294.

Umamaheswari, S., Vijayalakshmi, G.S. 2003. Vermicomposting of paper mill sludge using an African earthworm species *Eudrillus eugeniae* (KINBERG) with a note on its physico-chemical features. *Pollution Research* 22:339–341.

Ushakumari, K., Prabhakumari, P., Padmaja, P. 1999. Efficiency of vermicompost on growth and yield of summer crop okra (*Abelmoschus esculentus* Moench). *Journal of Tropical Agriculture* 37:87–88.

Van, G.C.A.M., Breemen, V.V.E.M., Baerselman, R. 1992. Influence of environmental conditions on the growth and reproduction of the earthworm *Eisenia andrei* in an artificial soil substrate. *Pedobiology* 36:109–120.

Vijaya, K.S., Seethalakshmi, S. 2011. Contribution of parthenium vermicompost in altering growth, yield and quality of *Alelmoschus esculentus* (I) Moench. Advance Biotechnology 11:44–47.

Vijayakumari, B., Hiranmai Yadav, R. 2005. Bioconversion of *Parthenium hysterophorus* as organic manure for chilli (*Capsicum annuum* L.). *Indian Journal of Environment and Ecoplantation* 10:27–29.

Vijayakumari, B., Hiranmai Yadav, R. 2008a. Biodiversity conservation. In: *To Fresh, Composted, Vermicomposted Parthenium hysterophorous and Poultry Manure.* M.S. Binoj Kumar and P.K. Gopalakrishnan (eds.). Scientific Publishers, India. ISBN:81-7233-489-3.

Vijayakumari, B., Hiranmai Yadav, R. 2008b. Effect of in situ green manuring on the quality parameters of radish (*Raphanus sativus* var. pusa chetki). *Research on Crops* 9:86–89.

Vijayakumari, B., Hiranmai Yadav, R. 2009a. Effect of fresh, composted and vermicomposted parthenium and poultry droppings on yield and quality of radish (*Raphanus sativus* var. Pusa Chetki). *Environmental Science: An Indian Journal* 4:Online.

Vijayakumari, B., Hiranmai Yadav, R. 2009b. Influence of green manure (chick pea) on the yield and quality of radish (*Raphanus sativus* var. Pusa Chetki). *Research & Reviews in Bio-Sciences* 3:21–23.

Vijayakumari, B., Hiranmai Yadav, R. 2012. Influence of fresh, composted and vermicomposted *parthenium* and poultry manure on the growth characters of sesame (*Sesamum indicum* var. VRI1). *Journal of Organic Systems* 7:14–19.

Vijayakumari, B., Hiranmai Yadav, R. 2013. Influence of fresh, composed, vermicomposed *Parthenium hysterophorus* and poultry manure on the quality parameters of chickpea (*Cicer arietinum* var. Co. 1). *The* 15th Biennial Conference *of The* Crop Science Society *of* Ethiopia *on* Crop Research *and* Development Towards Quality Products *to* Enhance Benefit *and* Competitiveness, held at Ethiopian Institute of Agricultural Research, Addis Ababa, Ethiopia, from 9th–10th May, 2013, pp. 18–19.

Vijayakumari, B., Hiranmai Yadav, R., Raja, A., Vergeese, X. 2009. Influence of fresh, composted and vermicomposted *Parthenium hysterphorus* and poultry droppings on quality parametres of radish. *Journal of Applied Sciences & Environmental Management* 13:79–82.

Vijayakumari, B., Jaime, A., Teixeirada, S., Yadav, R.H. 2011. Influence of fresh and decomposed *Parthenium* and poultry droppings on sesame yield. *The Asian and Australasian Journal of Plant Science and Biotechnology* 5:54–61.

Vishan, I., Sivaprakasam, S., Kalamdhad, A. 2017. Isolation and identification of bacteria from rotary drum compost of water hyacinth. *International Journal of Recycling of Organic Waste from Agriculture* 6:245–253.

Wareing, P.F. 1982. Plant growth substances. In: *Proceedings of the 11th International Conference on Plant Growth Substances.* P.F. Wareing (ed.). Academic Press, London.

Williams, C.M., Barker, J.C., Sims, J.T. 1999. Management and utilization of poultry wastes. *Review of Environment and Contamination Technology* 162:105–157.

Wu, L., Ma, L.Q. 2001. Effects of sample storage on biosolids compost stability and maturity evaluation. *Journal of Environment Quality* 30:222–228.

Wu, L., Ma, L.Q., Martinez, G.A. 2000. Comparison of methods for evaluating stability and maturity of biosolids compost. *Journal of Environment Quality* 29:424–429.

Yadav, A., Garg, V.K. 2010. Bioconversion of food industry sludge into value added product (vermicompost) using epigeic earthworm *Eisenia foetida. World Review of Science and Technology and Sustainable Development* 7:225–238.

Yadav, A., Garg, V.K. 2011. Vermicomposting – An effective tool for the management of invasive weed *Parthenium hysterophorus. Bioresource Technology* 102:5891–5895.

Yadav, A., Gupta, R., Garg, V.K. 2013. Organic manure production from cow dung and biogas plant slurry by vermicomposting under field conditions. *International Journal of Recycling of Organic Waste in Agriculture* 2:1–7.

Zaved, H.K., Rahman, M.M., Rahman, M.M., Rahman, A., Arafat, S.M.Y., Rahman, M.S. 2008. Isolation and characterization of effective bacteria for solid waste degradation for organic manure. *Science and Technology Journal* 8:2.

7 Beneficial Soil Microflora for Enhancement in Crop Production and Nutritive Values in Oil-Containing Nuts

Rishabh Chitranshi and Raj Kapoor

CONTENTS

7.1 INTRODUCTION

Proper and healthy nutrition is a basic need for today. Nuts appear to be an important healthful snack around the globe. Their oils are excellent sources of natural fats, proteins, and vitamins, which also contain a wide range of micronutrients and phytochemicals (phenolic acids, flavonoids, phytoestrogens, and carotenoids) with multi-faceted benefits (Orem et al. 2013). Moreover, tree nuts have natural bioactive and health-promoting components. Tree nuts can be defined as hard-shelled fruits that are dry in nature and possess edible kernels. Seeds are embryonic plants that are enclosed in a seed coat. Due to their nutritional properties, high energy value (23.4 to 26.8 kJ/g), and a wide variety of flavors, nuts are an important element of a diet for many cultures. In addition, many scientific studies have analyzed and confirmed the beneficial effects of nuts and their oils on various human health-related issues (Nile and Park 2014). Nuts can be eaten raw or utilized as an ingredient in many dishes in several other ways, such as in snack foods, cold soups, pastries, or cakes. They can also serve as dietary supplements, and oil can be used for cooking, as food additives, etc. These nuts are also used in the cosmetics industry, and may be consumed as a form of nutritional therapy in different human traditions around the world (Slavin and Lloyd 2012; Vayalil 2012).

Due to being a natural and healthier alternative of other traditional oil products, in recent years, oils containing nuts have not been restricted within the food sector; these nuts are used in a variety of other products, besides confectionery products. Hence, there is no contention about the fact that the nuts and seeds sector has been growing on account of the ever-expanding food and beverage sectors, amongst other industries globally. According to a report, the expected growth of oil containing nuts worldwide is at a CAGR of 1.7%, along with an increase at a CAGR of 10.0% for seeds till the end of 2021. The projected global value of the nuts market is US$1,279.44 bn by 2021.

Today's agriculture is mostly based on technology, so advances in the strategies used for cultivating healthy nuts and oil seeds are undoubtedly a serious issue. In recent agro-economic trends, where a nature-friendly soil is needed for the cultivation of healthy and nutritive crops, the use of beneficial soil microbes could also be a better approach. However, by using chemical pesticides, the gap between yield and productivity is filled to a great extent. But dependence on chemical pesticides has a great many side effects, mainly those of environmental concerns, but also health issues.

This chapter focuses on innovative uses of soil-based beneficial micro-flora as a promising approach for increasing the productivity of oil-containing nuts without any associated ecological harm for a sustainable environment and agriculture.

7.2 PRODUCTION AND ENVIRONMENTAL CLIMATE

The commercially successful cultivation of many fruit and nut trees requires winter chilling, which is specific for every tree cultivar. In order to avoid frost damage of the sensitive tissue in the cold winters of their regions of origin, trees from temperate or cold climates have evolved a period of dormancy during the cold season. After a certain duration of cold conditions (chilling), endodormancy is broken and the tree is ready to resume growth in spring. Chilling requirements vary substantially between species and cultivars from different parts of the world, and the commercial production of temperate tree crops requires selecting appropriate cultivars for the climatic conditions of the planned production site (Luedeling 2012).

7.3 OIL-CONTAINING NUTS

7.3.1 PEANUTS

The peanut is believed to be native to Brazil, Peru, Argentina, and Ghana, from where it was introduced into Cuba, Jamaica, and various West Indies islands (Paranthaman et al. 2009). The plant was introduced by the Portuguese into Africa, from where it was introduced into North America. It was introduced into India during the first half of the sixteenth century from one of the Pacific islands of China, where it was introduced earlier from either Central America or South America (Hammons 1994). The major peanut-producing countries of the world are India, China, Nigeria, Senegal, Sudan, Burma, and the US. Out of a total area of 18.9 million hectares and a total production of 17.8 million tons in the world, these countries account for 69% of the area and 70% of the production (FAO 2011). India occupies a significant position in the world, in regard to both the area and the production of groundnuts. About 7.5 million hectares are used for Indian production annually; the production is about 6 million tons (Madhusudhana 2013). 70% of the area and 75% of the production are concentrated in the four states of Gujarat, Andhra Pradesh, Tamil Nadu, and Karnataka. Andhra Pradesh, Tamil Nadu, Karnataka, and Orissa have irrigated area forms which constitute about 6% of the total groundnut area in India (Sujatha et al. 2016) (Table 7.1).

7.3.2 SESAME

Sesame (*Sesamum indicum* L.) seeds are tropical crops worldwide and known as one of the leading crops processed for oil production. It is known as til (Hindi), hu ma (Chinese), sésame (French), goma (Japanese), gergelim (Portuguese), and ajonjolí (Spanish) (Arawande and Alademeyin 2018). According to early Hindu legends, sesame seeds firstly originated in India, representing a symbol of immortality (Cerletti and Duranti 1979). From India, sesame seeds were introduced throughout the Middle East, Africa, and Asia. India is one of the major creators of oilseed crops and the country consumes a significant amount of sesame oil as a cooking medium (Nagaraj 2009). Sesame has been cultivated mainly for its high content of edible oil and protein (Anilakumar et al. 2010).

TABLE 7.1
Nutritional Value of Peanuts

Content	Percentage
Protein	25.2
Oil	48.2
Starch	11.5
Soluble sugar	4.5
Crude Fiber	2.1
Moisture	6.0

Source: agropidia

TABLE 7.2
Nutritional Value of Sesame Seeds

Nutrient Quantity	Percentage
Moisture	4.0–5.3
Protein	18.3–25.4
Oil	43.3–44.3
Ash	5.2–6.2
Glucose	3.2
Fructose	2.6

Sesame seeds are majorly grown in states Andhra Pradesh, Madhya Pradesh Maharashtra, Gujarat, Rajasthan, West Bengal, Tamil Nadu, and Telangana. The country has increased sesame seed production with a wide range of varieties and seed grades (Anilakumar et al. 2010). With the capability of producing thousands of tons of sesame seeds in a year, India leads the market globally. However, the actual trade is dependent upon the crop quality, monsoon, and many other factors. The Indian crop of sesame seeds was short by 60% in 2018, as against 418000 MT in 2017. The harvest yield of sesame seeds was estimated at around 178000 MT during 2018, due to unfortunate weather conditions in the Eastern and Western regions, which are known for their production of sesame seeds (Table 7.2).

7.3.3 ALMONDS

Almonds are considered one of the best dry fruits for the human body because of their nutritive values (Fernandes et al. 2010). Almonds are also known as one of the most vastly grown fruits of the world. The Middle East, Africa, and Europe are major participants in the production of almonds (N.I.I.R. 2005). Australia, Greece, Iran, Italy, Morocco, Spain, Syria Turkey, Tunisia, and the United States are the top 10 almond-producing countries globally. The United States is the leading producer, with a yield of 898167 metric tons/year. A large amount is cultivated in California (Freeman et al. 2000). Spain is known as a commercial producer for this crop. Catalonia, Valencia, Aragon, Murcia, and Andalusia are known for the cultivation of almonds (Pérez-Campos et al. 2011). Two major varieties – Desmayo Largueta and Marcona – are produced in Spain. Italy ranks third, with 100664 metric tons/year, predominately in the Southern part of the country, including Sicily (De Herralde et al. 2003).

7.3.4 WALNUTS

The walnut (*Juglans sp.*) is a temperate nut fruit. India cultivates different shapes, sizes, and categories of walnuts. The categories of Indian walnuts are thin-shelled, paper-shelled, medium-shelled, and hard-shelled. Jammu and Kashmir, Arunachal Pradesh, Himachal Pradesh, and Uttaranchal are the major walnut-producing states in India (Table 7.3).

TABLE 7.3
Nutritional Value of Walnuts

Calories	183
Fat	18 g
Sodium	0.6 mg
Carbohydrates	3.8 g
Fiber	1.9 g
Sugars	0.7 g
Protein	4.3 g

USDA for one ounce (28 g)

7.4   NUTRITIVE VALUES

Dry nuts are a good source of nutrition, as they contain protein, fat, minerals and vitamins of the B group, and fatty acids. Fatty acids found in dry nuts can be helpful for high blood fat content, if included in a balanced diet (Ros 2010). Moreover, they are rich in concentrated calories, and therefore not to be consumed in large quantities. Nuts can be consumed in their natural form (Saturni et al. 2010). Dry nuts have a lot of nutritional and medicinal properties, because an ample amount of nutrients are present in them (Balch 2006). Though dry fruits are expensive, and regarded as delicacies, the health benefits that they possess justify their cost. Dried fruit contains no fat, cholesterol, or sodium.

7.4.1   PROTEINS AND AMINO ACIDS

Peanut is a leguminous plant so it has a higher protein content than any other nut. All amino acids available in peanuts are the biggest source of the protein called "arginine" (USDA 2014). The protein digestibility-corrected amino acid Score (PDCAAS) shows that peanut proteins are nutritionally comparable to meat and eggs for human health (Massey 2003). Peanut proteins are essentially plant-based proteins, which carry some additional components that have numerous health benefits for mankind. These proteins have excellent water retention and high solubility.

Sesame (*Sesamum indicum* L.) seeds have been grown in tropical regions throughout the world (Anilakumar et al. 2010). An adequate amount of essential amino acids such as methionine, cysteine, and tryptophan, which are limiting amino acids, are found in sesame seeds (Fasuan et al. 2018).

Walnuts do not contain all of the amino acids the body requires. While the nuts have a high concentration of amino acids such as tryptophan, valine, leucine, and threonine, they lack others like methionine, alanine, and proline. Because of this, walnuts are considered an incomplete protein.

7.4.2   VITAMINS

Dry nuts are also termed "superfoods" because they are superior to other food items in their nutritional value (Wolfe 2010). A single bite of these nuts can supply abundant nutrition in the form of vitamins and minerals. Dry nuts have ample amounts of calcium, copper, iron, magnesium, potassium, phosphorus, riboflavin, vitamins A-C-E-K-B6, and zinc (Shear and Faust 1980). Consequently, they have many health benefits, such as protection from anemia, heart disease, cholesterol, immunity, amongst others.

7.4.3 ANTIOXIDANTS

Antioxidants are a class of molecules that inhibit the oxidation of other molecules. Antioxidants are able to slow or even reverse the effects of oxidative damage. This accounts for the fact that they can provide a protective mechanism to reduce the risk of chronic disease (Fraga 2007). Coumaric acid and resveratrol and antinutrients like phytic acid are some of the antioxidants found in peanut seeds (Olatidoye et al. 2019). While in sesame seeds, there are powerful antioxidants: IP-6 (AKA: Phytate; one of the most powerful antioxidants yet found, and one of the most potent natural anti-cancer substances. Especially abundant in grain and sesame, there are lignans, sesamin, sesaminol, sesamolinol, sesamolin, pinoresinol, vitamin E, lecithin, myristic acid, and linoleate are found (Al-Mashadani et al. 2010). Apart from that, almonds contain relatively high levels of vitamin E, which includes tocopherol in its composition. About 28.4 g of plain almonds provides approx. 7.27 mgs of vitamin E, which is around half the daily requirement of the human body (Hollis and Mattes 2007). Walnuts are an excellent source of antioxidative polyphenols, containing double the amount of any other dry nuts. Hence, they can help fight oxidative damage in your body due to "bad" LDL cholesterol, which promotes atherosclerosis (Dhiman et al. 2014).

7.5 BENEFICIAL SOIL MICROBES FOR CROP PRODUCTION

In recent agrological trends, microbes like bacteria and fungus are very effective for agriculture. Numerous bacterial genera such as *Pseudomonas*, *Rhizobium*, *Bacillus*, *Azotobacter*, *Azospirillum*, *Streptomyces*, *and Thiobacillus* are currently used, and have proven their ability in respective fields (Arora et al. 2016). Moreover, *Trichoderma*, *Ampelomyces*, *Candida*, and *Coniothyrium* are some of the fungal genera which are very popular from an agricultural point of view (Bhardwaj et al. 2014). Although plant-growth-promoting rhizobacteria (PGPR) are naturally present in soil, for better growth and enhancement in crop production target-based microorganisms are used, which is totally dependent upon soil physicochemical properties and environmental conditions (Saharan and Nehra 2011). Bioformulations can be an alternative solution to harmful chemicals. In addition, they are economical, environmentally friendly, and nontoxic natural products (Arora and Mishra 2016).

7.5.1 MECHANISM OF ACTION

In an era of sustainable agricultural production, considerable attention has been given to the immense potential of using fluorescent pseudomonads as biocontrol agents against soil-borne fungal pathogens. Efforts are now being made to use the fluorescent pseudomonads for enhancing crop growth and yield in a sustainable manner (Dey et al. 2004). Beneficial free-living soil bacteria are isolated from the rhizosphere; this has been shown to improve plant health or increase yield, and are usually referred to as interactions in the rhizosphere. These microbes play a pivotal role in transformation, mobilization, solubilization, etc. (Jeffries et al. 2003).

Pseudomonas spp. has gained major attention in the agricultural industry because of its widespread application in various biotechnological processes. An important ubiquitous member of this group, *Pseudomonas aeruginosa* is an opportunistic pathogen of plants and humans (Walker et al. 2004; De Bentzmann and Plésiat 2011). More recently, Pandey et al. (2013) reported that *Pseudomonas* strains were plant-growth-promoting Endorhizospheric bacteria-inhabiting sunflower. Microorganisms have the capability to improve plant-growth-promoting activity. Plant-growth-promoting rhizobacteria (PGPR) are a group of bacteria that actively colonize plant roots, and increase plant growth by the production of various plant growth hormones, P-solubilizing activity, N2 fixation, and biological activity (Deshwal and Kumar 2013; Singh 2013).

Fluorescent pseudomonads are characterized by their production of yellow-green pigments, termed "pyoverdines" or "pseudobactins"; they are fluorescent under UV irradiation, and function as a siderophore. Pseudomonads produce HCN which control the growth of root rot pathogens (Vanitha and Ramjegathesh 2014). Weller et al. (1995) similarly observed that pseudomonads exert beneficial effects on plants by the production of diverse microbial metabolites like HCN. Deshwal and Kumar (2013) mentioned that *Pseudomonas* strains isolated from Mucuna-produced HCN. Gupta et al. (2006) isolated the IAA-producing fluorescent pseudomonads in the potato rhizosphere. Reetha et al. (2014) reported that IAA-producing rhizobacteria enhanced the root length, which is one of the plant-growth-promoting activity rhizobacteria. Rhizobacteria also produce gibberellic acid (Salamone et al. 2001), cytokinins, and ethylene (Glick et al. 1995). Deshwal et al. (2011) reported that *Pseudomonas* strains improve plant growth in soybean crops.

The genus *Pseudomonas* is one of the leading bacteria which inhibit the growth of pathogenic fungus in agriculture. Lanteigne et al. (2012) isolated HCN-producing *Pseudomonas* and observed the biological control activity of *Pseudomonas*. A similar observation was reported by Ramyasmruthi et al. (2012). Plant-growth-promoting *Pseudomonas* strains produced IAA, HCN and siderophore activity (Bhakthavatchalu et al. 2013).

7.5.2 Application

An ample amount of beneficial soil microorganisms is already present in soil, and are responsible for driving nutrients, soil restoration and fertility, plant health and the ecosystem, along with primary production (Chaparro et al. 2012). A number of beneficial microorganisms (actinomycetes, bacteria, diazotrophic, fungi, rhizobia, mycorrhizal), are symbiotically associated with plant roots to promote nutrient uptake and availability by producing plant growth hormones, and sometimes act as a biocontrol agent of plant pests, parasites, or diseases (Arora 2015). Many of these microorganisms are naturally occurring in soil. Nevertheless, under special conditions, the populations of these organisms may be increased by either inoculation, or by applying various management techniques for sustainable agriculture.

7.5.3 FUTURE APPROACHES

Due to the explosion in demand for food which increases on a daily basis, there is a need for healthy plant resources. There are several kinds of pathogens that attack food plants in fields. It is therefore necessary to control the pathogenic attack by using some measures. Chemical-based fungicides and pesticides are widely used in crop protection but they also have some limitations, such as toxicity resentence, polluted environment, etc. So in recent agro-economic trends, it is necessary to discover some environmentally friendly alternatives for disease control or pest management for oil-containing nuts. Biological control is a better alternative for recent agriculture systems. In this system, chemical-based pesticides are replaced by microbial pesticides commonly known as bio-pesticides which are natural products, being plant-incorporated protectants (transgenic crops). PGPR is the most valuable biocontrol agent in today's agriculture. It seems to be applicable here due to its action mechanism for strategical crop protection, and improving the efficacy of the particular biocontrol agent. The microbes used in biocontrol treatment are principally nontoxic and environmentally friendly, have no pollutants, and no other relevant side effects, in comparison to chemical-based pest or disease control methods. Biocontrol agents also appear to be nontoxic to mankind as well. Hence, after this study it was observed that biological treatments of disease management and crop protection have a healthy future for sustainable agriculture, with lower investment and less harmful effects.

7.6 COMMERCIAL USES

The commercialization of beneficial soil microbes or PGPRs is subject to institute-industry linkage. Moreover, the commercial success of PGPRs requires a consistent and broad spectrum of action with longer shelf life, safety, and stability, low capital costs, with easy availability of career materials. It is essential that suitable carrier materials are used to maintain cell viability under adverse environmental conditions. PGPR as a bio-formulation is a partial or complete substitute for chemical fertilizer and pesticides to offer a sustainable and healthy environment to increase crop production (Gupta et al. 2015).

The design of such a formulation is a crucial issue, because an effective formulation with an effective bacterial strain can determine the success or failure of that particular biological agent (Bashan et al. 2014.). It is also very important to deliver the same concentration of microbial cells from the laboratory or production unit to soil or the rhizosphere; for that purpose, carrier materials are needed for the delivery (Arora et al. 2010). Carrier materials are generally intended to provide a protective niche to microbial inoculants in the soil by physical or natural methods (Gentili et al. 2006) Therefore, in the bioformulation of PGPR, a carrier should be favorable for the survival of bacteria with high water holding and retention capacity, chemical and physical uniformity, an almost neutral pH, be nontoxic in nature and biodegradable, environmentally friendly, and cost-effective.

7.7 CONCLUSION

Nuts are an important agro-economic source worldwide. Despite often being marginal in economic productive terms in national agro-food industries, nut farms are of vital importance for the areas in which cultivation of these products is strongly rooted. The use of beneficial soil microbes is very effective and an environmentally friendly alternative for chemical-based fungicides and pesticides for the preservation and control of oil and tree nuts. A considerable amount of research is ongoing for the analysis of PGPRs and their effects as an antagonistic agent. Pseudomonads are the most popular member in this group for PGP activities and biocontrol with effective crop enhancement. It appears that the sustainable use of these microbes in both target-based and broad-spectrum bioformulations seem to be beneficial for the enhancement of healthy nut crops in a natural environment.

REFERENCES

Al-Mashadani, H.A., Al-Daraji, H.J., Al-Hayani, W.K. 2010. Effect of feeding diets containing sesame oil or seeds on productive and reproductive performance of laying quail. *Al-Anbar Journal of Veterinary Sciences* 3:56–67.

Anilakumar, K.R., Pal, A., Khanum, F., Bawa, A.S. 2010. Nutritional, medicinal and industrial uses of sesame (*Sesamum indicum* L.) seeds – An overview. *Agriculturae Conspectus Scientificus* 75(4):159–168.

Arawande, J.O., Alademeyin, J.O. 2018. Influence of processing on some quality identities of crude sesame (*Sesamum indicum*) seed oil. *International Journal of Food Nutrition and Safety* 9:59–74.

Arora, N.K. (ed.) 2015. *Plant Microbes Symbiosis: Applied Facets* (Vol. 147). New Delhi: Springer.

Arora, N.K., Khare, E., Maheshwari, D.K. 2010. Plant growth promoting rhizobacteria: Constraints in bioformulation, commercialization, and future strategies. In: *Plant Growth and Health Promoting Bacteria* (97–116). Berlin, Heidelberg: Springer.

Arora, N.K., Mehnaz, S., Balestrini, R. (eds.) 2016. *Bioformulations: For Sustainable Agriculture* (1–283). New Delhi: Springer.

Arora, N.K., Mishra, J. 2016. Prospecting the roles of metabolites and additives in future bioformulations for sustainable agriculture. *Applied Soil Ecology* 107:405–407.

Balch, P.A. 2006. *Prescription for Nutritional Healing*. London: Penguin.

Bashan, Y., de-Bashan, L.E., Prabhu, S.R., Hernandez, J.P. 2014. Advances in plant growth-promoting bacterial inoculant technology: Formulations and practical perspectives (1998–2013). *Plant and Soil* 378:1–33.

Bhakthavatchalu, S., Shivakumar, S., Sullia, S.B. 2013. Characterization of multiple plant growth promotion traits of *Pseudomonas aeruginosa* FP6, a potential stress tolerant biocontrol agent. *Annals of Biological Research* 4:214–223.

Bhardwaj, D., Ansari, M.W., Sahoo, R.K., Tuteja, N. 2014. Biofertilizers function as key player in sustainable agriculture by improving soil fertility, plant tolerance and crop productivity. *Microbial Cell Factories* 13:66.

Board, N.I.I.R. 2005. *Cultivation of Fruits, Vegetables and Floriculture*. New Delhi: NIIR Project Consultancy Services.

Cerletti, P., Duranti, M. 1979. Development of lupine proteins. *Journal of the American Oil Chemists' Society* 56:460–463.

Chaparro, J.M., Sheflin, A.M., Manter, D.K., Vivanco, J.M. 2012. Manipulating the soil microbiome to increase soil health and plant fertility. *Biology and Fertility of Soils* 48:489–499.

De Bentzmann, S., Plésiat, P. 2011. The *Pseudomonas aeruginosa* opportunistic pathogen and human infections. *Environmental Microbiology* 13:1655–1665.

De Herralde, F., Biel, C., Save, R. 2003. Leaf photosynthesis in eight almond tree cultivars. *Biologia Plantarum* 46:557–561.

Deshwal, N., Sharma, A.K., Sharma, P. 2011. Review on hepatoprotective plants. *International Journal of Pharmaceutical Sciences Review and Research* 7:15–26.

Deshwal, V.K., Kumar, P. 2013. Production of plant growth promoting substance by Pseudomonads. *Journal of Academia and Industrial Research* 2:221–225.

Dey, R.K.K.P., Pal, K.K., Bhatt, D.M., Chauhan, S.M. 2004. Growth promotion and yield enhancement of peanut (*Arachis hypogaea* L.) by application of plant growth-promoting rhizobacteria. *Microbiological Research* 159:371–394.

Dhiman, P., Soni, K., Singh, S. 2014. Nutritional value of dry fruits and their vital significance – A review. *Pharma Tutor* 2:102–108.

Fasuan, T.O., Gbadamosi, S.O., Omobuwajo, T.O. 2018. Characterization of protein isolate from *Sesamum indicum* seed: In vitro protein digestibility, amino acid profile, and some functional properties. *Food Science and Nutrition* 6:1715–1723.

Fernandes, D.C., Freitas, J.B., Czeder, L.P., Naves, M.M.V. 2010. Nutritional composition and protein value of the baru (*Dipteryx alata* Vog.) almond from the Brazilian Savanna. *Journal of the Science of Food and Agriculture* 90:1650–1655.

Forest Europe (Organization), Liaison Unit Oslo. 2011. *State of Europe's Forests, 2011: Status and Trends in Sustainable Forest Management in Europe*. Ministerial Conference on the Protection of Forests in Europe, Forest Europe, Liaison Unit Oslo.

Fraga, C.G. 2007. Plant polyphenols: How to translate their in vitro antioxidant actions to in vivo conditions. *IUBMB Life* 59:308–315.

Freeman, S., Minz, D., Jurkevitch, E., Maymon, M., Shabi, E. 2000. Molecular analyses of *Colletotrichum* species from almond and other fruits. *Phytopathology* 90:608–614.

García de Salamone, I.E., Hynes, R.K., Nelson, L.M. 2001. Cytokinin production by plant growth promoting rhizobacteria and selected mutants. *Canadian Journal of Microbiology* 47:404–411.

Gentili, A.R., Cubitto, M.A., Ferrero, M., Rodriguéz, M.S. 2006. Bioremediation of crude oil polluted seawater by a hydrocarbon-degrading bacterial strain immobilized on chitin and chitosan flakes. *International Biodeterioration and Biodegradation* 57:222–228.

Glick, R.E., Schlagnhaufer, C.D., Arteca, R.N., Pell, E.J. 1995. Ozone-induced ethylene emission accelerates the loss of ribulose-1, 5-bisphosphate carboxylase/oxygenase and nuclear-encoded mRNAs in senescing potato leaves. *Plant Physiology* 109:891–898.

Gupta, G., Parihar, S.S., Ahirwar, N.K. 2015. Plant growth promoting rhizobacteria (PGPR): Current and future prospects for development of sustainable agriculture. *Microbial and Biochemical Technology* 7:96–102.

Gupta, M., Graham, J., McNealy, B.R.I.A.N., Zarghami, M., Landin-Olsson, M.O.N.A., Hagopian, W.A., Palmer, J., Lernmark, Å., Sanjeevi, C.B. 2006. MHC class I chain-related gene-A is associated with IA2 and IAA but not GAD in Swedish type 1 diabetes mellitus. *Annals of the New York Academy of Sciences* 1079:229–239.

Hammons, R.O. 1994. The origin and history of the groundnut. In: *The Groundnut Crop* (24–42). Dordrecht: Springer.

Hollis, J., Mattes, R. 2007. Effect of chronic consumption of almonds on body weight in healthy humans. *British Journal of Nutrition* 98:651–656.

Jeffries, P., Gianinazzi, S., Perotto, S., Turnau, K., Barea, J.M. 2003. The contribution of arbuscular mycorrhizal fungi in sustainable maintenance of plant health and soil fertility. *Biology and Fertility of Soils* 37:1–16.

Lanteigne, C., Gadkar, V.J., Wallon, T., Novinscak, A., Filion, M. 2012. Production of DAPG and HCN by *Pseudomonas* sp. LBUM300 contributes to the biological control of bacterial canker of tomato. *Phytopathology* 102:967–973.

Luedeling, E. 2012. Climate change impacts on winter chill for temperate fruit and nut production: A review. *Scientia Horticulturae* 144:218–229.

Madhusudhana, B. 2013. A survey on area, production and productivity of groundnut crop in India. *IOSR Journal of Economics and Finance* 1:1–7.

Massey, L.K. 2003. Dietary animal and plant protein and human bone health: A whole foods approach. *The Journal of Nutrition* 133:862S–865S.

Nagaraj, G. 2009. *Oilseeds: Properties, Processing, Products and Procedures.* New India: New India Publishing.

Nile, S.H., Park, S.W. 2014. Edible berries: Bioactive components and their effect on human health. *Nutrition* 30:134–144.

Olatidoye, O.P., Alabi, A.O., Sobowale, S.S., Balogun, I.O., Nwabueze, B.C. 2019. Effect of cooking methods on antinutrient content and phenolic acid profiles of groundnut varieties grown in Nigeria. *International Journal of Food and Nutrition Research* 1:27–37.

Orem, A., Yucesan, F.B., Orem, C., Akcan, B., Kural, B.V., Alasalvar, C., Shahidi, F. 2013. Hazelnut-enriched diet improves cardiovascular risk biomarkers beyond a lipid-lowering effect in hypercholesterolemic subjects. *Journal of Clinical Lipidology* 7:123–131.

Pandey, R., Chavan, P.N., Walokar, N.M., Sharma, N., Tripathi, V., Khetmalas, M.B. 2013. Pseudomonas stutzeri RP1: A versatile plant growth promoting endorhizospheric bacteria inhabiting sunflower (*Helianthus annus*). *Journal of Biotechnology* 8:48–55.

Paranthaman, R., Vidyalakshmi, R., Murugesh, S., Singaravadivel, K. 2009. Production of Tannase by various fungal cultures in solid state fermentation of ground nut shell. *International Journal of Biotechnology and Biochemistry* 5:125–135.

Pérez-Campos, S.I., Cutanda-Pérez, M.C., Montero-Riquelme, F.J., Botella-Miralles, O. 2011. Comparative analysis of autochthonous almond (*Prunus amigdalus* Bastch) material and commercial varieties in the Castilla-La Mancha Region (Spain). *Scientia Horticulturae* 129:421–425.

Ramyasmruthi, S., Pallavi, O., Pallavi, S., Tilak, K., Srividya, S. 2012. Chitinolytic and secondary metabolite producing *Pseudomonas fluorescens* isolated from Solanaceae rhizosphere effective against broad spectrum fungal phytopathogens. *Asian Journal of Plant Science and Research* 2:16–24.

Reetha, S., Bhuvaneswari, G., Thamizhiniyan, P., Mycin, T.R. 2014. Isolation of indole acetic acid (IAA) producing rhizobacteria of *Pseudomonas fluorescens* and *Bacillus subtilis* and enhance growth of onion (*Allim cepa*. L). *International Journal of Current Microbiology and Applied Sciences* 3:568–574.

Ros, E. 2010. Health benefits of nut consumption. *Nutrients* 2:652–682.

Saharan, B.S., Nehra, V. 2011. Plant growth promoting rhizobacteria: A critical review. *Life Sciences and Medicine Research* 21:30.

Saturni, L., Ferretti, G., Bacchetti, T. 2010. The gluten-free diet: Safety and nutritional quality. *Nutrients* 2:16–34.

Shear, C.B., Faust, M. 1980. Nutritional ranges in deciduous tree fruits and nuts. *Horticultural Reviews* 2:142–163.

Singh, J.S. 2013. Plant growth promoting rhizobacteria. *Resonance* 18(3):275–281.

Slavin, J.L., Lloyd, B. 2012. Health benefits of fruits and vegetables. *Advances in Nutrition* 3:506–516.

Sujatha, E.R., Dharini, K., Bharathi, V. 2016. Influence of groundnut shell ash on strength and durability properties of clay. *Geomechanics and Geoengineering* 11:20–27.

USDA Foreign. 2014. *Oilseeds: World Markets and Trade*. USA: USDA Foreign. Accessed 10 December 2014.

Vanitha, S., Ramjegathesh, R. 2014. Bio control potential of *Pseudomonas fluorescens* against coleus root rot disease. *Journal of J Plant Pathology & Microbiology* 5(1):1–4.

Vayalil, P.K. 2012. Date fruits (*Phoenix dactylifera* Linn): An emerging medicinal food. *Critical Reviews in Food Science and Nutrition* 52:249–271.

Walker, D.J., Clemente, R., Bernal, M.P. 2004. Contrasting effects of manure and compost on soil pH, heavy metal availability and growth of *Chenopodium album* L. in a soil contaminated by pyritic mine waste. *Chemosphere* 57:215–224.

Weller, D.M., Thomashow, L.S., Cook, R.J. 1995. Biological control of soil-borne pathogens of wheat: Benefits, risks and current challenges. *Plant and Microbial Biotechnology Research Series* 5:1.

Wolfe, D. 2010. *Super Foods: The Food and Medicine of the Future*. Berkeley, CA: North Atlantic Books.

8 Fungal Amylases for the Detergent Industry

Robinka Khajuria and Shalini Singh

CONTENTS

8.1 INTRODUCTION

Commercially, amylases are one of the most important enzymes with an approximately 25% share of the global enzyme market. Amylases are starch-hydrolyzing enzymes that are derived from different sources including plants, animals, and microorganisms (Niyonzima and More 2014). Microbial and plant-based amylases have been used for centuries in the brewing industry. Many continental foods preparations make use of fungal amylases (Rizak et al. 2019). The history of amylases began in 1811, when the first starch-degrading enzyme was discovered by Kirchhoff. It was much later, in 1930, that Ohlsson suggested the classification of starch digestive enzymes in malt as α-and β-amylases, according to the anomeric type of sugars produced by the enzyme reaction (Gupta et al. 2003). The first enzyme produced commercially was an amylase secreted by a fungal source in 1894, and was used for the treatment of digestive disorders(Crueger and Crueger 1984). Industrially, α-amylases and γ-amylases are the most important amylases with application in various sectors including food, pharmaceuticals, textiles, and detergents. The major advantage of using microorganisms for the production of amylases over plant and animal sources is the relative ease of economical bulk production capacity, and the ability to manipulate microbes to produce enzymes with desired characteristics. Microbial amylases have successfully replaced chemical hydrolysis of starch in starch-processing industries. Fungal enzymes, in particular, are preferred over other microbial sources due to their generally regarded as safe (GRAS) status, which is widely accepted (Karim et al. 2017).

Fungi belonging to *Aspergillus spp., Rhizopus spp.* and *Penicillium spp.* can produce different cell wall degrading enzymes that breakdown large polysaccharides into simple reducing sugars which are consumed for growth and multiplication (Ajayi and Adedeji 2014). Microbial production can be carried out using submerged and solid-state fermentation. Filamentous fungi are the best adapted for solid-state fermentation. The hyphal mode of fungal growth and their good tolerance to low water activity and high osmotic pressure conditions make fungi efficient and competitive in natural microflora for the bioconversion of solid substrates. Amylases have been produced by submerged fermentation. In recent years, however, the solid state fermentation (SSF) processes have been increasingly applied for the production of this enzyme. Solid state fermentation compared to submerged fermentation is simpler, requires slower capital, has superior productivity, reduces energy requirements, makes use of simple fermentation media and absence of rigorous control of fermentation parameters, uses less water and produces lower wastewater, has easier control of bacterial contamination, and requires a low cost for downstream processing (Saranraj and Stella 2013). The alkaline amylases and other detergent enzymes are now key ingredients in the manufacture of detergents in developed countries. Duramyl and PurafectOxAm are two detergent amylases manufactured by Novozyme and Genencore International, respectively, and are available in the market (Niyonzima and More 2014). The demand for detergent amylases with better properties is constantly increasing, since the current detergent amylases do not have all the desired characteristics. This chapter discusses in detail the various aspects associated with fungal amylases.

8.2 MODE OF ACTION

The amylase family is the largest family of glycoside hydrolases and includes enzymes with different modes of action. Based upon the mode of action, amylases are divided into two categories:

1. Endoamylases: These enzymes catalyze hydrolysis in a random manner. They act in the interior of the starch molecule and form linear and branched oligosaccharides of different chain lengths as product.
2. Exoamylases: These enzymes hydrolyze starch from the non-reducing end successively, leading to short-end products (El-Enshasy et al. 2013).

In another classification devised by Sivaramakrishnan et al. (2006), the amylases were divided into four main groups based on the nature of the bond cleaved and their catalytic activities:

1. Endoamylases: cleave internal α-1,4 bonds forming in α-anomeric products.
2. Exoamylases: cleave α-1,4 or α-1,6 bonds of the external glucose residues, yielding α- or β-anomeric products.

3. Debranching enzymes: hydrolyze α-1,6 bonds, exclusively leaving long linear polysaccharide.
4. Transferases cleave α-1,4-glycosidic bond of the donor molecule and transfer part of the donor molecule to a glycosidic acceptor, hence forming a new
   glycosidic bond.

However, it has been reported that some of the amylases can perform two types of
catalytic reactions at different rates. For instance,γ-amylases (discussed in the next
section) cleave α-1,4 linkages and α-1,6-glycosidic linkages. Nevertheless, the rate of
hydrolyzing latter bonds is slower than the former. However, these enzymes can lead
to complete degradation of starch to glucose (El-Enshasy et al. 2013).

8.3 TYPES OF AMYLASES

Among the various types of amylases, the three main industrially important amylases are α-amylase, β-amylase, and γ-amylase. These enzymes vary from each other
in their structures, catalytic activities, and production processes. The following section briefly discusses these four amylases.

8.3.1 α-AMYLASE (ENDO-1,4-D-GLUCAN GLUCOHYDROLASE EC)

These are calcium metallo-enzymes that require calcium ion for functioning. They
act randomly at locations along the starch chain and break down long-chain carbohydrates, forming maltotriose and maltose from amylose and maltose, glucose and
"limit dextrin" from amylopectin. α-amylases belong to the family of endoamylases
that leave the 1,4-D-glycosidic linkages between adjacent glucose units. α- amylases
are often divided into two categories according to the degree of hydrolysis of the
substrate. Saccharifying α-amylases hydrolyze 50 to 60% and liquefying α-amylases
cleave about 30 to 40% of the glycosidic linkages of starch. Since they act randomly,
α-amylases are faster-acting than β-amylases. α-Amylaseis a major digestive enzyme
in animals and in humans, both saliva and pancreatic amylases are α-amylases.
Alpha amylases are one of the most widely used enzymes with applications in various industries such as food, baking, brewing, detergent, textile, paper, and distilling
(Tiwari et al. 2015; Saranraj and Stella 2013; Gupta et al. 2003; Pandey et al. 2000).

8.3.2 β-AMYLASE (1,4-α-D-GLUCAN MALTOHYDROLASE;
GLYCOGENASE; SACCHAROGEN AMYLASE, EC)

This is another form of amylase synthesized by bacteria, fungi, and plants. It acts
on the non-reducing end and catalyzes the hydrolysis of the second α-1,4 glycosidic
bond, cleaving off two glucose units (maltose) at a time.ß-amylases convert the anomeric configuration of the liberated maltose from α to ß. During the ripening of fruit,
β-amylase breaks starch into maltose, resulting in the sweet flavor of ripeβ-amylase
being present in an inactive form prior to germination, whereas α-amylase and proteases appear once germination has begun. Animal tissues do not contain β-amylase,
although it may be present in microorganisms contained within the digestive tract.

8.3.3 γ-Amylase

In addition to cleaving the last (1-4)glycosidic linkages at the non-reducing end of amylose and amylopectin, yielding glucose, γ-amylase will cleave (1-6) glycosidic linkages. Unlike the other forms of amylase, γ-amylase is most efficient in acidic environments and has an optimum pH of 3.This enzyme is widely distributed among plants, animals, and other types of microorganisms. However, it is characterized by an odd feature for re-polymerization of the degraded mono-sugar-forming maltose, isomaltose, when glucose exceeds 30–35% (Tiwari et al. 2015; Saranraj and Stella 2013; Gupta et al. 2003; Pandey et al. 2000).

8.4 AMYLASE STRUCTURE

Among the three industrially important amylases, α- and β-amylase adopt the structure of a TIM barrel fold. The catalytic domain in these enzymes consists of an $(\alpha/\beta)_8$ barrel made up of eight parallel β-strands surrounded by eight α-helices. γ-amylase, on the other hand, has the structure of $(\alpha/\alpha)_6$ barrel, consisting of an inner barrel that is composed of six α-helices that are surrounded by six more. The strands and helices of the $(\beta/\alpha)_8$-barrel domain and $(\alpha/\alpha)_6$ barrel are connected by loop regions of various lengths (Machovic and Janecek 2006).

Due to high industrial importance, the tertiary structures of α-amylases have been studied most extensively. Regardless of the origin of the enzymes, all α-amylases comprise of one central domain, A and two other domains, B and C. Domain A is the central domain of a $(\beta/\alpha)_8$ barrel, while Domain B and C are located at the opposite sides of Domain A in the TIM barrel. Domain A contains the active site residues, while Domain B with its irregular β-rich structure contains the binding cleft that is functionally important for substrate specificity. Domain C is made up of the C-terminal part of the sequence and is a β-sandwich domain containing a Greek key motif as shown in Figure 8.1. The active site is known to contain a pair of carboxylic acids, one acting as a general acid and a general base, while the other acts as a nucleophile for stabilization of an oxo-carbonium ion and the diffusion of the leaving group (Saranraj and Stella 2013; Souza and Magalhães 2010; Taniguchi and Honnda 2009).

The molecular weights of α-amylases vary from 10 kDa to 210 kDa (Gupta et al. 2003). The lowest value molecular weight enzyme (10 kDa) was isolated from *Bacillus caldolyticus* whereas the highest (210 kDa) was obtained from the thermophilic, photosynthetic bacterium *Chloroflexus aurantiacus*. However, the molecular weights of most microbial amylases of industrial importance lie between 50 kDa and 60 kDa (Saranraj and Stella 2013; Gupta et al. 2003). Amylases are not only identified by their catalytic domain but also other domains that bind with solid starch for hydrolysis and saccharification process. Like the catalytic domain classification, the CAZy database has also classified the carbohydrate-binding modules (CBMs) into 53 families, according to ligand specificity (Lin et al. 2009). Out of these, the START-binding families are divided into seven CBM families (CBM20, CBM21, CBM25, CBM26, CBM34, CBM41, and CBM45) (Machovic and Janecek 2006). The SBD is a functional domain that can bind granular starch and thus increase the

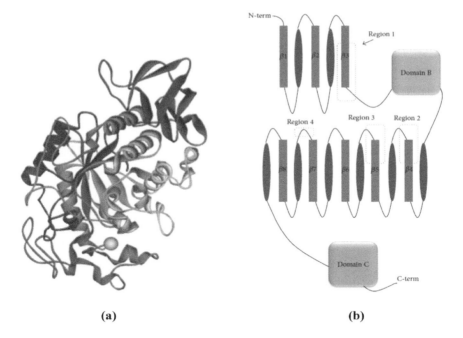

(a)   (b)

FIGURE 8.1   (a) Three-dimensional structures of α-amylase (RCSB PDB accession code 1SMD). (b) Topology of α-amylases. (Adapted from Gopinath et al. 2017.)

local concentration of the substrate at the enzyme active site and may also disrupt the structure of the starch surface, thereby enhancing the amylolytic rate(Santiago et al. 2005).

8.5   SOURCES OF AMYLASES

Over the decades, various amylase-producing fungi, bacteria, fungi, and other microbes have been isolated and characterized. The advantage of microbial amylases is that both bacteria and fungi secrete them extracellularly and hence are commercially important. Among fungi, *Aspergillus niger*, *Aspergillus oryzae*, *Thermomyces lanuginosus*, and *Penicillium expansum* have been reported to produce high levels of amylase. Additionally, many species of *Mucor* and white rot fungi *Gandoerma*have also been reported to produce amylases extracellularly. Among yeast, strains of *Filobasidiuim capsuligenum* are capable of amylase production (El-Fallal et al. 2012; Nguyen et al. 2015). Table 8.1 shows various fungi reported to produce amylases and their optimum pH and temperature conditions.

Among bacteria, *Bacillus spp.* is known to produce a variety of extracellular amylases that are of immense importance for commercial applications. Some of the members of *Bacillus spp.*that are commercially used for amylase production include *B. cereus*, *B. circulans*, *B. subtilis*, and *B. licheniformis*. Bacteria belonging mainly to the genus Bacillus are known for the production of thermostable α-amylase. Also to be noted is that most of the Bacillus such as *B. amyloliquefaciens* and *B. stearothermophilus* produce liquefying amylases with pH optima between 5 and 7.5. They

TABLE 8.1
Fungal Amylases: Sources and Properties

Fungus	MW	Optimum pH	Optimum Temperature
α-Amylase			
Aspergillus sp.	56	5.5	40
A. flavus	52.5	6.0	55
A. fumigatus	65	5.5	40
Aspergillus niger	43	9.5	30
Aspergillus oryzae	52	4-5	50
Filobasidium capsuligenum	56	5.6	45
Lipomyces kononenkoae	76	4.5-5.0	70
Rhizopus sp.	64	4.0-5.6	60-65
Scytalidium sp.	87	6.5	50
Thermomyces lanuginosus	45	4.6-5.2	60
γ-Amylase			
Acremonium sp. YT-78	74	5	50
Aspergillus awamori	83.7	4.5	60
A. awamori	110,86		
A. awamori var. kawachi	57-90	3.8-4.5	
A. niger	74,96	4.2,4.5	60,65
A. niger	61-112	4.4	60
A. oryzae	38,76	4.5	50-60
A. saitri	90	4.5	
A. terreus	70	5	60
Chalara paradoxa	68	5	45
Corticium rolfsii	69,70,78,79		
Neurospora crassa	82	5.4	60
Paecilomyces variotii	69	5	
Pyricularia oryzae	94	4.5	50-55
Rhizopus sp.	58.6-74	4.5-5	
R. niveus		4.5-6	40
Schizophyllum commune	66	5	70
T. lanuginosus	37	4.9,6.6	65-70

Source: Adapted from Nguyen et al. 2015

have also been reported to produce alkaline amylases that are all of the saccharifying type. Amylases from *Bacillus sp.* strain 707 and *B. licheniformis* TCRDC-B13 are the only exceptions to this. Bacillus species have also been reported to produce thermostable β-amylases. A variety of ruminal bacteria such as *Bacteroides ruminicola, Ruminobacter amylophilus, Butyrivibrio fibrisolvens, Selenomonas ruminantium,* and *Streptococcus bovis* also exhibit the ability to utilize starch as a growth substrate and are present in the rumen. Genes encoding intracellular α-amylases have been reported for *Escherichia coli* and *Streptococcus bovis.* α-Amylases are secreted by several species of Streptomyces; for example, *S. albus, S. griseus,*

S. thermocyaneoviolaceus. Hyperthermophilic microorganisms viz., *Thermotoga maritima, Geobacillus thermoleovorans* also possess the ability to produce thermostable starch-hydrolyzing enzymes. Among hyperthermophilic Archaea, deep-sea *Thermococcale* and *Sulfolobus* species have been reported to produce α-amylases (Rizk et al. 2019; Nguyen et al. 2015; Karimet al. 2017; El-Fallal et al. 2012)

8.6 INDUSTRIAL PRODUCTION

The two most common methods used for commercial production of amylases are submerged fermentation (SmF) and solid-state fermentation (SSF). SmF uses liquid medium like molasses and nutrient broths, and the product is released in the fermentation medium. The rate of substrate consumption is very high in SmF and hence the process requires regular supply of the substrate. This fermentation method is used when the product is being produced by bacteria as they require high moisture content for growth(Couto and Sanroman 2006). The advantages of SmFare that it can support the use of genetically modified organisms that have specific nutrient requirements, ease of media sterilization, product recovery, and controlling process parameters such as pH, temperature, moisture, oxygen transfer, etc. In SSF, solid nutrients such as bagasse, bran, wheat straw, corncob, paper pulp, etc. are used as substrate. The microorganisms consume these substrates very slowly and at a constant rate. SSF is used commonly for fungal species. SSF offers advantages like higher product recovery as it simulates the natural environment for microbial growth. Economically, solid-state fermentation is also more favored in developed countries since the solid substrate (agriculture material) acts as a source of carbon, nitrogen, minerals, and growth factors and has a capacity to absorb the water necessary for microbial growth (Suriya et al. 2016). It was in 1894 that Dr. Jhokichi Takamine filed a US patent on the process for the commercial production of amylase by *Aspergillus oryzae*. This method used solid-state culture on wheat bran and alcohol for enzyme extraction (Souza and Magalhães 2010). The fungal enzyme produced by this method was for the treatment of digestive disorders and was commercially available under the brand name of Taka-Diastase. As discussed, commercially amylases are produced from multiple sources, this section will however will only examine the production of fungal amylases.

With regards to fungal amylases, most work has researched the production of α-amylase by *Aspergillusoryzae*. It has been reported that pH of the media critically affects the morphology of *A.oryzae*. The fungus has been reported to be present in dispersed hyphal form at pH 3.0 to 3.5, pellet and dispersed hyphal fragment at pH 4 to 5, and pellet form at pH higher than 6. The fungi have optimum growth temperature of 35°C and an increase in the α-amylase production by the fungus is observed under glucose exhaustion (Saranraj and Stella 2013; Laderman et al. 2003; Aehle and Misset 1999). Different types of substrates such as coconut oil cake, sesame oil cake, groundnut oil cake, and olive oil cake have been used for the production of fungal amylases using SSF. Ramachandra et al. (2004) reported the production of alpha-amylase by *Aspergillus oryzae* using groundnut oil cake supplemented with lactose and ammonium nitrate at 30°C for 72 hrs. It was also observed that the enzyme activity increased 2.4-fold after partial purification of the enzyme using ammonium sulfate fractionation. Similarly, corncob leaf, rye straw, wheat straw, and wheat bran have been used as substrates in SSF for α-amylase production by *Penicillium chrysogenum* with highest enzyme production of 160V/ml in wheat bran (Balkan and Ertan 2006).In another study by Anto et al. (2006), wheat bran was used as a substrate for the production of γ-amylase by *Aspergillus* sp. isolated from local soil. Highest enzyme production of 264 ± 1.44 U/gds) was obtained at 28 ± 2°C, pH 5. The enzyme activity increased with the supplementation of organic nitrogen sources such as yeast extract and peptone. Fossi et al. (2005) reported the production of amylase with an optimum enzyme activity at 70°C and pH 5.5–6.5 from a yeast strain. *Penicillium fellutanum* isolated from coastal regions was reported to produce α-amylase for 96 h at 30°C and pH 6.5 using submerged fermentation (Kathiresan and Manivannan2006). α-amylases have also been produced by submerged fermentation from various amylolytic isolates viz., be *P. frequestans A. niger, A. fumigatus, A.s flavus,* and *Helminthosporium oxyspor. H. oxysporum* was reported to secrete 10.77 E.U of enzyme on cassava media and 10.42 E.U on yam peel media (Alva et al. 2007). Similarly, Gupta et al. (2013) used submerged fermentation for the production of α-amylase by *Aspergillus niger* at temperature optima range of 30–40°C andpH optima range of at 4–6.

Media composition and operation parameters have a significant effect on the production of α –amylase. One such parameter is carbon source; it has been found that enzyme production requires carbon source induction. It is therefore important to use a suitable carbon source such as starch or maltose in the cultivation medium (Abou Dobara et al. 2011; Asoodeh et al. 2010). Carlsen and Nielsen (2001) reported that glucose can act as an inducer and repressor, based upon concentrations in the case of enzyme production by *A. oryzae* in submerged fermentation. Glucose acts as enzyme inducer below a concentration of 10 mg/L and as a repressor at higher concentrations. In any case, it has now been established that catabolite repression can be eliminated by genetic manipulation in some strains (Rajagopalan and Krishnan 2008). Another controlling parameter is the nature and concentration of nitrogen source enhance. Organic nitrogen sources have been reported to enhance the production of α-amylase. Various other reports have established that yeast extract is the most preferred organic nitrogen source for α-amylase production as it is also a rich source of vitamins and growth factors (El-Fallal et al. 2012; Abou Dobara et al. 2011). Amino acids such as thiamine, cysteine, and pyridoxine have also been reported to enhance enzyme production (Rajagopalan and Krishnan 2008). Most α-amylase producer strains are aerobic microorganisms and require high aeration during the growth phase. Therefore, high dissolved oxygen tension was reported as a vital factor for α-amylase production enzyme production (El-Enshasyet al. 2013).In the case of fungal amylase production, the relation between nutritional facts, cultivation conditions, and their effects on morphology is critical and should be taken into consideration while optimizing the working conditions.

In the case of γ-amylases, higher yields are obtained using a semisolid fermentation system, rather than solid-state fermentation systems (Norouzian et al. 2006). Conventionally, γ-amylases were produced by the SmF system using fungal strains, such as *A niger, A. awamori,* and *Rhizopus* sp. Various authors have also reported the production of γ-amylases by *A. niger* using agro-industrial residues such as wheat bran, rice bran, rice husk, gram flour, wheat flour, corn flour, tea waste, and copra waste (El-Enshasy et al. 2013). *Aureobasidium pullulans,* a marine yeast, was reported to produce 58.5 U/mg of protein within 56 h of fermentation (Liu et al. 2008). Advances led to the use of different types of bioreactors such as rotary bioreactors, packed-bed column bioreactors, and airlift bioreactors for the production of γ-amylases. For instance, higher yields were achieved from *A. niger* in a glucose-limited chemostat culture. It has been seen that batch cultivation leads to the highest enzyme yields in the case of γ-amylase production by SmF (El-Enshasy et al. 2013).

As compared to α-amylases and γ-amylases, β -amylases are mainly produced by plants. However, various wild-type and recombinant microorganisms are used for the commercial production of β-amylases by SmF or SSF. Fungal β-amylases are mainly produced by fungi belonging to industrial production, which is conducted by using bacterial strains belonging to *Aspergillus* spp. and *Rhizopus* spp. Most of the industrialβ-amylases are produced by microbes belonging to *Bacillus* sp., such as *B.cereus, B.megatarium,* and *B.polymyxa.* Some microorganisms such as *B.subtilis* MIR-5 can produce α- and β-amylases simultaneously in batch and continuous culture (El-Enshasy et al. 2013). Just as in the case of α-amylases, β-amylase production is also highly regulated by the carbon source.

8.7   AMYLASES IN DETERGENT INDUSTRY

Enzymes have now become an essential ingredient of modern detergents. The biggest advantage that enzyme-based detergents offer over enzyme-free detergents is their mild nature. Conventional dishwashing detergents were harsh, injurious when ingested, and incompatible with delicate wooden and china dishware. These shortcomings of conventional detergents led to the search of milder and efficient solutions.α-amylases are one of the enzymes used in laundry detergents (Gupta et al. 2003).

They were used for the first time in 1975, and now 90% of all liquid detergents contain α-amylase. Initially, amylases added in detergents exhibited sensitivity to calcium with a decreased stability in a low calcium environment. Moreover, most wild-type α-amylases are sensitive to oxidants which are a component of detergent formulations. However, two of the major detergent enzyme suppliers, Novozymes and Genencore International, have used various strategies to improve the stability of α-amylases. One such strategy used other enzymes such as protease to enhance amylase stability against oxidants in household detergents.Additionally, protein engineering is being employed to improve the bleach stability of the amylases. For instance, the replacement of Met at position 197 by Leu in *B. licheniformis* amylase resulted in an amylase with improved resistance against oxidative compounds. Genencor International and Novoymes released these products in the market under the trade names Purafect OxAm and Duramyl respectively (Gopinath et al. 2017).

Starch in its native form is slowly degraded by α-amylases, but gelatinization and swelling are needed to make the starch susceptible to enzymatic breakdown. In the case of foods, different degree of gelatinization occurs during cooking. Therefore, in detergents for laundry and automatic dishwashing, amylases facilitate the removal of starch-containing stains, e.g., potato, gravy, chocolate, and baby food. Amylases also work by preventing swollen starch from adhering to the surface of laundry and dishes. Complexes or reaction products between protein, starch, and/or fat are usually found in prepared foods. In such cases, enzyme synergy effects make it possible to remove soil even more efficiently than with single enzyme systems (Olsen and Falholt 1998). Industrially important α-amylases are made from the *Bacillus* and *Aspergillus* species. During enzymatic liquefaction, the α-1,4-linkages in starch are hydrolyzed at random, thus reducing the viscosity of the gelatinized starch and increasing the solubility of attached starch, which is converted to water-soluble dextrins and oligosaccharides. Therefore, Termamyl®, Duramyl®, and BAN are often referred to as "liquefying amylases." Fungamyl is a fungal exo-amylase, which also hydrolyzes the α-1,4-linkages in liquefied starch. Fungamyl is relevant in detergents for use at low pH levels in industrial cleaning tasks. Duramyl® was developed to achieve increased oxidation stability in the humid environment of detergent powders containing moisture. In practice, amylases for detergent applications are selected on the basis of performance and stability tests in specific detergents. In laundry detergents, amylases may maintain or even contribute to increased whitening of dingy fabrics, and inhibit the graying of white fabrics resulting from a combination of starch and particulate soiling (Nguyen et al. 2015; Ajita and Krishna Murthy 2014).

8.8   CONCLUSION

Even though different types of α-amylase preparations are available with various manufacturers, enhancements in terms of both quality and quantity are still needed. Qualitative improvements in amylase gene and its protein can be achieved by recombinant DNA technology and protein engineering. Quantitative enhancement needs strain improvement through site-directed mutagenesis and/or standardizing the nutrient medium for the overproduction of active α-amylases. Another approach is to search for novel microbial strains, specifically fungal strains for the production of desired amylases.

REFERENCES

Abou Dobara, M.I., El-Sayed, A.K., El-Fallal, A.A., Omar, N.F. 2011. Production and partial characterization of high molecular weight extracellular α-amylase from *Thermoactinomyces vulgaris* isolated from Egyptian soil. *Polish Journal of Microbiology* 601:65–71.

Aehle, W., Misset, O. 1999. Enzymes for industrial applications. In: *Biotechnology*, 2nd ed. Germany: Wiley-VCH, pp. 189–216.

Ajayi, A.A., Adedeji, O.M. 2014. Modification of cell wall degrading enzymes from *Soursop Annona muricata* fruit deterioration for improved commercial development of clarified Soursop juice – A review. *Journal of Medicinal and Aromatic Plants* 41:1–5.

Ajita, S., Krishna Murthy, T.P. 2014. α-Amylase production and applications: A review. *Journal of Applied and Environmental Microbiology* 24:166–175.

Alva, S., Anupama, J., Savla, J., Kumudini, B.S., Varalakshmi, K.N. 2007. Production and characterization of fungal amylase enzyme isolated from *Aspergillus* sp. in solid state culture. *African Journal of Biotechnology* 6:576–581.

Anto, H., Trivedi, U., Patel, K. 2006. Amylase production by *Bacillus cereus* using solid SSF. *Food Technology and Biotechnology* 44:241–245.

Asoodeh, A., Chamanic, J., Lagzian, M. 2010. A novel thermostable, acidophilic α-amylase from a new thermophilic *"Bacillus sp. Ferdowsicous"* isolated from Ferdows hot mineral spring in Iran: Purification and biochemical characterization. *International Journal of Biological Macromolecules*463:289–297.

Balkan, B., Ertan, F. 2006. Production of amylase from *Penicillium chrysogenum* under SSF by using some agricultural by-product. *Food Technology and Biotechnology* 45:439–442.

Carlsen, M., Nielsen, J. 2001. Influence of carbon source on amylase production by *Aspergillus oryzae*. *Applied Microbiology and Biotechnology* 57:346–349.

Couto, S.R., Sanroman, M.A. 2006. Application of solid-state fermentation to food industry – A review. *Journal of Food Engineering* 763:291–302.

Crueger, W., Crueger, A. 1984. Enzymes. In: *Bio-Technology. A Text book of Industrial Microbiology*. Sunderland, MA: Sinauer Associated Inc., pp. 161–186.

El-Enshasy, H.A., Fattah, Y.R.A., Othman, N.Z.A. 2013. Amylases: Characteristics, sources, production, and applications. In: *Bioprocessing Technologies in Biorefinery for Sustainable Production of Fuels, Chemicals, and Polymers*, S.H. Yang, H.A. El-Enshasy, N. Thongchul (Eds.). Hoboken, NJ: John Wiley & Sons, Inc.

El-Fallal, A., Dobara, M.A., El-Sayed, A., Omar, N. 2012. Starch and microbial α-amylases: From concepts to biotechnological applications. In: *Carbohydrates – Comprehensive Studies on Glycobiology and Glycotechnology*, C.F. Chang (Ed.). London: IntechOpen.

Fossi, B.T., Tavea, F., Ndjouenkeu, R. 2005. Production and partial characterization of amylase from ascomycete yeast strain isolated from starchy soils. *African Journal of Biotechnology* 41:14–18.

Gopinath, S.C.B., Anbu, P., Md.Arshad, M.K., Lakshmipriya, T., Voon, C.H., Hashim, U., Chinni, V.S. 2017. *Biotechnological Processes in Microbial Amylase: Production Microbial Enzymes and Their Applications in Industries and Medicine*. Article ID 1272193. doi:10.1155/2017/1272193.

Gupta, A., Gautam, N., Modi, D.R. 2013. Optimization of amylase production from immobilized cells of *Aspergillus niger*. *Journal of Biotechnology and Pharma Research* 11:1–8.

Gupta, R., Gigras, P., Goswami, V.K., Chauhan, B. 2003. Microbial amylase: A biotechnological perspective. *Process Biochemistry* 38:1599–1616.

Karim, K.M.R., Husaini, A., Tasnim, T. 2017. Production and characterization of crude glucoamylase from newly isolated *Aspergillus flavus* NSH9 in liquid culture. *American Journal of Biochemistry and Molecular Biology* 73:118–126.

Kathiresan, K., Manivannan, S. 2006. Amylase production by *Penicillium fellutanum* isolated from rhizosphere soil. *African Journal of Biotechnology* 510:829–832.

Laderman, K.A., Lewis, M.S., Griko, Y.V., Privalov, P.L. 2003. The purification and characterization of an extremely thermostable amylase from *Pyrococcus furiosus*. *Journal of Biological Chemistry* 268:24394–24401.

Lin, S.C., Lin, I.P., Chou, W.I., Hsieh, C.A., Liu, S.H., Huang, R.Y., Sheu, C.C., Chang, M.D.T. 2009. CBM 21 starch-binding domain: A new purification tag for recombinant protein engineering. *Protein Expression and Purification* 65:261–266.

Liu, Y.H., Lu, F.P., Li, Y., Wang, J.L., Gao, C. 2008. Acid stabilization of *Bacillus licheniformis* alpha-amylase through introduction of mutations. *Applied Microbiology and Biotechnology* 80:795–803.

Machovic, M., Janecek, S. 2006. Starch-binding domains in the post-genome era. *Cellular and Molecular Life Sciences* 63:2710–2724.

Nguyen, Q.D., Bujna, E., Styevkó, G., Rezessy-Szabó, J.M., Hoschke, A. 2015. Fungal biomolecules for the food industry. In: *Fungal Biomolecules Sources, Applications and Recent Developments*. V.K.Gupta, R.L.Mach, S.Sreenivasaprasad (Eds.). UK: John Wiley & Sons, Ltd.

Niyonzima, F.N., More, S.S. 2014. Detergent-compatible bacterial amylases. *Applied Biochemistry and Biotechnology* 174:1215–1232.

Norouzian, D., Akbarzadeh, A., Scharer, J.M., Moo-Young, M. 2006. Fungal glucoamylases. *Biotechnology Advances* 24:80–85.

Olsen, H.S., Falholt, P. 1998. The role of enzymes in modern detergency. *Journal of Surfactants and Detergents* 14:555–567.

Pandey, A., Nigam, P., Soccol, C.R., Soccol, V.T., Singh, D., Mohan, R. 2000. Advances in microbial amylases. *Biotechnology and Applied Biochemistry* 312:135–152.

Rajagopalan, G., Krishnan, C. 2008. Optimization of medium and process parameters for a constitutive α-amylase production from a catabolite derepressed *Bacillus subtilis* KCC103. *Journal of Chemical Technology & Biotechnology* 83:654–661.

Ramachandran, S., Patel, A., Nampoothiri, K.M., Pandey, A. 2004. Amylase from a fungal culture grown on oil cakes and its properties. *Brazilian Archives of Biology and Technology* 47:309–317.

Rizk, M.A., El-Kholany, E.A., Abo-Mosalum, E.M.R. 2019. Production of α-amylase by *Aspergillus niger* isolated from mango kernel. *Middle East Journal of Applied Science* 0901:134–414.

Santiago, M., Linares, L., Sánchez, S., Dodríguez-Sanoja, R. 2005. Functional characterization of the starch-binding domain of *Lactobacillus amylovorus* α-amylase. *Biologia Bratislava* 60:111–114.

Saranraj, P., Stella, D. 2013. Fungal amylase – A review. *International Journal of Microbiology Research* 42:203–211.

Sivaramakrishnan, S., Gangadharan, D., Nampoothiri, K.M., Soccol, C.R., Pandey, A. 2006. α-Amylases from microbial sources: An overview on recent developments. *Food Technology and Biotechnology* 44:173–184.

Souza, P.M., Magalhães, P.O. 2010. Application of microbial amylase in industry – A review. *Brazilian Journal of Microbiology* 41:850–861.

Suriya, J., Bharathiraja, S., Krishnan, M., Manivasagan, P., Kim, S.K. 2016. Marine microbial amylases: Properties and applications. *Advances in Food and Nutrition Research* 79:161–177.

Taniguchi, H., Honnda, Y. 2009. Amylases. In: *Encyclopedia of Microbiology*, M.Schaechter (Ed.), 3rd ed. Cambridge, MA, USA: Academic Press, pp. 159–173.

Tiwari, S.P., Srivastava, R., Singh, C.S., Shukla, K., Singh, R.K., Singh, P., Singh, R., Singh, N.L., Sharma, R. 2015. Amylases: An overview with special reference to alpha-amylase. *Journal of Global Biosciences* 41:1886–1901.

9 Plant Growth-Promoting Rhizobacteria (PGPR) Activity in Soil

A. Suresh, P.M. Sameera, J. Chapla, and P. Rajarao

CONTENTS

9.1 INTRODUCTION

Soil is considered a good host of microbial activities; the microbes occupy less than 5 percent of total space in soil. Soil is the natural growth medium for plants and microorganisms. The dynamic environment, which harbors the diverse groups of microbes, is confined to the aggregates with accumulated organic matter. The major microbial habitat around the plant root is called the rhizosphere and it was defined by Hiltner (1904) as the volume of soil, influenced by the presence of living plant roots, whose extension may vary with soil type, plant species, age and other factors (Foster 1988). The rhizosphere is a hot spot of microbial interactions as exudates released by plant roots are the main food sources for microorganisms and a driving force of their population density and activity. The ratio of the microbial load of the rhizosphere soil to the non-rhizosphere soil is generally known as the rhizosphere effect. Several factors such as soil type, soil moisture, pH, temperature, age and conditions of the plant are known to influence the effect of rhizosphere. The actual root surface colonized by microorganisms is often referred to as rhizoplane. A specific group of microorganisms inhabits this microhabitat. Most of the fungi inhibit the root surface in a mycelial state. Plants play an important role in selecting and enriching the types of bacteria by the constituents of their root exudates. Thus, depending on the nature and concentrations of organic constituents of exudates, and also the corresponding ability of the bacteria to utilize these as sources of energy, the bacterial community develops in the rhizosphere Curl and Truelove 1986). There is a continuum of bacterial presence in the soil rhizosphere, rhizoplane and internal plant tissues Hallmann et al. 1997).

Microorganisms are the main foundation for the maintenance of soil ecosystems and microbial diversity. Microorganisms are bacteria, actinomycetes and fungi, occupying an important niche in every ecosystem. They are essential in N, P, K and S cycles in nature, but also play an important part in the decomposition of organic matter. Microorganisms such as bacteria are the most common and grow rapidly, having the ability to utilize a wide range of substances, such as carbon or nitrogen sources. The most prominent group of bacteria in the rhizosphere are the non-sporulating gram-negative rod-shaped bacteria. Though fungi and actinomycetes are present, their populations are negligible compared to bacteria.

9.2   PLANT GROWTH-PROMOTING RHIZOBACTERIA (PGPR)

Numerous species of soil bacteria which flourish in the rhizosphere of plants may be grown in or around plant tissues, and stimulate growth with a plethora of mechanisms. These bacteria are collectively known as plant growth-promoting rhizobacteria (PGPR). The abbreviation PGPR has become common usage ever since it was first used by Kloepper and Schroth (1978).

Rhizobacteria that exert beneficial effects on plant development via direct or indirect mechanisms have been defined as plant growth-promoting rhizobacteria. Rhizosphere bacteria that favorably affect plant growth and yield of crops are referred to as plant growth-promoting rhizobacteria (PGPR).

Soil is the natural growth medium for living plants and microorganisms. A soil-plant ecosystem depends on microbial activities which in turn contribute towards improving soil health, environmental quality and crop production. Bacteria are the most dominant group of microorganisms in the soil and probably equal to one half of the microbial biomass present. The most abundant genera of bacteria present in the soil are *Acetobacter, Actinoplanes, Agrobacterium, Azospirillum, Azotobacter, Bacillus, Cellulomonas, Clostridium, Flavobacterium, Pasteuria, Rhizobium* and *Bradyrhizobium,* fluorescent pseudomonads, *Micrococcus, etc.,* which are large groups among PGPRs.

Thus, the plant growth-promoting microorganisms (PGPMs) are defined by three intrinsic characters: (i) ability to colonize the root, (ii) survive and multiply in micro-habitats associated with the root, in competition with other microbiota, to express their plant growth and protection activities and (iii) promote plant growth Gamalero et al. 2004).

In recent years it has been proven that root colonization indeed is required for some biocontrol mechanisms, such as antibiosis and competition for nutrients and niches (CNN) (Uren NC 2007). Beneficial effects of rhizospheric bacteria have often been based upon increased plant growth, faster seed germination, better seedling emergences, enhanced nodulation and nitrogen fixation in leguminous crops and suppression of disease. Then, PGPR have been further divided into subsets like emergence-promoting rhizobacteria (EPR); nodulation-promoting rhizobacteria (NPR) and disease-suppressing rhizobacteria (DSR). Beneficial plant-microbe interactions in the rhizosphere are the determinants of plant health and soil fertility (Jaffries et al. 2003). They include different PGPR bacteria like *Azotobacter, Azospirillum, Bacillus, Pseudomonas, Acetobacter, Burkholderia* and Bacilli (Glick 1995).

9.3 MECHANISM OF GROWTH PROMOTION OF PGPR

Plant growth-promoting bacteria are rhizospheric bacteria with the ability to stimulate and enhance plant growth through different mechanisms (Glick et al. 1999). Plant growth-promoting bacteria may be important for plant nutrition by increasing N and P uptake by the plants and playing a significant role as PGPR in the biofertilization of crops (Cakmakci et al. 2005). These are thought to improve plant growth by colonizing the root system and pre-empting the establishment of suppressing deleterious rhizosphere microorganisms on the roots (Figure 9.1).

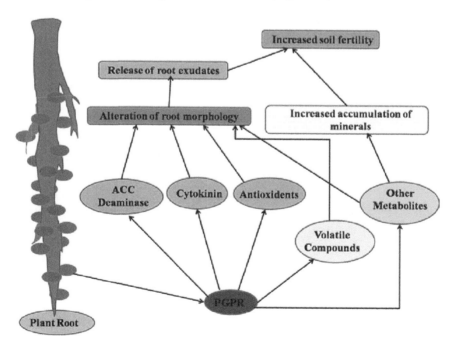

FIGURE 9.1 Soil fertility enhancement mechanism through PGPRs.

These mechanisms include the production of plant growth hormones, auxin (Khakipour et al. 2008) and 1-aminoacyclopropane-1-carboxylic acid deaminase (ACC) increased solubility of immobile nutrients such as P and Fe (by producing siderophores), fixation of atmospheric N and neutralizing the unfavorable effects of pathogens on plant growth (Jalili et al. 2009).

Plant hormones are very effective on plant growth and development, and among them is IAA (auxin), one of the most important growth regulators (Stepanova et al. 2008), often produced by rhizospheric bacteria. Production of auxin at the rates higher than the plant's requirement produces extra amounts of ACC, which is a prerequisite for ethylene production and is catalyzed by ACC-oxidase (Table 9.1). However, in some plants, ethylene can increase seed germination and interrupt seed dormancy (Glick et al. 1994). Under soil stress conditions, there is an increase in the ethylene level and a decrease in the plant growth. PGPR are able to alleviate such stresses on plant growth by producing ACC-deaminase enzyme, which turns ACC into ammonium and α-ketobutyric acid (Penrose and Glick 2003).

PGPR can affect plant growth in two different ways: direct and indirect. The direct promotion of plant growth by PGPR for the most part entails either providing the plant with a compound that is synthesized by the bacterium, or facilitating the uptake of certain nutrients from the environment. The indirect promotion of plant growth is when PGPR prevent the deleterious effect on one or more phytopathogenic organisms. These two mechanisms by which PGPR promote plant growth have been reviewed (Lata et al. 2002). It has been suggested that the production of plant hormone-enhanced nutrient uptake and suppression of phytopathogenic microorganisms such as auxins (Asghar et al. 2002), cytokinins (Arkhipova et al. 2005) and gibberellins (Joo et al. 2004) as well as through the solubilization of phosphate minerals. Indirect growth promotion occurs through the elimination of pathogens by the production of ß 1,3-glucanase (Fridlender et al. 1993), antibiotics (Raaijmakers et al. 1997), cyanide (Owen and Zdor 2001) and siderophores (Pidello 2003). Many PGPR stimulate the growth of plants by helping to control pathogenic organisms (Zehnder et al. 2001) (Table 9.1).

TABLE 9.1

Influence of Inoculations of PGPRS on Growth of Chili Cultivar under Greenhouse Conditions (45 DAS)

Isolate	Plant Height (cm)	Shoot Dry Weight (grams)	Root Length (cm)	Root Dry Weight (cm)
PGPR1	15.5	10.02	3.7	2.10
PGPR2	6.4	7.05	3.3	1.0
PGPR3	11.5	7.6	6.6	2.50
PGPR1004	8.5	6.89	2.9	1.57
Control	6.3	5.86	4.8	1.02
LSD(0.05)	0.003	0.020	0.016	0.010

Values are mean of three replicates and significant at $P < 0.05$

PGPR enhance the nutrient status of host plants, and these improvements may be categorized into:

1. Biological nitrogen fixation
2. Increase of the availability of nutrients in the rhizosphere
3. Enhancement of the root surface areas
4. Beneficial symbiosis of the host
5. Combination of modes of action

Many PGPR increase the availability of nutrients for the plant in rhizosphere. It involves the solubilization of unavailable nutrients or siderophores production, which facilitates the transport of certain nutrients. Bacterial-associated root increases in height are reported in response to PGPR. Most incidents noted an increase in root length and root surface area (Singh and Purohit 2008).

The PGPR and the mechanisms involved by which it promotes plant growth are ambiguous and not fully understood, but are thought to include the following characters, termed "PGPR traits" (Cattelan et al. 1999):

i) The ability to produce or change the concentration of plant hormones like indoleacetic acid (Mardukhova et al. 1984; Abbas et al. 2009; Karnwal 2009), gibberellic acid (Mahmoud et al. 1984; Gutierrez et al. 2001), cytokinins (Tein et al. 1979; Saleena et al. 2001) and ethylene (Arshad and Frankenberger 1991; Glick et al. 1995).
ii) Asymbiotic nitrogen fixation (Boddey and Dobereiner 1995; Kennewdy et al. 1997) and symbiotic nitrogen fixation.
iii) Antagonism against phytopathogenic microorganisms by the production of siderophores (Scher and Baker 1982; Alexander and Zuberer 1991), β-1,3-glucanase (Fridlender et al. 1993) chitinases (Renwick et al. 1991), antibiotics (Shanahan et al. 1992) and cyanide (Flaishman et al. 1996; Souza and Raaijmakers 2003; Sunish Kumar et al. 2005).
iv) Solubilization of mineral phosphates and other nutrients (Sperber 1958a, b; Chen et al. 2006; Pandey et al. 2006).

The microorganisms that fall under the broad category of PGPR are: *Rhizobia*, *Azotobacters*, *Azospirillum*, *Acetobacter*, phosphate solubilizers, *Mycorrhizae*, *Pseudomonas*, especially fluorescent pseudomonads.

9.4 MECHANISMS OF DIRECT GROWTH PROMOTIONS

9.4.1 PRODUCTION OF PLANT GROWTH REGULATORS

The mechanism most commonly invoked to explain the various effects of PGPR on plants is the production of phytohormones, the most common and well-characterized is IAA, which is known to stimulate both rapid and long-term response in plants. Many rhizosphere bacteria produce IAA in culture media, especially in the presence of tryptophan.

Naturally occurring substances with indole nucleus possessing growth-promoting activity are referred to as auxins. In many host-parasite interactions, definitive alterations in auxins (indole-3-yl-acetic acid, IAA) have been reported (Mahadevan 1984).

Chemical structure of Indole-3-acetic acid

Plant extracts, plant growth hormones or other natural products recently have recently been substituted for the chemical control of disease (Bekheit 2002).Indole-3-acetic acid (IAA) is well known for being a bioactive growth regulator, controlling stem elongation, geotropism, apical dominance, root initiation, etc.It has also been established that it is able to induce a concomitant resistance against phytopathogen attack through the regulation of defense mechanisms in plants (Mayda et al. 2000). IAA, in controlling diseases, not only induces resistance in a host but may also extend to the pathogen itself (Sharaf and Farrag 2004). The inhibitory activity was detected for several pathogenic fungi like *Gaeumannomyces graminis* var. *triticii, Rhizoctonia cerealis, Helminthosporium sativum* and *Phytophthora capsici.*

There are numerous soil microorganisms involved in the synthesis of auxins in pure culture and soil (Barazani and Friedman 1999). Some microorganisms produce auxins in the presence of a suitable precursor such as L-tryptophan. The concentration effects of auxins on plant seedlings are that low concentration may stimulate growth, while high concentrations may be inhibitory (Arshad and Frankenberger 1991). Different plant seedlings respond differently to variable auxin concentrations (Sarwar and Frankenberger 1994) and also different types of microorganisms.

Microorganisms inhabiting rhizospheres of various plants are likely to synthesize and release auxin as secondary metabolites because of rich supplies of substrates exuded from the roots compared with non-rhizospheric soils. Plant morphogenetic effects may also be a result of different ratios of plant hormones produced by roots as well as by rhizosphere bacteria. Diverse soil microorganisms including bacteria, fungi and algae are capable of producing physiologically active quantities of auxins which may exert pronounced effects on plant growth. Plant hormones are very effective for plant growth and development, and among them auxin (IAA) is one of the most important growth regulators often produced by rhizospheric bacteria (Mayak et al. 2004). Production of auxin at a rate higher than the plant requirement results in extra amounts of 1-aminoacyclopropane-1-carboxylic acid deaminase (ACC), which is a prerequisite for ethylene production and is catalyzed by ACC-oxidase.

Plant growth-promoting rhizobacteria (PGPR) producing plant growth regulators play an important role in plant growth promotion. The effect of these plant growth regulators on the plant are concentration-dependent. Moreover, IAA of *pseudomonas* origin induces some resistance in plants, for instance it stimulates resistance in *Phaseolus vulgaris* against *Colletotrichum vulgari s*(Huges and Dickerson 1990; Singh et al. 2018, 2017a, b, c, 2019; Tiwari et al. 2018, 2019a, b; Kour et al. 2019a).

The inhibitory activity of IAA was also detected for several pathogenic fungi like *Gaeumannomyca graminis var tritici*, *Rhizoctonia cerealis*, *Helminthosporium sativum* and *Phytophthiriacapsici* (Lu et al. 2000). The PGPR-producing IAA hormones are known to have a dual role in influencing plant growth by involvement in the biocontrol, together with glutathione-s-transferases in defense-released plant reactions and inhibit the germination of spore and growth of mycelium of different pathogenic fungi. IAA hormone, when supplied to excised potato leaves, eventually reduced the severity of the disease provoked by *Phytophthora infestans* (Martinez et al. 2001). Hence plant growth hormones are seen as potential biocontrol agents.

The ability of pseudomonad-produced auxin can very much affect plant growth (Khakipour et al. 2008), as it has some very important functions in plants such as hormonal adjustment, plant cell division, development and nodule formation. Since tryptophan is necessary for auxin production, its production at the control medium can be related to bacterial cell degradation (Frankenberger and Brunner 1983). According to Benizri et al. (1998), although there was not any tryptophan in corn root exudates, the *Pseudomonas fluorescens* strain M31 was able to produce auxin. Table 9.1 shows PGPR strains involved in phytohormone production in different plants (Jha and Bhattacharyya 2012).

9.4.2 PRODUCTION OF GIBBERELLIC ACID

Gibberellins are cyclic diterpens containing 20-carbon in their structure. They contain the skeleton enantiomers of gibberallane, which proves to be advantageous, as it enables the use of a uniform numbering system. Similar to other cyclic diterpens, so far more than 110 gibberellins have been identified.

Chemical structure of Gibberellic acid (GA_1)

Phytohormone production including gibberellins (Fulchieri et al. 1993; Lucangeli and Bottini 1997) is one mechanism that has been proposed. Gibberellic acids are a class of phytohormones with many demonstrated effects on a number of physiological processes (Davies 1995). Among 130 gibberellic acids identified thus far from plants, fungi and bacteria, GA_1, GA_3 and GA_4 are the three most common directly effective gibberellic acid shoot elongation promotions.

Gibberellic acid, which comes from a naturally produced growth hormone, is a member of a type of plant hormone called gibberellins, which regulate the growth rate of plants. Gibberellins are involved in several plant development processes and promote a number of desirable effects including stem elongation, uniform flowering, reduction in the time of flowering and increase in the flower number and size (Jaleel and Gopi 2009). Gibberellic acid is known to increase the antioxidant metabolism and alkaloid production in *Catharanthus roseus* (Jaleel et al. 2009).

Leaf shape and plant structure also be affected by altered levels of gibberellic acids and the application of exogenous GA_3 (Aguirre and Blanco 1992). Since GA_1 is thought to stimulate cell elongation, thus higher GA_1 content in tissues may also be associated with stem elongation and subsequent bloating (Jaleel et al. 2009). Stem swelling and alterations in leaf anatomy with gibberellic acid applications were reported in mustard (Xu et al. 2008; Singh et al. 2016).

Gibberellins are economically and industrially important products. They are commonly used in agriculture, viticulture, gardens and horticulture (Bandelier and Renaud 1997). Gibberillins are naturally present in plants in which they act as growth regulators. At an industrial scale, they are produced primarily by submerged fermentations using *Gibberella fujikorii*. They can also be obtained from several bacterial sources such as *Azotobacter*, *Pseudomonas* and *Azospirillum* (Basiacik and Nilufer 2004). Gibberellins are produced by microorganisms as typical secondary metabolites; upon exhaustion of nitrogen sources, exponential growth ceases and secondary metabolism is triggered (Gelmi and Perez 2000). More than ninety different types of gibberellins are known to occur in higher plants and microorganisms (Mander and Owen 1996).

9.4.3 NITROGEN FIXATION

Some diazotrophic PGPR supply a portion of the fixed nitrogen required to their plant hosts. Nitrogen-fixing organisms can contribute to nutrition for nitrogen and increased efficiency in the use of the nitrogen plant *Rhizobium*.

9.4.4 UPTAKE OF MINERALS

Several reports suggested that PGPR stimulates plant growth by facilitating the absorption of minerals into the plant, particularly phosphate. PGPR was found to be involved in inorganic phosphate solubilization, most of which were *Pseudomonas* and *Bacillus* species. Toro et al. (1997) studied the effects of mycorrhizal and nodulated *Pueraria phaseoloids* on the yield and nutrition exerted by rhizobacterium P-solubilizing. Significant nutrient stimulation (N, P, K and Ca) uptake was observed with *Azospirillum* sp. fungus-bacteria combinations. A majority of agricultural soils contain large reserves of phosphorus, of which a considerable part is accumulated as a result of regular application. The phenomenon of fixation and precipitation of P is soil dependent on pH. In acidic soils, P is precipitated as Al and Fe phosphates, whereas in calcareous soil, high Ca concentration results in P precipitation. The soil is actually a habitat for a range of organisms that use a variety of solubilization reactions to release soluble phosphorus from insoluble phosphates. The potential of these phosphate-solubilizing microorganisms has been used as bioinoculants for crops grown in less nutrient-present soils and modified to tricalcium phosphate with rock phosphate.

9.4.5 ROOT COLONIZATION

The distribution of bacteria in the rhizosphere can be considered from two angles: from the outside to the inside of the root, or from the root base (seed) to the root tip longitudinally. Distribution results from several dynamic phenomena that occur during the establishment of rhizosphere bacterial populations, such as post-inoculation bacterial migration, attachment to the roots and the ability of the roots to survive and proliferate. It should be remembered before interpreting distribution patterns that the types of measurement procedures and statistical analyzes used partly affect the results (Kloepper and Beauchamp 1992).

It is assumed that PGPR stimulation of plant growth generally requires binding of the bacterium to the plant root. The successful use of either rhizobial or PGPR inoculants in agriculture depends on the inoculation of viable bacteria to the root zones, most commonly achieved by inoculating seeds with the preparation of dormant bacterial cells through coated seeds or bulk inoculants.

9.5 INDIRECT GROWTH PROMOTION MECHANISMS

9.5.1 PRODUCTION OF SIDEROPHORES

The availability of iron is extremely limited in the rhizosphere, which is one of the most important nutrients required for the growth of nearly all living organisms. It has been found that Fe-binding ligands called siderophores with high affinity to sequester Fe from the microenvironment have been secreted to survive environmental organisms. One way the plant growth-promoting rhizobacteria can prevent phytopathogens from prolifying and facilitate plant growth by producing and secreting siderophore with a very high affinity to Fe. The secret siderophore molecules bind most of the Fe^{3+} available in the rhizosphere, effectively preventing any pathogens in their immediate vicinity from proliferating due to lack of Fe. As a mechanism for promoting plant growth and the biological control of pathogens, Kloepper et al. (1980) first demonstrated the importance of siderophore production.

The scarcity of bioavailable iron in soil habitats and on plant surfaces foments a furious competition (Loper and Henkels 1997). Under iron-limiting conditions, plant growth-promoting bacteria produce low molecular weight compounds called siderophores to competitively acquire ferric ion (Whipps 2001). Although various bacterial siderophores differ in their abilities to sequester iron, in general they deprive pathogenic fungi of this essential element since the fungal siderophores have lower affinity (Loper and Henkels 1999). Some PGPB strains go one step further and draw iron from heterologous siderophores produced by cohabiting microorganisms (Lodewyekx et al. 2002).

9.5.2 ANTIBIOTICS

The synthesis of antibiotics is one of the most effective mechanisms a PGPR can use to prevent the proliferation of phytopathogens. Most antibiotics belong to the nitrogen class that contains heterocycles such as phenazines and antibiotics of the type pyrrolintrin. Howell and Stipanovic (1980) reported *P.fluorescences* Pf-5 as the purified antibiotics pyoluteorin and pyrrolnitrin. The cotton damping could be suppressed by *Pythium ultimum* or *Rhizoctonia solani* such as strain production.

The usefulness of *Bacillus* as a source of antagonism to many plant pathogens is well known. Several potent strains from different *Bacillus* species have been tested for their ability to control multiple diseases on a wide variety of plant species. *Bacillus* has ecological benefits as it produces endospores that are tolerant to extreme conditions such as heat and desiccation. *Bacillus* species and actinomycetes share several characteristics that make them attractive to biological control agents including their abundance in the soil by producing various biologically, potentially active metabolites against a range of fungi.

The field of biological control of soil-borne plant pathogenic fungi was revolutionized by fluorescent pseudomonads. They have emerged over the past three decades as the largest and potentially most promising plant growth group promoting rhizobacteria involved in plant disease biocontrol. For several compelling reasons, fluorescent pseudomonads have received the greatest attention as they colonize roots readily in nature. Simple nutritional requirements and the ability to use many carbon sources that exude from roots and compete with indigenous microflora may explain their ability to colonize the rhizosphere. Pseudomonads are also suitable for genetic manipulation. These features make them useful vehicles for supplying the rhizosphere with antimicrobial and insecticidal compounds and plant hormones. Suresh et al. (2016) examined the characteristics of fluorescent pseudomonads such as antibiotic production, hydrogen cyanide, siderophore involved in plant pathogen suppression.

9.5.3 AMMONIA AND CYANIDE PRODUCTION

In 1988 Howell et al. reported volatile compounds such as ammonia produced by *Enterobacter cloacae* were involved in the suppression of *Pythium ultimum*, induced by the damping off of cotton. Hydrocyanic acid (HCN) is produced by many rhizobacteria and postulated to play a role in the biological control of pathogens. Cyanide production (Fig 9.2) in one study was thought to be detrimental to plant growth

FIGURE 9.2   Production of siderophore on CAS agar plate.

(Baker and Schippers 1987). However, in a standardized gnotobiotic system, cyanide has been shown to be involved in the suppression of the black root rot (Voisard et al. 1989) and several other pathogens like *Gaeumanomyces graminis*, causing disease in cereals (Paszkowski 1998). HCN is produced in certain conditions as well as specific growth stages of bacteria. The contribution of this compound in the disease-controlling ability of the producer strain varies among different species and strains (Anith et al. 1999). This volatile substance is studied specifically in pseudomonads for their biocontrol ability. Indeed, many rhizosphere-inhabiting fluorescent pseudomonads exhibit antagonistic effects towards fungal pathogens of plant roots, thereby protecting the plant from disease (Schisler et al. 1997). Among the different mechanisms involved in disease suppression, the production by fluorescent pseudomonads of antimicrobial secondary metabolites such as hydrogen cyanide (HCN) or 2,4-diacetylphloroglucinol (Phl) (Keel et al. 1992) is recognized to be significant for effective biocontrol.

9.5.3.1 Competition

Fluorescent pseudomonads are reported to be nutritionally versatile and grow rapidly in the rhizosphere, thereby excluding the other organisms from reaching the niche.

9.5.4 Lytic Enzymes

Some PGPR strains have been found to produce enzymes that can lyse fungal cells. These enzymes are able to digest and lyse *Fusarium solani* mycelia, thereby preventing the fungus from causing crop loss owing to root rot. Chitinase produced by *S. plymuthica* C48 inhibited spore germination and germ-tube elongation in *Botrytis cinerea* (Frankowski et al. 2001). The ability to produce extracellular chitinases is considered crucial for *Serratia marcescens* to act as an antagonist against *Sclerotium rolfsii* (Ordentlich et al. 1988), and for *Paenibacillus* sp. strain 300 and *Streptomyces* sp. strain 385 to suppress *Fusarium oxysporum* f. sp. *cucumerinum*. It has been also demonstrated that extracellular chitinase and laminarinase synthesized by *Pseudomonas stutzeri* digest and lyse mycelia of *F. solani* (Lim et al. 1991). Although chitinolytic activity appears less essential for PGPB such as *S. plymutica* IC14 when used to suppress *S. sclerotiorum* and *B. cinerea*, the synthesis of proteases and other biocontrol traits are involved (Kamensky et al. 2003). The β-1,3-glucanase synthesized by *Paenibacillus* sp. strain 300 and *Streptomyces* sp. Strain 385 lyse fungal cell walls of *F. oxysporum* f. sp. *Cucumerinum* (Singh et al. 1999). *B. cepacia* synthesizes β-1,3-glucanase that destroys the integrity of *R. solani*, *S. rolfsii* and *Pythium ultimum* cell walls (Fridlender et al. 1993). Similar to siderophores and antibiotics, the regulation of lytic enzyme production (proteases and chitinases in particular) involves the GacA/GacSor GrrA/GrrS regulatory systems and colony phase variation (Lugtenberg et al. 2001).

9.5.5 Induction of Systemic Resistance

In many plants, long-lasting and broad spectrum systemic resistance to disease-causing agents including fungal pathogens can be induced by treating the plant or seed with a PGPR. In this case the PGPR appear to turn on the synthesis of some antipathogenic metabolites within the plants in a mechanism that does not involve any direct interaction between the PGPR and the pathogen. Ramamoorthy et al. (2001) identified how many strains of pseudomonads can indirectly protect the plants by inducing systemic resistance against various diseases.

9.5.6 Promotion of Symbiosis/Enhancement of Legume Nodulation

The symbiosis between microorganisms and plants is also influenced by free-living bacteria and thus indirectly stimulates plant growth. Some plant growth-promoting bacterial inocula may interact positively with various symbiotic plant microorganisms such as *Rhizobium*, *Bradyrhizobium*, *Frankia* and mycorrhizal fungi.

9.5.7 Phosphate Solubilization

Phosphorous is considered an essential micronutrient and a great portion of phosphorous from chemical fertilizers becomes insoluble by its conversion to calcium or magnesium salts in soils, and so becomes unavailable to plants. Soil microorganisms transform the insoluble forms of phosphorous into soluble forms and thus influence the subsequent availability of phosphate to plant roots. (Richardson et al. 2001). Phosphate-solubilizing microorganisms have been employed in agriculture and horticulture and have been considered highly significant due to their potential in ecological amelioration. It is believed that microbial-mediated solubilization of insoluble phosphates in soil is through the release of organic acid microbial metabolites (Rodriguez et al. 2004). However, in addition to acid production, other mechanisms can cause phosphate solubilization (Nautiyal et al. 2000). Phosphate solubilization has been reported to depend on the structural complexity and particle size of phosphates and the quantity of organic acid secreted by microbes (Gaur 1990).

Different soil microorganisms including fluorescent pseudomonads play an important role in solubilizing inorganic phosphates that are chemically fixed. This is accomplished mainly by the secretion of organic and inorganic acids by microorganisms. These acids form a chelate with calcium ion in addition to the lowering of pH. Carbon and nitrogen sources greatly influence this process. The secretion of organic acids is intimately related to substrate metabolism by the organisms. The carbon and nitrogen sources are an important parameter for the active proliferation of organisms and the production of organic and inorganic acids (Bagyaraj et al. 2000). Recently, the solubilization of tricalcium and rock phosphates by *Pseudomonas fluorescens* and the influence of different carbon and nitrogen sources have been studied (Patel and Dare 2003).Scientific reports on phosphate solubilization by fluorescent pseudomonads in general are scanty. But we can observe an increase in the surface area covered by the root system and mineral solubilization ability of fluorescent pseudomonads, thus facilitating an increased nutrient uptake, which may increase seedling biomass (Nautiyal 1999).

Phosphate-solubilizing bacteria have been reported for promoting plant growth and enhancing yield (Kapoor et al. 1989). The secretion of organic acids and phosphatase enzymes are common mechanisms that facilitate the conversion of insoluble forms of phosphorous to plant available forms (Richardson 2001). The solubilization of insoluble phosphorous to accessible forms like orthophosphate is one of the important traits of plant growth-promoting rhizobacteria (PGPR).

9.6 PLANT GROWTH PROMOTION AND YIELD

There is an enormous body of work on the application of bacteria for the improvement of plant performance but few bacteria like *Azotobacter, Azospirillum* and *Pseudomonas* have been developed as commercial products. The organisms for potential use in agriculture are bacteria belonging to the genera *Pseudomonas* and *Bacillus* sp.

The effects of PGPR on crop growth have been reviewed (Goel et al. 2001). Inoculation of three PGPR isolates each belonging to *Proteus vulgaris, Klebsiella planticola* and *Bacillus subtillis* markedly increased the seed yield of sunflower and maize.

Microbial formulations are carrier-based preparations containing beneficial microorganisms in a viable state intended for seed or soil application. They are designed to improve soil fertility and help plant growth by increasing their numbers and thus their biological active root environment (Bashan 1989). The inoculation of seeds (Table 9.1) with PGPR is known to increase nodulation, nitrogen uptake and the growth and yield response of crop plants (Dorosinsky and Kadyrov 1975). Phosphate-solubilizing bacteria are also known to enhance phosphorus uptake, resulting in better growth and a higher yield of crop plants (Johri et al. 2003).

Dobbelare et al. (2002) assessed the inoculation effect of PGPR *Azospirillum brasilense*on the growth of spring wheat. They observed that inoculated plants resulted in better germination, early development and flowering and also an increase in the dry weight of the root system and upper plant parts. Similarly, promotion in growth parameters and yields of various crop plants in response to inoculation with PGPR were reported by Gravel et al. (2007). Inoculation of maize seeds with *Pseudomonas* strains under experiment conditions resulted in a more visible increase in shoot development, especially during the establishment of the plant. Khalid et al. (2004) showed that responses of wheat growth to inoculation with rhizobacteria depended on the plant genotype and PGPR strains, as well as the environmental conditions.

Burd et al. (2000) reported that plant growth-promoting rhizobacteria might enhance plant height and productivity by synthesizing phytohormones, increasing the local availability of nutrients and facilitating the uptake of nutrients by the plants by decreasing heavy metal toxicity in the plants' antagonizing plant pathogens. The enhancing effect of seed inoculation with rhizobacteria on shoot dry weight and yield of maize were reported by Pandy et al. (1998) and Shaharoona et al. (2006). Such an improvement might be attributed to the nitrogen-fixing and phosphate-solubilizing capacity of bacteria, as well as the ability of these microorganisms to produce growth-promoting substances (Salantur et al. 2006).

The inoculation of three PGPR isolates each belonging to *Bacillus, Pseudomonas* and *Rhizobium* increased yields of sunflower and maize over uninoculated plants grown in fields. Increases in plant height and root and shoot biomass were reported in different isolates of PGPR belonging to fluorescent pseudomonads (Suresh et al. 2016). Similarly, treatment of sunflower seeds with fluorescent pseudomonads resulted in yield increases of 30% in field trails. PGPR are potent inoculants but are not commercialized due to lack of consistency under field conditions. Researches indicate that a combination of PGPR strains (2 or more) which have diverse modes of plant growth promotion or antagonism against soil-borne pathogens are more effective than single strain inoculums. IAA-producing *Bacillus* isolates promoted root growth or nodulation when co-inoculated with Rhizobium species on a *Phaseolus vulgaris* contender in growth chambers. Gupta et al. (1998) reported that two strains of Enterobacter and one strain each of Pseudomonas fluorescens and Bacillus sp. have been found to promote the growth of green gram. Field studies also indicated that PGPR strains in conjugation with Rhizobium or Azospirillum increased the grain yield of chickpea and wheat.

The effects of PGPR on crop growth have been reviewed (Goel et al. 2001). Inoculation of three PGPR isolates each belonging to *Proteus vulgaris*, *Klebsiella planticola* and *Bacillus subtilis* markedly increased the seed yield of sunflower, maize and cotton over uninoculated plants grown in field (Malik et al. 1998). In an exploratory experiment, the inoculation of *Kurthia* spp. a gram-positive, ammonifying, urea hydrolyzing and thermo-tolerant organism isolated from the rhizosphere of wheat resulted in the improvement in growth and yield of rapeseed (*Brassica campestris* var *toria*) in pot culture studies under alluvial soil conditions (Malik et al. 1999).

9.7 CONCLUSIONS

Only those microorganisms which grow in the rhizosphere act as bio-control agents. The rhizosphere functions as a first line of defense for the roots of plants against the attacks on the soil-borne pathogens. Thus, there is a strong need to screen the rhizospheric bacteria having plant growth-promoting ability and to develop the PGPRs. Not only this, the bioinoculants with PGP and bio-control activity will be successful only when rhizosphere competence is able to exert the desired effect on the plants. It is necessary to introduce education to farmers in this regard. In doing so, the negative effects from the use of chemical fertilizers would be replaced by encouraging in the use of PGPRs in crop fields, resulting in high yields, the maintenance of soil fertility and sustainable development, all achieved in a natural manner.

REFERENCES

Anith, K.N., Tilak, K.V.B.R., Manomohandas, T.P. 1999. Analysis of mutation affecting antifungal property of a fluorescent *Pseudomonas* sp. During cotton – Rhizoctonia interaction. *Indian Phytopathology* 52:366–369.

Arkhipova, T.N., Veselov, S.U., Melentiev, A.I., Martynenko, E.V., Kudoyarova, G.R. 2005. Ability of bacterium *Bacillus subtilis* to produce cytokinins and to influence the growth and endogenous hormone content of lettuce plants. *Plant and Soil* 272:201–209.

Arshad, M., Frankenberger, W.T.J. 1991. Microbial production of plant hormones. *Plant and Soi l*133:1–8.

Asghar, H.N., Zahir, Z.A., Aeshad, M., Khaliq, A. 2002. Relationship between in vitro production of auxins by rhizobacteria and their growth promoting activities in *Brassica juncea* L.*Biology and Fertility Soils* 35:231–237.

Basiacik, S.K., Nilufer, A. 2004. Optimization of carbon-nitrogen ratio for production of Gibberillic acid by *Pseudomonas* sp. *Polish Journal of Microbiology* 53:117–120.

Bekheit, H.K.M. 2002. Biological control in Egypt. Advance and contrains. In: 2nd International Conference Plant Protection Research Institute, Cairo, Egypt, 21–24, December, pp. 112–124.

Cakmakci, R., Donmez, D., Aydin, A., Sahin, F. 2005. Growth promotion of plants by plant growth promoting rhizobacteria under green house and two different field soil conditions. *Soil Biology and Biochemistry* 38:1482–1487.

Cattelan, A.J., Hartel, P.G., Furhmann, F.F. 1999. Screening for plant growth promoting rhizobacteria to promote early soybean growth. *Soil Science Society of American Journal* 63:1670–1680.

Curl, E.A., Truelove, B. 1986. *The Rhizosphere Biology*, Springer Verlag, Berlin, pp. 288.

Foster, R.C. 1988. Microenvironments of soil microorganisms. *Biology Fertility Soils* 6:189–203.

Fridlender, M., Inbar, J., Chet, I. 1993. Biological control of soilborne plant pathogens by a β-1,3-glucanase-producing *Pseudomonas cepacia*. *Soil Biology Biochemistry* 25:1211–1221.

Gamalero, E., Lingua, G., Capri, F.G., Fusconi, A., Berta, G., Lemanceau, P. 2004. Colonization pattern of primary tomato roots by *Pseudomonas fluorescens* A6RI characterized by dilution plating, flow cytometry, fluorescence, confocal and scanning electron microscopy. *FEMS Microbiology Ecology* 48:79–87.

Glick, B.R., Karaturovic, D.M., Newell, P.C. 1995. A novel procedure for rapid isolation of plant growth promoting pseudomonads. *Canadian Journal of Microbiology* 43:533–536.

Glick, B.R., Patten, C.L., Holguin, G., Penrose, D.M. 1999. *Biochemical and Genetic Mechanisms Used by Plant Growth Promoting Bacteria*. Imperial College Press, London, pp. 267.

Hallmann, J., Quandt-Hallmann, A., Mahaffee, W.F., Kloepper, J.W. 1997. Bacterial endophytes in agricultural crops. *Canadian Journal of Microbiolog y*43:895–914.

Hiltner, L. 1904. Uber neuere Erfahrungen und Probleme auf dem Gebiete der Bodenbakteriologie unter bessonderer Berücksichtigung der Grundung und Brache. *Arb Dtsch Landwirtsch Ges Berl* 98:59–78.

Huges, R.K., Dickerson, A.G. 1990. Auxin regulation of the response of *Phaseolus vulgaris* to a fungal elicitor. *Plant Cell Physiology* 31:667–675.

Jaffries, P., Gianinazzi, S., Perotto, S., Turnau, K., Barea, J.M. 2003. The contribution of arbuscular mycorrhizal fungi in sustainable maintenance of plant health and soil fertility. *Biology Fertility Soil* 37:1–16.

Jalili, F., Khavazi, K., Pazira, E., Nejati, A., Asadi Rahmani, H., Rasuli Sadaghiani, H., Miransari, M. 2009. Isolation and characterization of ACC deaminase producing fluorescent pseudomonads, to alleviate salinity stress on canola (*Brassica napus* L.) growth. *Journal of Plant Physiology* 166:667–674.

Joo, G.J., Kim, Y.M., Lee, I.J., Song, K.S., Rhee, I.K. 2004. Growth promotion of red pepper plug seedlings and the production of gibberellins by *Bacillus cereus, Bacillus macroides* and *Bacillus pumilus*. *Biotechnology Letters* 26:487–491.

Kamensky, M., Ovadis, M., Chet, I., Chernin, L. 2003. Soil-borne strain IC14 of *Serratia plymuthica* with multiple mechanisms of antifungal activity provides biocontrol of *Botrytis cinerea* and *Sclerotinia sclerotiorum* diseases. *Soil Biology Biochemistry* 35:323–331.

Keel, C., Schnider, V., Maurhofer, M., Voisard, C., Laville, J., Burger, U., Wirthner, P., Hass, D., Defago, G. 1992. Suppression of root diseases by *Pseudomonas fluorescens* CHAO: Importance of the bacterial secondary metabolite 2,4 diacetyl phloroglucinol. *Molecular Plant Microbe Interactions* 5:4–9.

Khakipour, N., Khavazi, K., Mojallali, H., Pazira, E., Asadirahmani, H. 2008. Production of auxin hormone by fluorescent pseudomonads. *American – Eurasian Journal of Agricultural and Environmental Science* 4:687–692.

Kloepper, J.W., Leong, J., Teintze, M., Schroth, M.N. 1980. *Pseudomonas* siderophores: A mechanism explaining disease suppressive soils. *Current Microbiology* 4:317–320.

Kloepper, J.W., Schorth, M.N. 1978. Plant growth-promoting rhizobacteria on radishes. In: *Proceedings of the 4th International Conference on Plant Pathogenic Bacteria*. Vol. 2, Station de Pathologie Vegetable et Phytobacteriology, INRA, Angers, France, pp. 879–882.

Kour, D., Rana, K.L., Yadav, N., Yadav, A.N., Rastegari, A.A., Singh, C., Negi, P., Singh, K., Saxena, A.K. 2019a. Technologies for biofuel production: Current development, challenges, and future prospects. In: Rastegari, A.A. et al. (eds.), *Prospects of Renewable Bioprocessing in Future Energy Systems, Biofuel and Biorefinery Technologies*, Vol. 10. Springer, Berlin, pp. 1–50.

Lata Saxena, A.K., Tilak, K.V.B.R. 2002. Biofertilizers to augment soil fertility and cro production. In: Krishna, K.R. (ed.), *Soil Fertility and Crop Production*. Science Publishers, USA, pp. 279–312.

Lim, H.S., Kim, Y.S., Kim, S.D. 1991. *Pseudomonas stutzeri* YPL-1genetic transformation and antifungal mechanism against *Fusarium solani* an agent of plant root rot. *Applied Environmental Microbiology* 57:510–516.

Loper, J.E., Henkels, M.D. 1999. Utilization of heterologus siderophores enhances levels of iron available to *Pseudomonas putida* in the rhizosphere. *Applied Environmental Microbiology* 65:5357–5363.

Lugtenberg, B.J.J., Dekkers, L., Bloemberg, G.V. 2001. Molecular determinants of rhizosphere colonization by *Pseudomonas*. *Anneal Review Phytopathology* 39:461–490.

Mahadevan, A. 1984. *Growth Regulators, Microorganisms and Diseased Plants*. Oxford & IBH, New Delhi.

Martinez, N.G.M.A., Madrid, E.A., Botin, R., Lamattina, L. 2001. Indoleacetic acid attenuates disease severity in potato-phytophthora infestans interaction and inhibits the pathogen growth in vitro. *Plant Physiology Biochemistry* 39:815–823.

Ordentlich, A., Elad, Y., Chet, I. 1988. The role of chitinase of *Serratia marcescens* in biocontrol of *Sclerotium rolfsii*. *Phytopathology* 78:84–88.

Owen, A., Zdor, R. 2001. Effect of cynogenic rhizobacteria on the growth of velvet leaf (*Abutilontheo phrasti*) and corn (*Zea mays*) in autoclaved soil and the influence of supplemental glycine. *Soil Biology Biochemistry* 33:801–809.

Paszkowski, W.L. 1998. The effect of HCN excreted by *Pseudomonas fluorescens* on the growth of some phytopathogenic fungi (in Polish). *Zesz. Nau. Akad. Roln We Wroclawiu Konference* 332:85–92.

Penrose, D.M., Glick, B.R. 2003. Methods for isolating and characterizing ACC deaminase containing plant growth promoting rhizobacteria. *Physiology Plant* 18:10–15.

Pidello, A. 2003. The effect of *Pseudomonas fluorescens* strains varying in pyoverdine production on the soil redox status. *Plant and Soil* 253:373–379.

Ramamoorthy, V., Viswanathan, R., Raguchander, T., Prakasam, V., Samiyappar, R. 2001. Induction of systemic resistance by plant growth promoting rhizobacteria in crop plants against pests and diseases. *Crop Protection* 20:1–11.

Singh, C., Chowdhary, P., Singh, J.S., Chandra, R.2016. Pulp and paper mill wastewater and coliform as health hazards: A review. *Microbiology Research International*4:28–39.

Singh, C., Tiwari, S., Boudh, S., Singh, J.S. 2017a. Biochar application in management of paddy crop production and methane mitigation. In: Singh, J.S., Seneviratne, G. (eds.), *Agro-Environmental Sustainability: Managing Environmental Pollution*, 2nd ed. Springer, Switzerland, pp. 123–146.

Singh, C., Tiwari, S., Gupta, V.K., Singh, J.S. 2018. The effect of rice husk biochar on soil nutrient status, microbial biomass and paddy productivity of nutrient poor agriculture soils. *Catena* 171:485–493.

Singh, C., Tiwari, S., Singh, J.S. 2017b. Impact of rice husk biochar on nitrogen mineraliza-tion and methanotrophs community dynamics in paddy soil. *International Journal of Pure and Applied Bioscience* 5:428–435.

Singh, C., Tiwari, S., Singh, J.S. 2017c. Application of biochar in soil fertility and envi-ronmental management: A review. *Bulletin of Environment, Pharmacology and Life Sciences* 6:07–14.

Singh, C., Tiwari, S., Singh, J.S. 2019. Biochar: A sustainable tool in soil 2 pollutant biereme-diation. In: Bharagava, R.N., Saxena, G. (eds.), *Bioremediation of Industrial Waste for Environmental Safety.* Springer, Berlin, pp. 475–494.

Stepanova, A.N., Robertson-Hoyt, J., Yun, J., Benavente, L.M., Xie, D.Y., Dolezal, K., Jurgens, S.G., Alonso, J.M. 2008. IAA-mediated auxin biosynthesis is essential for hormone crosstalk and plant development. *Cell* 133:177–191.

Suresh, A., Munjam, S., Praveen Kumar, V., Krishna Reddy, V., Ram Reddy, S. 2016. Fluorescent pseudomonads as prospective bioinoculants for sunflower (*Helianthus annus* L.) *Octa Journal Biosciences* 4:28–32.

Tiwari, S., Singh, C., Boudh, S., Rai, P.K., Gupta, V.K., Singh, J.S. 2019a. Land use change: A key ecological disturbance declines soil microbial biomass in dry tropical uplands. *Journal of Environmental Management* 242:1–10.

Tiwari, S., Singh, C., Singh, J.S. 2018. Land use changes: A key ecological driver regulat-ing methanotrophs abundance in upland soils. *Energy, Ecology, and the Environment* 3:355–371.

Tiwari, S., Singh, C., Singh, J.S. 2019b. Wetlands: A major natural source responsible for methane emission. In: Upadhyay, A.K. et al. (eds.), *Restoration of Wetland Ecosystem: A Trajectory Towards a Sustainable Environment.* Springer, Berlin, pp. 59–74.

10 Nisin in Food Packaging

Shalini Singh and Robinka Khajuria

CONTENTS

10.1 INTRODUCTION

In ancient times, food was eaten fresh, because no particular facilities were available to store it for periods of time. With changing times and requirements, food packaging or food storage became a necessity. Humans developed various ways to protect their food materials from spoilage, owing to contact of foods with different spoiling factors (Sacharow and Griffin 1970; Kelsey 1989). The discovery of microorganisms, including those responsible for food spoilage, along with the beneficial effects of food sterilization to destroy such microbes, further added to an understanding of food characters, and the need to maintain its quality for different durations. Food packaging, one of the final steps in food processing before storage and consumption, thus, became a crucial step to incorporate antimicrobial effects for controlling post-processing contamination (Salim et al. 2017). The prevention of food spoilage addresses deterioration by microbial, biochemical, physical, textural, and chemical agents (Cutter 2002). Natural materials such as tree parts, papyrus, gourds, shells, animal parts, etc. are examples of natural materials used for preventing food spoilage in those times. Also, ancient practices to ferment, dry, smoke, sugar/salt, the addition of beeswax, honey, oil, etc., for various foods further helped in preserving foods. With time, improvements and variations to the early practices were also seen and found to be effective for storing different types of foods (Sacharow and Griffin 1970; Kelsey 1989; Cutter 2002a, b).

10.2 FOOD PACKAGING FORMULATIONS

Food packaging has been drawing a lot of attention worldwide and efforts are continuously being made to innovate and improve food packaging systems. A system that strives to preserve the quality and safety of the product it contains from the time of manufacture to the time it is used by the consumer, and protects the food from damage due to physical, chemical, or biological hazards (Dallyn and Shorten 1988; Hotchkiss 1995), makes for an ideal packaging formulation.

The selection of food packaging materials play a crucial role in the success of food packaging, as it provides a barrier to spoilage agents, and even adds additional functions to the package, which prolongs shelf life, enhances safety, as well as improving the nutritional value of foods (Appendini and Hotchkiss 2002) so that foods reach consumers in a safe and satisfactory condition.

It is thus important that the right package is selected for the right food, through sensible consideration of various influencing factors. Several important aspects to be considered are a slow but sustained delivery of antimicrobials; the evaluation of a microbe for which antimicrobial is to be targeted; additional attributes aimed for the end-use properties of food; compatibility between the antimicrobial agent to be delivered; and the packaging system used, especially with careful consideration of the potential chances of microbial resistance to appear. The most common food packaging materials are glass, wood, metal, plastics, paper, and other flexible packages such as coatings and adhesives (Kelsey 1989). Materials like glass, plastic bottles, pottery items, wood boxes, etc., belong to the rigid category of food packaging materials while plastic films, paper, foils, etc., are flexible packaging materials (Raheem 2012). These materials, singly or in combination with other preservation techniques, offer the necessary prevention of food spoilage (Cutter 2002; Suppakul et al. 2003). With time, both rigid as well as flexible systems have improved (Sacharow and Griffin 1970) Plastic materials are moldable, heat sealable, easy to print, and can be integrated into production processes where the package is formed, filled, and sealed in the same production line (Marsh and Bugusu 2007). At the same time, they exhibit variable permeability to agents like light, gases, etc. Paper and paperboard are sheet materials, basically made of cellulose. They are biodegradable and have good printability and are commonly used in corrugated boxes, milk cartons, folding cartons, bags and sacks, and wrapping paper. The use of rubber and adhesive components, polyester, polypropylene, polyolefins, polyvinyl, polyethylene, vinylidene, vinyl chloride, surlyn, and nylon, etc., has further improved packaging formulations by providing improved strength, permeability, and sealability to the package formulations. Such flexible systems also interact positively with gaseous environments to extend the shelf life of many foods, including meat and poultry (Sacharow and Griffin 1970; Cutter 2002; Stollman et al. 1994). Polymers such as low-density poly ethylene (LDPE) constitute the majority of primary packages for foods and beverages and have been commonly used in active polymer packaging (Rooney 1995a). Besides the synthetically derived packaging materials mentioned above, flexible packaging can encompass the use of edible films, gels, or coatings made from polysaccharides, proteins, lipids, or composites of any or all three. The benefits of using edible films as packaging materials are threefold: These films may resist the migration of outer moisture into the food during storage (Ooraikul 1991); films may serve as gas and solute barriers; and films complement other types of packaging by improving the quality and shelf life of foods (Ooraikul 1991).

A number of innovations such as modified atmosphere packaging (MAP), intelligent and/or active packaging are significantly contributing to the cause (Brennan and Grandison 2012). Variants like controlled atmosphere, vacuum, and modified atmosphere packaging use different gases and flushing systems to alter the headspace within a package (Ooraikul 1991; Genigeorgis 1985; Hintlian and Hotchkiss 1986; Farber 1991; Ooraikul and Stiles 1991; Labuza et al. 1992; Church and Parsons 1995; Garcia et al. 1995).

Modified atmosphere packaging (MAP) involves the addition of a single or a mixture of gases after removal of air from the pack (Blakistone 1999). Active packaging is a type of modified atmosphere packaging, which aims to improvise upon shelf life/improved safety/various improved properties of food by modifying the conditions existing in the food package, usually done with the addition of additives into the packages (Brennan and Grandison 2012; Alvarez 2000; Debeaufort et al. 2000; Quintavalla and Vicini 2002; Vermereinen et al. 2002; Salim et al. 2017; Brody et al. 2001; Appendini and Hotchkiss 2002; Han 2003; Cha and Chinnan 2004; Devlieghere et al. 2004; Dobiáš et al. 1998, 2000; Brody et al. 2001). These systems are developed on the basis of compatibility of the system with the food properties (Rooney 1995b). Selective permeability, temperature modulation, gas control, alcohol or ethylene scavenging systems, moisture control, controlled release of food colors, flavors, removal of odors, inclusion of antimicrobial agents, etc., are some of the important variants of active packaging systems (Labuza and Breene 1989; Rooney 1995a; De Kruijf et al. 2002). The quality of packaged food during transportation and storage can further be monitored through advanced intelligent packaging systems (Ahvenainen 2003).

Antimicrobial packaging has drawn a lot of attention worldwide, as awareness for minimally processed and chemical-free food products is increasing like never before (Imran et al. 2014; Mauriello et al. 2005). The antimicrobials may be directly coated on food material or incorporated into packaging materials (Hoffman et al. 2001; Kamper and Fennema 1984; Wilson 2007; Quintavalla Vicini 2002). As stated above, a number of factors such as the compatibility of food with antimicrobial agents, food composition, mode of action of antimicrobial substances, diffusion kinetics, polymer use, etc., shall all be considered while choosing the antimicrobial packaging system (Appendini and Hotchkiss 2002). Such compounds, generally in conjunction with other inhibitors or conditions, are used for the control of food spoilage and such use of multiple interventions is sometimes called 'hurdle technology' (Leistner 2000; Leistner and Gorris 1995).

At the same time, issues like loss in antimicrobial activity in the case of preservatives such as organic acids or their respective acid anhydrides, spice extracts, chelating agents, metals, enzymes, bacteriocins, etc., might arise when such agents are directly added to food (Teerakarn et al. 2002). The problem can be minimized by a 'controlled release packaging system', a form of active packaging utilizes packaging as a delivery vehicle to efficiently bring the actives in specifically controlled rates over prolonged periods to the food product to further improve its quality and safety (Cha and Chinnan 2004; Sung et al. 2013; Lacoste et al. 2005; Han 2000; Guerra et al. 2005; Han and Floros 1997; Davies et al. 1999; Hoffman et al. 2001; Cutter 2002; Quintavalla and Vicini 2002). This enables control of the release of antimicrobials and makes food safer to consume. Sometimes, the antimicrobial packaging films are categorized as those where volatile antimicrobial agents are added to sachets and pads, those where antimicrobial agents are incorporated into polymers, those where polymer surfaces are coated with antimicrobials, those where antimicrobials are immobilized by ionic or covalent linkages to polymers, or those where the use of polymers which are inherently antimicrobial is made (Appendini and Hotch-Kiss 2002; Sung et al. 2013).

The development of edible packaging films, coated with antimicrobials, is a new and attractive biodegradable active packaging concept that aims to improve the environmental friendliness of the packaging (Wong et al. 1994). Viable edible films and coatings produced from whey proteins (Ramos et al. 2012; An et al. 2000; Kamper and Fennema 1984, 1985; Fennema and Kester 1991; Han 2000) is one such example. Foods like meat and meat products have been seen to successfully benefit from edible antimicrobial packaging films Meyer et al. 1959; Siragusa and Dickson 1992, 1993; Baron 1993; Cutter and Siragusa 1996; Fang et al. 1996). The use of antibacterial nanoparticles in packaging films (Sorrentino et al. 2007) is an example of improved packaging film systems. Nanocomposites have been successfully used in vacuum-packaged meat, fish, poultry or cheese and offer improved strength, barrier properties, etc., to the packages (Sorrentino et al. 2007).

10.3 NATURAL ANTIMICROBIALS FOR FOOD PACKAGING

Though a number of preservatives are now available for food preservation, the growing preference of consumers for natural foods containing fewer synthetic additives has made the use of natural or food-grade antimicrobials in food popular (Sohaib et al. 2016; Pekcan et al. 2006; Davidson 1997).

Naturally occurring antimicrobials include compounds from microorganisms, plants, and animals.

Examples include organic acids or their respective acid anhydrides, spice extracts, chelating agents, metals, enzymes, bacteriocins, etc., which are quite popular in food preservation (Dainelli et al. 2008; Han and Floros 2000; Davies et al. 1999; Hoffman et al. 2001; Cutter 2002; Quintavalla and Vicini 2002).

Some of the active packaging systems include O_2 or CO_2 scavengers, ethylene and moisture absorption systems, CO_2 or ethanol emitting systems, and antimicrobials antioxidants releasing or containing systems. Eugenol, cinnamaldehyde, pectins, carageenan, starch derivatives, alginic acid, cellulose, collagen, chitosan, gelatin, antioxidants, fragrances (vanillin, clove, orange or citric extract, proteins (conalbumin, casein, whey protein, gelatin/collagen, fibrinogen, soy protein, wheat gluten, corn zein, or egg albumen), polysaccharides, lipid (fats, waxes, or oils) and seaweed extracts, grape seed extracts, spice extracts (thymol, p-cymene, cinnamaldehyde), enzymes (peroxidase, lysozyme), natural essential oils of plants such as, rose-mary, lemongrass, etc., can all be incorporated in antimicrobial edible films, where apart from acting as antimicrobial agents, they also provide selective desirable qualities to foods ((De Kruijf et al. 2002; Han 2002; Krochta et al. 1994; Jung 2000; Gennadios et al. 1997; Han 2000; Cutter 2002; Ozdemir and Floros 2004; Appendini and Hotchkiss 2002; Klangmuang and Sothornvit 2016; O'Callaghan and Kerry 2014; Takala et al. 2013; Wang et al. 2017; Mulla et al. 2017; Kuorwel et al. 2013; Kapetanakou et al. 2014; Siripatrawan and Vitchayakitti 2016; Dotto et al.

2014; Dutta et al. 2009; El-Saharty and Bary 2002; Siripatrawan and Noipha 2012; Yoshida et al. 2010; Ouattara et al. 2000; Aider 2010; Cruz-Romero et al. 2013; Soysal et al. 2015; Guo et al. 2014; Siripatrawan and Noipha 2012).

Organic acids like lactic acid, diacetic acid, have been used to reduce pathogens on beef, cheese, and processed meats (FDA 2000) and their activity was found to be better in an immobilized state (Siragusa and Dickson 1992, 1993). Other organic acids such as propionic, benzoic, sorbic, and lauric are also promising agents (Han 2000; Cutter 2002; Cruz-Romero et al. 2013; Branen et al. 2001; Perez et al. 2014; Rodriguez-Martinez et al. 2016; Hauser and Wunderlich 2011; Silveira et al. 2007; Dobias et al. 2000; Zhou et al. 2007; Sohaib et al. 2016).

Materials like ascorbic acid, photo-sensitive dyes, iron powder, are packed to scavenge oxygen and thereby prevent the growth of aerobic bacteria and molds (Floros et al. 1997). Lysozyme was found to inhibit bacteria in meat and that it inflicted wine malolactic fermentation (FDA 2000; Buonocore et al. 2003). Similarly, potassium sorbate's antibacterial and antimycotic effect has been studied on cheeses (Szente and Szejtli 2004). Milk protein-based film containing pimento and oregano was found to improve the shelf life of beef during storage at 4o C, where, oregano was most effective against *Pseudomonas* sp and *E. coli* O157: H7 (Oussalah et al. 2004). Baron and Sumner (1993) found potassium sorbate and lactic acid to inhibit *S. typhimurium* and *E. coli* O157:H7 on poultry. Juhl et al. (1994) proposed a number of agents like antioxidants, fragrances, colorants, antimycotic agents, or biocides to be used in thermoplastic polymer.

10.4 BACTERIOCINS FOR FOOD PACKAGING

A subgroup of naturally occurring antimicrobials is the bacteriocins, proteins/peptides produced by gram-positive and gram-negative species of microorganisms (Sohaib et al. 2016). However, only bacteriocins produced by food-grade lactic acid bacteria are of particular interest to the food industry as biopreservative lactic acid bacteria, e.g., *Lactococcus*, *Lactobacillus*, and *Pedicoccus* species, and a few other bacteria (Deegan et al. 2006; Imran et al. 2014; Sebti et al. 2003; Bierhalz et al. 2012; Fucinos et al. 2015; Duran et al. 2016) have been drawing great attention as a promising and natural antimicrobial agent for food preservation, amongst other areas of application. Only a few bacteriocins like nisin have regulatory approval for application to foods. The use of bacteriocins and other biologically derived antimicrobials in packaging material has been attracting increasing interest recently (Wilhoit 1996a, b; Ming et al. 1997; Sirgausa 1992), especially as many researchers (Crandall and Monville 1998; Dean and Zottola 1996) have realized that the existing methods of preservation may not be sufficient to preclude foodborne listeriosis. Several reports have thus addressed the use of bacteriocins to suppress effectively the growth of gram-positive bacteria (Degnan et al. 1993; Delves-Broughton et al. 1996; Ryser et al. 1994; Siragusa and Cutter 1993). Interestingly, a controlled release application using bacteriocins has been found to control resistant bacterial strains (Benoit et al. 1998). In such applications, the antimicrobial film must contact the surface of the food so that bacteriocins can diffuse to the surface, which then facilitates the gradual release of bacteriocins to the food surface (Appendini and Hochkiss 2002).

Bacteriocins like nisin and lacticin have also been used in edible food packaging films (Han 2000; Cutter 2002).

A polymer-based film solution coating has proved to be the most acceptable method in terms of stability for attaching a bacteriocin to a plastic packaging film (An et al. 2000). Incorporating bacteriocins into food packaging films to control spoilage of food pathogenic microorganisms has been an area of research for the last decades. Two procedures have been commonly used to make packaging films with bacteriocins. Two packaging methods, heat-press and casting, were used to deliver nisin into films made from soy protein and corn zein protein. Both the procedures produced excellent films by inhibiting the growth of the microbe *L. plantarum* (Appendini and Hotchkiss 2002). Immobilization of organic acids in edible coatings based on calcium alginate gel or whey protein has also been used to control *Listeria monocytogenes* on beef tissue (Cagri et al. 2001; Sirgausa and Dickinson 1992). Studies on meat and meat products (Bell and Lacy 1986; Mauriello et al. 2005) indicated that the surface addition of bacteriocins facilitates distribution through the product, and reduces interaction with meat components, thereby improving the antimicrobial effect. As with other antimicrobial food preservants, including both natural as well as synthetic/chemical, bacteriocins are found to be influenced by a number of factors which dictate their efficiency in the control of food spoilage. pH is one of the examples of external factors influencing the effect of bacteriocins in food. Certain studies indicate that food-borne pathogens resist the action of bacteriocins more in acidic pH (Van Schaik et al. 1999).

The emergence of varieties of spoilage/pathogenic microorganisms resistant to baacteriocins is also a matter of concern, as such varieties are on the increase. This is observed in cases where a single compound is being used to control microbial population, while no such observation has been reported in cases where combinations of bacteriocins or combinations of bacteriocins with other preservation methods or combination of bacteriocins with other antimicrobials, have been used (Mulet-Powell et al. 1998; Schillinger et al. 1998). Such observations have been reported for nisin, pediocin, and bavaricin by some researchers, but variable findings have been reported too (Crandall and Montville 1998; Rasch and Knøchel 1998). Interestingly, acquired resistance to a bacteriocin does not appear to impart resistance to other antimicrobials or preservative treatments nor any natural advantage for a population in the absence of the inhibitor. Nevertheless, more research is necessary (Sofos et al. 1998). The coating of pediocin onto cellulose casings and plastic bags has inhibited *L. monocytogenes* when checked in meats and poultry at 4° C (Natrajan and Sheldon 2000a). Leucocin- and sakacin-resistant *L. monocytogenes* strains have also been reported (Dykes and Hastings 1998), while Gravesen et al. (2002) reported a pediocin-resistant *L. monocytogenes* strain.

10.5   CHARACTERISTIC FEATURES OF NISIN

Nisin, a bacteriocin produced by some strains of *Lactococcus lactis* subsp. lactis, is commercially available for use in foods as an antimicrobial agent (Delves-Broughton 1990; Guinane et al. 2005). It is known to be non-toxic and approved as safe (**GRAS**) by the American Food and Drug Administration and World Health Organization in

1969 (Sindt 2001). A safe and natural preservative, widely used in the food industry, nisin is a heat-stable antimicrobial peptide with 34 amino acids and a molecular weight of 3.5 kDa (Mulders et al. 1991; Liu and Hansen 1990). It is classified as a lantibiotic, one of the two classes of bacteriocins (Class 1 and II), that contain amino acids such as lanthionine in their structure (Zendo et al. 2008; Cotter et al. 2005). The presence of unusual amino acid residue like dehydrobutyrine, dehydroalanine, lanthionine, and β-methyl-lanthionine, gives a unique structural and functional character to nisin (Dean and Zottola 1996; Schillinger et al. 1998; Winkowski et al. 1996). Largely reported to be active against gram-positive bacteria, especially against spore-forming bacteria, nisin is found to exhibit good antimicrobial effects against food-borne pathogens such as *Listeria*, *Staphylococcus*, *Bacillus*, lactic acid bacteria, *Clostridium*, *Listeria*, and *Streptococcus* (Guerra et al. 2005; Brewer et al. 2002; Lopez Pedemonte et al. 2003). Immediate digestibility by the enzyme alpha-chymotrypsin, heat stability at a low pH, and absence of color and flavor are some of the other important features of nisin that add to its popularity as an antimicrobial preservative for improving the shelf life of foods (Pongtharangkul and Demirci 2004).

Nisin characteristically gets adsorbed on various surfaces, and added to packaging films (McAuliffe et al. 2001). It binds to the anionic phospholipids of the microbial cell membrane and inserts into the cell by forming a pore, and further disrupting the breached cytoplasmic membrane. This causes the release of intracellular components and eventual depletion of the proton motive force (Crandall and Montville 1998). Gram-negative bacteria, on the other hand, are better protected against the action of nisin, owing to the protective effects of the outer membrane (Olasupo et al. 2003). Still, certain reports mention ease of action of nisin against gram-negative bacteria one the bacterial cell wall is weakened by specific agents like lyozyme, NaCl, or EDTA (Carneiro de Melo et al. 1998; Stevens et al. 1991), before the addition of nisin.

Numerous studies have thus incorporated nisin in various packaging formulations, and have shown that nisin is able to exhibit antimicrobial effects in such formulations (Chollet et al. 2008; Coma et al. 2001; Gadang et al. 2008; Ko et al. 2001; Kristo et al. 2008; Natrajan Sheldon 2000a; Neetoo et al. 2008; Nguyen et al. 2008; Teerakarn et al. 2002). In France, the use of nisin in the preservation of cheese was allowed without any restrictions (Ripoche et al. 2006). No reports have indicated any toxic effects on humans and other animals, even when used in higher amounts (Hagiwara et al. 2010).

Nisin activity may be affected by many factors, such as concentration, the target microorganisms, interaction with food components, fat content and phosphate type, processing, and storage conditions of food (Chollet et al. 2008; Davies et al. 1999; Soriano et al. 2004). Nutrient rich-media of foods or fatty acids in foods can affect nisin activity (Appendini and Hotchkiss 2002). While a low pH of the medium supports nisin activity through maintaining its solubility, higher pHs are not very helpful (Mauriello et al. 2005; Van Schaik et al. 1999). Still, pH-dependent response is not conclusively established for different foods. Along with other important factors like nisin concentration and film types used, nisin efficiency is also influenced by the diffusion rate of nisin from films into the food surface, which in turn, is influenced by many factors, as mentioned above (Ripoche et al. 2006). The action of nisin, as with other bacteriocins is dependent on its concentration, the presence of enzymes, and other additives that might be present in the food material to be preserved (Ripoche et al. 2006). Therefore, it is imperative to evaluate its effectiveness in in vitro conditions.

10.6 NISIN FOR FOOD PACKAGING

As nisin is reported to act against many pathogenic bacteria (Han 2005), studies have been undertaken to inhibit the growth of bacteria like *Bacillus cereus*, *Lactobacillus plantarum*, *Staphylococcus aureus*, *Listeria monocytogenes*, *Lactobacillus plantarum*, *Micrococcus luteus*, *Micrococcus flavus*, and *Brochothrix thermospacta* (Imran et al. 2014; Sebti et al. 2003; Periago and Moezelaar 2001; Pranoto et al. 2005; Delves-Broughton et al. 1996; Singh et al. 2001; Mahadeo and Tatini 1994; Ming et al. 1997).

The direct incorporation of nisin in packaging films is well studied (Quintavalla and Vicini 2002; Cooksey 2005; Chollet et al. 2008; Kristo et al. 2008; Manikantan and Varadharaju 2012; Padgett et al. 2000), and the effect of the type of packaging film on the activity of nisin in foods investigated, where it was found that the degree of cross-linking of agar-based film dictated the efficiency of nisin to migrate and control food spoilage microbes in broiler skin (Natrajan and Sheldon 2000b). A patented application of nisin, along with a number of antimicrobials, was proposed by Juhl et al. (1994) in a thermoplastic polymer. Its combination with chelators is well documented too (Teerakarn et al. 2002). The antimicrobial influence in sausages revealed its importance in preserving the quality of foods, even in cooked foods (Reunanen and Saris 2004).

The successful incorporation of nisin in a variety of edible films (Teerakarn et al. 2002; Padget et al. 2000; Scannel et al. 2000; Hong et al. 2005; Guerra et al. 2005; Luchansky et al. 2004; Sebti et al. 2002, 2003; Grower et al. 2004; Franklin et al. 2004; Guiga et al. 2009; Chollet et al. 2009; Guiga et al. 2010; Nguyen et al. 2008; Buonocore et al. 2003; Siragusa et al. 1999; Natrajan and Sheldon 1995; Cutter et al. 2001; Kim et al. 2002; Daeschel et al. 2001; Lee et al. 2003; Choi et al. 2001; Limjaroen et al. 2003, etc.), is enough proof of its efficacy as an antimicrobial agent in different matrices. Edible packaging using nisin, in combination with lysozyme and EDTA, was evaluated and found to inhibit the growth of *Brochothri thermosphacta*. Interestingly, *Listeria monocytogenes* was not found to be affected by the antimicrobial packaging formulation, while a strain of *E. coli* was partially influenced under the given study (Rodrigues and Han 2000). A similar combination, minus lysozyme, was tested by Economou et al. (2009) to improve the shelf life of raw poultry too, while Gill and Holley (2000a, b) showed the use of nisin in combination with lysozyme to control the population of *Lactobacillus* and *Leuconostoc* in sausage. Dawson et al. (2005) studied nisin in combination with silica and corn starch powders.

A combination of nisin with chitosan in a packaging film was successfully tested by Mauriello et al. (2005) against milk-borne microorganisms and Micrococcus sp., while Lee et al. (2003) used the same combination to store milk. Nisin in low density polyethylene packaging films could reduce the growth of gram-positive bacteria like *Staphylococcus aureus* and *Listeria* sp., Cooksey (2000). Appendini and Hotchkiss (2002) demonstrated that the same levels of nisin incorporated in cast films exhibited larger inhibitory zones than the heat-press films and also that when EDTA was incorporated into the films, the inhibitory effect of nisin against the bacteria *Escherichia coli* was increased. Cao-Hoang et al. (2010) presented the effectiveness of the nisin-coated films against *Listeria innocua* in cheese to improve the shelf life of processed cheeses, while it is commonly used in the preservation of processed cheese spreads (Hurst 1981).

Coma et al. (2001) incorporated nisin in Hypromellose films and tested against *Listeria innocua* and *Staphylococcus aureus*. The effect of nisin with other natural antimicrobials was studied against gram-positive and gram-negative bacteria, where thymol and eugenol were evaluated along with nisin (Tippayatum and Chonhenchob 2007). Interestingly, the essential oils tested yielded positive results towards the test organisms, but nisin could inhibit the activity of all but one (*E.coli*). Nisin, in calcium alginate matrix, was found to reduce the population of bacteria, and even sustained bacteriocin release from the matrix in beef as well as pork (Cutter and Siragusa 1996, 1997; Fang et al. 1996). Such gelatin-based matrices have been tested for nisin in fibrinogen/thrombin and gelatin for raw meat and ham and sausages, respectively, too (Cutter and Siragusa 1998; Gill 2000).

A large number of investigations have tested the antimicrobial activity of nisin in meat and meat products (Chung et al. 2005; Cooksey 2005; Geonaras et al. 2006; Grisi and Gorlach-Lira 2005; Jofré et al. 2008; López-Mendoza et al. 2007; Mangalassary et al. 2008; Sivarooban et al. 2007, 2008; Theivendran et al. 2006). While Ruiz et al. (2009, 2010) tested the effect of nisin in different concentrations, on turkey ham samples for fixed as well as variable time periods of storage; they reported satisfactory control of *L. monocytogenes*, inoculated into the food samples. Similarly, a combination of nisin with lacticin was tested in pork and evaluated for their action against *Listeria innocua* for a number of days. Though both the antimicrobial agents were effective, nisin better controlled the growth of the pathogen (Soriano et al. 2004). The effect of nisin on *L.monocytogenes* has been examined on meats, and chicken samples by others too (Chung et al. 1989; Kara et al. 2014; Günes and Çibik 2002; Solomakos et al. 2008). Nisin, evaluated at different concentrations, inhibited the growth of *L.monocytogenes* in fermented sausage (Hampikyan and Ugur 2007), while the effect was found to be concentration dependent on the same organism by Ko et al. (2001) and Pawar et al. (2000). Similar observations were made when comparing the action of nisin with sakacin and when used in combination with nitrate, where higher concentrations performed better to control microbial growth, as compared to the use of nisin in a lower concentration (Aasen et al. 2003; Reunanen and Saris 2004). Various combinations of nisin with a chelating agent in different packaging film materials were also tested for meats against *Brochothrix termosphacta*, where a clear influence of packaging film on the efficacy of antimicrobials used was observed. A film made of polyethylene, for example, was a better performer than others when nisin and EDTA were bound to it (Cutter et al. 2001), Natrajan and Sheldon (1995, 2000a, b) also studied the effect of nisin broiler skin when the latter was coated with *Salmonella Typhimurium*. A comparison between gel-immobilized nisin and free nisin was made on meat spoilage organism, *Brochothrix thermosphacta*, and it was found that an immobilized system was better in the overall control of spoilage organism than the non-immobilized system (Cutter and Sirgusa 1996). A similar observation was again made on beef when an alginate immobilized system performed better than the non-immobilized one, and other packaging systems, on the same organism (Cutter and Sirgausa 1997; Siragusa et al. 1999). Hoffman et al. (2001) studied nisin, lauric acid, and EDTA and showed reduced *L. monocytogenes* growth. Cellulose-based films in conjunction with nisin have been evaluated for many types of foods including meat and meat products (Nguyen et al. 2008; Ming et al. 1997; Scannell et al. 2000; Franklin et al. 2004; Luchansky and Call 2004) against different food-borne microflora including serious pathogens like *L. monocytogenes*. A combination of nisin-coated films and low temperature storage was evaluated for its role in controlling the microbial spoilage in beef. It was found that the combination effectively controlled the growth of bacteria on the given food (Ercolini et al. 2010). Siragusa et al. (1999) also supported the use of nisin-coated packaging films for the control of *Lactobacillus helveticus* and *B. thermosphacta* in beef carcasses. Millette et al. (2007) promoted the use of nisin in palmitoylated-based alginate film containing nisin to control *S. aureus* on round beef steak. Neetoo et al. (2008) used nisin-coated plastic films to control *Listeria monocytogenes* on salmon.

The effect of temperature was highlighted by Raju et al. (2003) and Cutter and Siragusa (1998) when they explained the nisin to lose its antimicrobial activity in meat and fish. A reduction in the microbial load in fresh veal meat (Guerra et al. 2005), a combination of lactic acid and nisin in beef carcasses (de Martinis et al. 2002) have also been reported to support the use of nisin in food preservation. It is important to note that even when many findings confirm the successful use of nisin in the preservation of meat and meat products, some reports highlight that nisin might lose some of its activity in meat due to the presence of interfering factors like phospholipids, and high pH, in meat and meat products (O'Sullivan et al. 2002; Liu and Hansen 1990; Aasen et al. 2003; Rose et al. 2002; Yonema et al. 2004).

Dawson et al. (1996) and Padgett et al (2000) showed the use of a combination of nisin and lysozyme in edible films for preventing the growth of *E. coli* and *Lactobacillus plantarum*, while Hoffman et al. (2001) tested the use of nisin with EDTA, lauric acid in different combinations, against *Listeria monocytogenes*. Studies on nisin as an antimicrobial agent in packaging films have been reported by Scannell et al. (2000) too, on different foods, including ham and cheese. Activity against *Staphylococcus aureus* and *Listeria monocytogenes* have been supported by Cooksey (2000) and Chi-Zhang et al. (2004), while Mauriello et al. (2005) showed its antimicrobial efficiency against *Micrococcus luteus* and microflora of milk. Malic acid, natamycin, and nisin were further evaluated by Pintado et al. (2010) to control pathogenic and spoilage organisms in cheese. Growth reduction for Staphylococcus aureus in milk by using high intensity pulsed electric fields and nisin was reported by Sobrino-Lopez et al. (2006).

The use of nisin has not only been studied against many foodborne pathogenic bacteria but their spores as well, and the efficiency of nisin against both have been demonstrated (Ray 1992; Orr et al. 1998), and L. *plantarum* in peptone water to below detection level (Padgett et al. 2000). Few studies indicate that there might be no direct effect of nisin on microbial growth. Instead, it simply discourages the attachment of the microbe to the matrix provided (Bowers and McGuire 1995; Carballo and Arajjo 2005). Mahdavi et al. (2007) studied the effect of nisin on bio-film-forming foodborne bacteria.

The antimicrobial effect of nisin in different food materials is dictated by many factors. Investigations also suggest that the release of immobilized nisin from meat components is influenced by the method adopted for the release (Cutter and Sirgausa 1997; Tramer and Fowler 1964; Bell and DeLacy 1986) reported supporting results to the afore-mentioned observation. Guerra et al. (2005) suggested that the introduction of nisin to the meat surface was a better method than the introduction of nisin directly into the meat mass. The presence of proteins and fats in food (Jung et al. 1992), interference of salt in foods like meat with release/adsorption of nisin into the food surface (Bell and DeLacy 1985) are other examples. The influence of packaging material can be highlighted through the work of Papadokostaki et al. (1997). For instance, one finding showed that linear low-density polyethylene tends to repel nisin coating, thereby influencing the actual nisin activity in food. Physical conditions like temperature, pH, and film thickness were found to have a significant effect on nisin activity when rested in whey protein edible films (Rossi-Márquez et al. 2009).

Proposals to overcome the loss of nisin activity due to variable reasons have been discussed by researchers. One of the ways to overcome the activity loss/minimize activity loss is to include a carrier molecule, which helps in the diffusion of nisin into the food matrix. The concept was applied by Soto et al. (2016) with amaranth protein-pullulan nanofibers, by Dheraprasart et al. (2009) with gelatin electrospun fiber, by Wang et al. (2015) with poly (vinyl alcohol)/wheat gluten/zirconia nanofibers), as a few examples of successful utilization of a carrier in nisin-based systems.

Resistance towards bacteriocin, sometimes when used singly in food preservation is known as described above. It is interesting to note that the way through which nisin is released into food also dictates the appearance of resistance among test organisms. It was shown by Chi-Zhang et al. (2004) that a rapid release of nisin from the film matrix made the bacteria more resistant towards nisin, in comparison to the slow release of the antimicrobial. Enzymatic action, alteration of membrane susceptibility, random mutation appearance in select bacterial strains, etc., are some important reported mechanisms that resistant bacterial strains use to overcome the antibacterial action of nisin (Montville et al. 2001; Hoover and Hurst 1993; Mazzotta et al. 1997; Ming and Daeschel 1993; Schillinger et al. 1998; Crandall and Montville 1998). Mazzotta et al. (2000) demonstrated that nisin-resistant strains of *L. monocytogenes* and *C. botulinum* were not as resistant as wild-type strains to other traditional food antimicrobials including sodium chloride, sodium nitrite, and potassium sorbate. Spores of certain strains of *Clostridium botulinum* have also been reported to develop resistance against nisin (Mazzotta and Montville 1999).

10.7 FUTURE OF NISIN IN FOOD PACKAGING

With the rapid rise in applications and acceptance of antimicrobial active packaging technology and fast-growing popularity of natural, safe, and environment-friendly preservants, the world is bound to explore more of bacteriocins, including more extensive utilization of nisin than is already in the market for food preservation. As the success of such agents would lie in their efficiency and durability, nisin, or bacteriocins for that matter, need to outperform limitations associated with their use.

Moreover, biopolymers as antimicrobial carriers have also received great interest due to environmental concerns. The antimicrobial carriers, apart from being as environment-friendly as possible, need to improve upon their limitations of degradation rates, and mechanical properties for the overall success of nisin as a food preservant. Thorough understanding of films, therefore, is needed for better development of mechanical and barrier features. Laboratory-level successful application of other bacteriocins shall be investigated with dedication for a possible increase in their commercial utilization. At the same time, the properties of nisin, and the compatibility of nisin with antimicrobials, as well as different packaging films, need to be extensively explored. A slow, sustained delivery of nisin in different formulations, and in different types of foods is a matter of concern and needs to be addressed upon for improving the delivery of nisin into the food material. A large number of studies have focussed on the use of nisin in meat and meat products, but its use in other foods shall be investigated more for large scale applications for better exploitation of nisin's antimicrobial efficiency. As the nature of foods also plays a crucial role in influencing the activity of nisin, the ways through which nisin can act better in porous or irregular foods need to be studied. A lot of work has been done on developing films, but actual preservation efficiency studies in real time applications are still far less. The stability of nisin is still questionable in certain food types and hence, extensive exploration of possible stabilization of nisin in such foods is necessary too. The future of bacteriocins like nisin is promising in food applications. Overcoming glitches in its extensive utilization in food preservation will go a long way in using the full potential of these 'microbial gifts' to the human race.

REFERENCES

Aasen, I.M., Markussen, S., Moretro, T., Katla, T., Axelsson, L., Naterstad, K. 2003. Interactions of the bacteriocins sakacin P and nisin with food constituents. *International Journal of Food Microbiology* 87:35–43.

Ahvenainen, R. 2003. Active and intelligent packaging: an introduction. In: *Novel Food Packaging Techniques*, R. Ahvenainen (Ed.). Woodhead Publishing Ltd., Cambridge, UK, pp. 5–21.

Aider, M. 2010. Chitosan application for active bio-based films production and potential in the food industry: Review. *LWT – Food Science and Technology* 43:837–842.

Alvarez, M.F. 2000. Review: Active food packaging. *Food Science & Technology International* 6:97–108.

An, D.S., Kim, Y.M., Lee, S.B., Paik, H.D., Lee, D.S. 2000. Antimicrobial low density polyethylene film coated with bacteriocins in binder medium. Food Science and Biotechnology 9:14–20.

Appendini, P., Hotchkiss, J.H. 2002. Review of antimicrobial food packaging. *Innovative Food Science Emerging Technology* 3:113–126.

Baron, J.K. 1993. *Inhibition of Salmonella typhimurium and Escherichia coli O-157:H7 by an Antimicrobial-Containing Edible Film*. M.S. Thesis, University of Nebraska, Lincoln, NE.

Baron, J.K., Sumner, S.S. 1993. Antimicrobial containing edible films as inhibitory system to control microbial growth on meat products. *Journal of Food Protection* 56:916.

Bell, R.G., DeLacy, K.M. 1985. The effect of nisin-sodium chloride interactions on the outgrowth of *Bacillus licheniformis* spores. *Journal of Applied Bacteriology* 59:127–132.

Bell, R.G., DeLacy, K.M. 1986. Factors influencing the determination of nisin in meat products. *Journal of Food Technology* 21:1–7.

Benoit, M.A., Mousset, B., Delloye, C., Bouillet, R., Gillard, J. 1998. Antibiotic-loaded plaster of Paris implants coated with polylactide-co-glycolide as a controlled release delivery system for the treatment of bone infections. *International Orthopaedics* 21:403–408.

Bierhalz, A.C.K., da Silva, M.A., Kieckbusch, T.G. 2012. Natamycin release from alginate/pectin films for food packaging applications. *Journal of Food Engineering* 110:18–25.

Blakistone, B.A. 1999. *Principles and Applications of Modified Atmosphere Packaging of Foods.* Aspen Publication, Chapman and Hall, New York, NY.

Bowers, C.K.J., McGiure, M.A.D. 1995. Suppression of *Listeria monocytogenes* colonization following adsorption of nisin onto silica surfaces. *Applied and Environment Microbiology* 61:992–997.

Branen, A.L., Davidson, P.M., Salminen, S., Thorngate, J. 2001. *Food Additives.* CRC Press, Boca Raton, FL.

Brennan, J.G., Grandison, A.S. 2012. *Food Processing Handbook.* John Wiley & Sons, Hoboken, NJ.

Brewer, R., Adams, M.R., Park, S.F. 2002. Enhanced inactivation of *Listeria monocytogenes* by nisin in the presence of ethanol. *Letters in Applied Microbiology* 34:18–21.

Brody, A.L., Strupinsky, E.R., Kline, L.R. 2001. *Active Packaging for Food Applications.* Technomic Publishing Company, Lancester, 224 pp. ISBN:1-58716-045-5.

Buonocore, G.G., Del Nobile, M.A., Panizza, A., Corbo, M.R., Nicolais, L. 2003. A general approach to describe the antimicrobial agent release from highly swellable films intended for food packaging applications. *Journal of Controlled Release* 90:97–107.

Cagri, A., Ustunol, Z., Ryser, E.T. 2001. Antimicrobial, mechanical and moisture barrier properties of low pH whey protein based edible films containing paminobenzoic or sorbic acids. *Journal of Food Science* 66:865–870.

Cao-Hoang, L., Chaine, A., Grégoire, L., Waché, Y. 2010. Potential of nisin-incorporated sodium caseinate films to control Listeria in artificially contaminated cheese. *Food Microbiology* 27:940–944.

Carballo, J., Arajjo, A.B. 2005. Influence of surface characteristics of food contact materials on bacterial attachment. *Biomicro World.* International Conference of Biofilms in Spain.

Carneiro de Melo, A.M.S., Cassar, C.A., Miles, R.J. 1998. Trisodium phosphate increases sensitivity of gram-negative bacteria to lysozyme and nisin. *Journal of Food Protection* 61:839–844.

Cha, D.S., Chinnan, M.J. 2004. Biopolymer based anti-microbial packaging: A review. *Critical Reviews in Food Science and Nutrition* 44:223–227.

Chi-Zhang, Y., Yam, K.L., Chikindas, M.L. 2004. Effective control of *Listeria monocytogenes* by combination of nisin formulated and slowly released into a broth system. *International Journal of Food Microbiology* 90:15–22.

Choi, J.O., Park, J.M., Park, H.J., Lee, D.S. 2001. Migration of preservative from antimicrobial polymer coating into water. *Food Science and Biotechnology* 10:327–330.

Chollet, E., Sebti, I., Martial-Gros, A., Degraeve, P. 2008. Nisin preliminary study as a potential preservative for sliced ripened cheese: NaCl, fat and enzymes influence on nisin concentration and its antimicrobial activity. *Food Control* 19:982–989.

Chollet, E., Swesi, Y., Degraeve, P., Sebti, I. 2009. Monitoring nisin desorption from a multi-layer polyethylene-based film coated with nisin loaded HPMC film and diffusion in agarose gel by an immunoassay (ELISA) method and a numerical modeling. *Innovative Food Science and Emerging Technologies* 10:208–214.

Chung, K., Dickson, J., Crouse, J.D. 1989. Effects of nisin on growth of bacteria attached to meat. *Applied Environmental Microbiology* 55:1329–1333.

Chung, Y.K., Vurma, M., Turek, E.J., Chism, G.W., Yousel, A.E. 2005. Inactivation of baro-tolerant *Listeria monocytogenes* in sausage by combination of high-pressure process-ing and food-grade additives. *Journal of Food Protection* 68:744–750.

Church, I.J., Parsons, A.L. 1995. Modified atmosphere packaging technology: A review. *Journal of Science and Food Agriculture* 67:143–152.

Coma, V., Sebti, I., Pardon, P., Deschamps, A., Pichavant, F.H. 2001. Antimicrobial edible packaging based on cellulosic ethers, fatty acids and nisin incorporation to inhibit *Listeria innocua* and *Staphylococcus aureus*. *Journal of Food Protection* 64:470–475.

Cooksey, K. 2000. Utilization of antimicrobial packaging films for inhibition of selected microorganism. In: *Food Packaging: Testing Methods and Applications*, S.J. Risch (Ed.). American Chemical Society, Washington, DC, pp. 17–25.

Cooksey, K. 2005. Effectiveness of antimicrobial food packaging materials. *Food Additives and Contaminants* 22:980–987.

Cotter, P.D., Hill, C., Ross, R.P. 2005. Bacteriocins: Developing innate immunity for food. *Nature Reviews Microbiology* 3:777–778.

Crandall, A.D., Montville, T.J. 1998. Nisin resistance in Listeria monocytogenes ATCC 700302 is a complex phenotype. *Applied Environment Microbiology* 64:231–237.

Cruz-Romero, M.C., Murphy, T., Morris, M., Cummins, E., Kerry, J.P. 2013. Antimicrobial activ-ity of chitosan, organic acids and nano-sized solubilisates for potential use in smart anti-microbially-active packaging for potential food applications. *Food Control* 34:393–397.

Cutter, C.N. 2002a. Incorporation of antimicrobials into packaging materials, fresh meat/packaging II. *Proceedings of the 55th Reciprocal Meat Conference*, American Meat Science Association, pp. 83–87.

Cutter, C.N. 2002b. Microbial control by packaging: A review. *Critical Reviews in Food Science and Nutrition* 42:151–161.

Cutter, C.N., Sirgausa, G.R. 1996. Reduction of *Brochothrix thermosphacta* on beef surfaces following immobilization of nisin on beef surfaces following immobilization of nisin in calcium alginate gels. *Letters in Applied Microbiology* 23:9–12.

Cutter, C.N., Sirgausa, G.R. 1997. Growth of *Brochothrix thermosphacta* in ground beef fol-lowing treatments with nisin in calcium alginate gels. *Food Microbiology* 14:425–430.

Cutter, C.N., Siragusa, G.R. 1998. Incorporation of nisin into a meat binding system to inhibit bacteria on beef surfaces. *Letters in Applied Microbiology* 27:19–23.

Cutter, C.N., Willet, J.L., Siragusa, G.R. 2001. Improved antimicrobial activity of nisin incorporated polymer films by formulation change and addition of food grade chelator. *Letters in Applied Microbiology* 33:325–328.

Daeschel, M.A., Mcguire, J., Almakhlafi, H. 2001. Antimicrobial activity of nisin adsorbed to hydrophilic and hydrophobic silicon surfaces. *Journal of Food Protection* 55:731–735.

Dainelli, D., Gontard, N., Spyropoulos, D., Zondervan-van den Beuken, E., Tobback, P. 2008. Active and intelligent food packaging: Legal aspects and safety concerns. *Trends in Food Science & Technology* 19:99–108.

Dallyn, H., Shorten, D. 1988. Hygiene aspects of packaging in the food industry. *International Biodeterioration* 24:3876–3892.

Davidson, P.M. 1997. Chemical preservatives and natural antimicrobial compounds. In: *Food Microbiology Fundamentals and Frontiers*, M.P. Dayle, L.R. Beuchat, T.J. Montville (Eds.). ASM Press, Washington, DC, pp. 520–556.

Davies, E.A., Milne, C.F., Bevis, H.E., Potter, R.W., Harris, J.M., Williams, G.C., Thomas, L.V., Delves-Broughton, J. 1999 Effective use of nisin to control lactic acid bacte-rial spoilage in vacuum-packed bologna-type sausage. *Journal of Food Protection* 62:1004–1010.

Dawson, P.L., Acton, J.C., Han, I.Y., Padgett, T., Orr, R., Larsen, T. 1996. Incorporation of antibacterial compounds into edible and biodegradable packaging films. *Research and Development Association* 48:203–210.

Dawson, P.L., Harmon, L., Sotthibandhu, A., Han, I.Y. 2005. Antimicrobial activity of nisin-adsorbed silica and corn starch powders. *Food Microbiology* 2:93–99.

De Kruijf, N., Van Beest, M., Rijk, R., Sipilainen-Malm, T., Losada, P.P., De Meulenaer, B., et al. 2002. Active and intelligent packaging: Applications and regulatory aspects. *Food Additives & Contaminants* 19:144–162.

De Martinis, E.C.P., Alves, V.F., Franco, B.D.G.M. 2002. Fundamentals and perspectives for the use of bacteriocins produced by lactic acid bacteria in meat products. *Food Reviews International* 18:191–208.

Dean, M., Zottola, E.A. 1996. Use of nisin in ice cream and effect on the survival of *Listeria monocytogenes*. *Journal of Food Protection* 59:476–480.

Debeaufort, F., Quezada-Gallo, J.A., Delpote, B., Voilley, A. 2000. Lipid hydrophobicity and physical state effects on the properties of bilayer edible films. *Journal of Membrane Science* 180:47–55.

Deegan, L.H., Cotter, P.D., Hill, C., Ross, P. 2006. Bacteriocins: Biological tools for bio-preservation and shelf-life extension. *International Dairy Journal* 16:1058–1071.

Degnan, A.J., Buyong, N., Luchansky, J.B. 1993. Anti-listerial activity of pediocin AcH in model food systems in the presence of an emulsifier or encapsulated within liposomes. *International Journal of Food Microbiology* 18:127–138.

Delves, B.J., Blackburn, P., Evans, R.E., Hugenholtz, J. 1996. Applications of the bacteriocin, nisin. *Antonie van Leeuwenhoek* 69:193–202.

Delves-Broughton, J. 1990. Nisin and its use as a food preservative. *Food Technology* 44:100–117.

Devlieghere, F., Vermeiren, L., Debevere, J. 2004. New preservation technologies: Possibilities and limitations. *International Dairy Journal* 14:273–285.

Dheraprasart, C., Rengpipat, S., Supaphol, P., Tattiyakul, J. 2009. Morphology, release characteristics, and antimicrobial effect of nisin-loaded electrospun gelatin fiber mat. *Journal of Food Protection* 72:2293–2300.

Dobiáš, J., Chudáčková, K., Voldřich, M., Marek, M. 2000. Properties of polyethylene films with incorporated benzoic anhydride and/or ethyl and propyl esters of 4-hydroxibenzoic acid and their suitability for food packaging. *Food Additives and Contaminants* 17:1047–1053.

Dobiáš, J., Voldřich, M., Marek, M., Čeřovský, M. 1998. Active food packaging – Immobilization of preservatives on/in packaging materials. *Lebensmittelchemie* 52:3–36.

Dotto, G.L., Buriol, C., Pinto, L.A.A. 2014. Diffusional mass transfer model for the adsorption of food dyes on chitosan films. *Chemical Engineering Research and Design* 92:2324–2332.

Duran, M., Aday, M.S., Zorba, N.N.D., Temizkan, R., Buyukcan, M.B., Caner, C. 2016. Potential of antimicrobial active packaging "containing natamycin, nisin, pomegranate and grape seed extract in chitosan coating" to extend shelf life of fresh strawberry. *Food and Bioproducts Processing* 98:354–363.

Dutta, P.K., Tripathi, S., Mehrotra, G.K., Dutta, J. 2009. Perspectives for chitosan based antimicrobial films in food applications. *Food Chemistry* 114:1173–1182.

Dykes, G.A., Hastings, J.W. 1998. Fitness costs associated with class IIa bacteriocin resistance in *Listeria monocytogenes* B73. *Letters in Applied Microbiology* 26:5–8.

Economou, T., Pournis, N., Ntzimani, A., Savvaidis, I.N. 2009. Nisin-EDTA treatments and modified atmosphere packaging to increase fresh chicken meat shelf-life. *Food Chemistry* 114:1470–1476.

El-Saharty, Y.S., Bary, A.A. 2002. High-performance liquid chromatographic determination of neutraceuticals, glucosamine sulphate and chitosan, in raw materials and dosage forms. *Analytica Chimica Acta* 462:125–131.

Ercolini, D., Ferrocino, I., La Storia, A., Mauriello, G., Gigli, S., Masi, P., Francesco Villani, F. 2010. Development of spoilage microbiota in beef stored in nisin activated packaging. *Food Microbiology* 27:137–143.

Fang, T.J., Lin, L.W., Lin, C.C. 1996. Immobilization of nisin in calcium alginate gels and its application to meat decontamination. *Book of Abstracts. Institute of Food Technologists Annual Meeting*, Abstract # 14D-7, p. 30.

Farber, J.M. 1991. Microbiological aspects of modified-atmosphere packaging technology – A review. *Journal of Food Protection* 54:58–70.

FDA. 2000. Sodium diacetate, sodium acetate, sodium lactate and potassium lactate: Use as food additives. Food and Drug Administration, Washington, D.C. *Federal Register* 65:17128–17129.

Fennema, O., Kester, J.J. 1991. Resistance of lipid to transmission of water vapor and oxygen. *Advances in Experimental Medicine and Biology* 302:703–719.

Floros, J.D., Dock, L.L., Han, J.H. 1997. Active packaging technologies and applications. *Food Cosmetics and Drug Packaging* 20:10–17.

Franklin, N.B., Cooksey, K.D., Getty, K.J.K. 2004. Inhibition of *Listeria monocytogenes* on the surface of individually packaged hot dogs with a packaging film coating containing nisin. *Journal of Food Protection* 67:480–485.

Fucinos, C., Miguez, M., Cerqueira, M.A., Costa, M.J., Vicente, A.A., Rua, M.L., Pastrana, L.M. 2015. Functional characterisation and antimicrobial efficiency assessment of smart nanohydrogels containing natamycin incorporated into polysaccharide-based films. *Food and Bioprocess Technology* 8:1430–1441.

Gadang, V.P., Hettiarachchy, N.S., Johnson, M.G., Owens, C. 2008. Evaluation of antibacterial activity of whey protein isolate coating incorporated with nisin, grape seed extract, malic acid, and EDTA on a Turkey frankfurter system. *Journal of Food Science* 73:389–394.

Garcia, T., Martin, R., Sanz, B., Hernandez, P.E. 1995. Review: Shelf life extension of fresh meat. 1. Modified atmosphere packaging, lactic acid bacteria, and bacteriocins. *Reviews Espanol Cient Technologia Alimentaria* 35:1–18.

Genigeorgis, C.A. 1985. Microbial and safety implications of the use of modified atmospheres to extend the storage life of fresh meat and fish. *International Journal of Food Microbiology* 1:237–251.

Gennadios, A., Hanna, M.A., Kurth, L.B. 1997. Application of edible coatings on meats, poultry and seafoods: A review. *Food Science and Technology – Lebensmittel-Wissenschaft & Technologie* 30:337–350.

Geonaras, I., Skandamis, P.N., Belk, K.E. Scanga, J.A., Kendall, P.A., Smith, G.C., Sofos, J. 2006. Postprocess control of *Listeria monocytogenes* on commercial frankfurters formulated with and without antimicrobials and stored at 10°C. *Journal of Food Protection* 69:53–61.

Gill, A.O. 2000. *Application of Lysozyme and Nisin to Control Bacterial Growth on Cured Meat Products*. M.S. Thesis, The University of Manitoba, Winnipeg, MB, Canada.

Gill, A.O., Holley, R.A. 2000a. Inhibition of bacterial growth on ham and bologna by lysozyme, nisin and EDTA. *Food Research International* 33:83–90.

Gill, A.O., Holley, R.A. 2000b. Surface application of lysozyme, nisin, and EDTA to inhibit spoilage and pathogenic bacteria on ham and bologna. *Journal of Food Protection* 63:1338–1346.

Gravesen, A., Jydegaard Axelsen, A.-M., Mendes da Silva, J., Hansen, T.B., Knøchel, S. 2002. Frequency of bacteriocin resistance development and associated fitness costs in *Listeria monocytogenes*. *Applied and Environment Microbiology* 68:756–764.

Grisi, T.C.S.L., Gorlach-Lira, K. 2005. Action of nisin and high pH on growth of *Staphylococcus aureus* and *Salmonella* sp. in pure culture and in the meat of land crab (*Ucides cordatus*). *Brazilian Journal of Microbiology* 36:151–156.

Grower, J.L., Cooksey, K., Getty, K.J.K. 2004. Development and characterization of an antimicrobial packaging film coating containing nisin for inhibition of *Listeria monocytogenes*. *Journal of Food Protection* 67:475–479.

Guerra, N.P., Macias, C.L., Agrasar, A.T., Castro, L.P. 2005. Development of a bioactive packaging cellophane using Nisaplin as biopreservative agent. *Letters in Applied Microbiology* 40:106–110.

Guiga, W., Galland, S., Peyrol, E., Degraeve, P., Carnet-Pantiez, A., Sebti, I. 2009. Antimicrobial plastic film: Physico-chemical characterization and nisin desorption modeling. *Innovative Food Science and Emerging Technologies* 10:203–207.

Guiga, W., Swesi, Y., Galland, S., Peyrol, E., Degraeve, P., Sebti, I. 2010. Innovative multilayer antimicrobial films made with Nisaplin or nisin and cellulosic ethers: Physico-chemical characterization, bioactivity and nisin desorption kinetics. *Innovative Food Science and Emerging Technologies*. doi:10.1016/j.ifset.2010.01.008.

Guinane, C.M., Cotter, P.D., Hill, C., Ross, R.P. 2005. A review: Microbial solution to microbial problems; lactococcal bacteriocins for the control of undesirable biota in food. *Journal of Applied Microbiology* 98:1316–1325.

Güne, E., Çibik, R. 2002 Deneysel olarak kontamine edilmiþ inegöl köftelerde antimikrobiyel bir protein olan nisinin L. *monocytogenes* üzerine inhibe edici etkisi. FEMS Symposium *on the* Versality *of* Listeria spp., October 10–11, İzmir, Turkey.

Guo, M., Jin, T.Z., Wang, L., Scullen, O.J., Sommers, C.H. 2014. Antimicrobial films and coatings for inactivation of *Listeria innocua* on ready-to-eat deli turkey meat. *Food Control* 40:64–70.

Hagiwara, A., Imai, N., Nakashima, H., Toda, Y., Kawabe, M., Furukawa, F., Delves-Broughton, J., Yasuhara, K., Hayashi, S.M. 2010. A 90-day oral toxicity study of nisin A, an anti-microbial peptide derived from *Lactococcus lactis subsp. lactis*, in F344 rats. *Food and Chemical Toxicology* 48:2421–2428.

Hampikyan, H., Ugur, M. 2007. The effect of nisin on L. *monocytogenes* in Turkish fermented sausages (sucuks). *Meat Science* 76:327–332.

Han, A., Floros, H. 2000. Simulating migration models and determining the releasing rate of potassium sorbate from antimicrobial plastic film. *Food Science and Biotechnology* 9:68–72.

Han, J.H. 2000. Antimicrobial food packaging. *Food Technology* 54:56–65.

Han, J.H. 2002. Protein-based edible films and coatings carrying antimicrobial agents. In: *Protein-based Films and Coatings*, A. Gennadios (Ed.). Technomic Publishing Co., Inc., Lancaster, PA, pp. 485–500.

Han, J.H. 2003. Antimicrobial food packaging. In: *Novel Food Packaging Techniques*, R. Ahvenainen (Ed.). Woodhead Publishing Limited, Cambridge, United Kingdom and CRC Press LLC, Boca Raton, FL, pp. 70–72.

Han, J.H. 2005. Antimicrobial packaging systems. In: *Innovations in Food Packaging*, J.H. Han (Ed.). Academic Press Inc., Cambridge, MA, pp. 80–201.

Han, J.H., Floros, J.D. 1997. Casting antimicrobial packaging films and measuring their physical properties and antimicrobial activity. *Journal of Plastic Film Sheeting* 13:287–298.

Hauser, C., Wunderlich, J. 2011. Antimicrobial packaging films with a sorbic acid based coating. *Procedia Food Science* 1:197–202.

Hintlian, C.B., Hotchkiss, J.H. 1986. The safety of modified atmosphere packaging: A review. *Food Technology* 40:70–76.

Hoffman, K.L., Han, I.Y., Dawson, P.L. 2001. Antimicrobial effects of corn zein films impregnated with sisin, lauric acid and EDTA. *Journal of Food Protection* 64:885–889.

Hong, S.I., Lee, J.W., Son, S.M. 2005. Properties of polysaccharide-coated polypropylene films as affected by biopolymer and plasticizer types. *Packaging Technology and Science* 18:1–9.

Hoover, D.G., Hurst, A. 1993. Nisin. In: *Antimicrobials in Foods*, P.M. Davidson, A.L. Branen (Eds.), 2nd ed. Marcel Dekker, Inc., New York, NY, pp. 369–394.

Hotchkiss, J.H. 1995. Safety considerations in active packaging. In: *Active Food Packaging*, M.L. Rooney (Ed.). Blackie Academic & Professional, New York, NY, pp. 238–255.

Hurst, A. 1981. Nisin. *Advances in Applied Microbiology* 27:85–123.

Imran, M., Klouj, A., Revol-Junelles, A.-M., Desobry, S. 2014. Controlled release of nisin from HPMC, sodium caseinate, polylactic acid and chitosan for active packaging applications. *Journal of Food Engineering* 143:178–185.

Jofré, A., Garrida, M., Aymerich, T. 2008. Inhibition of *Salmonella sp. Listeria monocytogenes* and *Staphylococcus aureus* in cooked ham by combining antimicrobials, high hydrostatic pressure and refrigeration. *Meat Science* 78:53–59.

Juhl, R.L., Lustig, S., Tijunelis, D. 1994. Transferable modifier-containing film. U.S. Patent 5288532.

Jung, D.S., Bodyfelt, F.W., Daeschel, M.A. 1992. Influence of fat and emulsifiers on the efficacy of nisin in inhibiting *Listeria monocytogenes* in fluid milk. *Journal of Dairy Science* 75:387–393.

Jung, H. 2000. Antimicrobial food packaging. *Food Technology* 54:56–65.

Kamper, S.L., Fennema, O. 1984. Water vapor permeability of an edible, fatty acid, bilayer. *Journal of Food Science* 49:1482–1485.

Kamper, S.L., Fennema, O. 1985. Use of edible to maintain water vapor gradients in foods. *Journal of Food Science* 50:382–384.

Kapetanakou, A.E., Agathaggelou, E.I., Skandamis, P.N. 2014. Storage of pork meat under modified atmospheres containing vapors from commercial alcoholic beverages. *International Journal of Food Microbiology* 178:65–75.

Kara, R., Yaman, H., Gök, V., Akkaya, L. 2014. The effect of nisin on *Listeria monocytogenes* in chicken burgers. *Indian Journal of Animal Research* 48:171–176.

Kelsey, R. 1989. *Packaging in Today's Society*, 3rd ed. Technomic Publishing Co., Inc., Lancaster, PA.

Kim, Y.M., Paik, H.D., Lee, D.S. 2002. Shelf-life characteristics of fresh oysters and ground beef as affected by bacteriocin-coated plastic packaging film. *Journal of the Science of Food and Agriculture* 82:998–1002.

Klangmuang, P., Sothornvit, R. 2016. Barrier properties, mechanical properties and antimicrobial activity of hydroxypropyl methylcellulose-based nanocomposite films incorporated with Thai essential oils. *Food Hydrocolloids* 61:609–616.

Ko, S., Janes, M.E., Hettiarachchy, N.S., Johnson, M.G. 2001. Physical and chemical of edible films containing nisin and their action against *Listeria monocytogenes*. *Journal of Food Science* 66:1006–1011.

Kristo, E., Koutsoumanis, K.P., Biliaderis, C.G. 2008. Thermal, mechanical and water vapor barrier properties of sodium caseinate films containing antimicrobials and their inhibitory action on *Listeria monocytogenes*. *Food Hydrocolloids* 22:373–386.

Krochta, J.M., Baldwin, E.A., Nisperos-Carriedo, M.O. (Eds.) 1994. *Edible Coatings and Films to Improve Food Quality*. Technomic Publishing Co., Inc., Lancaster, PA.

Kuorwel, K.K., Cran, M.J., Sonneveld, K., Miltz, J., Bigger, S.W. 2013. Migration of antimicrobial agents from starch-based films into a food simulant. *LWT – Food Science and Technology* 50:432–438.

Labuza, T.P., Breene, W.M. 1989. Applications of "active packaging" for improvement of shelf-life and nutritional quality of fresh and extended shelf-life foods. *Journal of Food Processing and Preservation* 13:1–69.

Labuza, T.P., Fu, B., Taoukis, P.S. 1992. Prediction for shelf life and safety of minimally processed CAP/MAP chilled foods: A review. *Journal of Food Protection* 55:741–750.

Lacoste, A., Schaich, K.M., Zumbrunnen, D., Yam, K.L. 2005. Advancing controlled release packaging through smart blending. *Packaging Technology and Science* 18:77–87.

Lee, C.H., An, D.S., Lee, S.C., Park, H.J., Lee, D.S. 2003. A coating for use as an anti-microbial and antioxidative packaging material incorporating nisin and α-tocopherol. *Journal of Food Engineering* 62:323–329.

Leistner, L. 2000. Basic aspects of food preservation by hurdle technology. *International Journal of Food Microbiology* 55:181–186.

Leistner, L., Gorris, L.G.M. 1995. Food preservation by hurdle technology. *Trends Food Science Technology* 6:41–46.

Limjaroen, P., Ryser, E., Lockhart, H., Harte, B. 2003. Development of a food packaging coating material with antimicrobial properties. *Journal of Plastic Film & Sheeting* 19:95–109.

Liu, W., Hansen, J.N. 1990. Some chemical and physical properties of nisin, a small-protein antibiotic produced by *Lactococcus lactis*. *Applied and Environmental Microbiology* 56:2551–2558.

López-Mendoza, M.C., Ruiz, P., Mata, C.M. 2007. Combined effects of nisin, lactic acid and modified atmosphere packaging in raw ground pork: Antimicrobials to control *Listeria* in meat. *International Journal of Food Science and Technology* 42:562–566.

Lopez-Pedemonte, T.J., Roig-Sagues, A.X., Trujillo, A.J., Capellas, M., Guamis, B. 2003. Inactivation of spores of *Bacillus cereus* in cheese by high hydrostatic pressure with the addition of nisin or lysozyme. *Journal of Dairy Science* 86:3075–3081.

Luchansky, J.B., Call, J.E. 2004. Evaluation of nisincoated cellulose casings for the control of *Listeria monocytogenes* inoculated onto the surface of commercially prepared frank-furters. *Journal of Food Protection* 67:1017–1021.

Mahadeo, M., Tatini, S.R. 1994. The potential use of nisin to control *Listeria monocytogenes* in poultry. *Letters in Applied Microbiology* 18:323–326.

Mahdavi, M., Jalali, M., Kermanshahi, R.K. 2007. The effect of nisin on biofilm form-ing foodborne bacteria using microtiter plate method. *Research in Pharmaceutical Sciences* 2:113–118.

Mangalassary, S., Han, I., Rieck, J., Acton, J., Dawson, P. 2008. Effect of combining nisin and/or lyzozyme with in-package pasteurization for control of *Listeria monocyto-genes* in ready-to-eat turkey bologna during refrigerated storage. *Food Microbiology* 25:866–870.

Manikantan, M.R., Varadharaju, N.A. 2012. Preparation and properties of linear low den-sity polyethylene based nanocomposite films for food packaging. *Indian Journal of Engineering and Material Sciences* 19:54–66.

Marsh, K., Bugusu, B. 2007. Food packaging–roles, materials and environmental issues. *Journal of Food Science* 72(3):39–55.

Mauriello, G., De Luca, E., La Storia, A., Villani, F., Ercolini, D. 2005. Antimicrobial activ-ity of a nisin-activated plastic film for food packaging. *Letters in Applied Microbiology* 41:464–469.

Mazzotta, A.S., Modi, K.D., Montville, T.J. 2000. Nisin resistant (NisR) Listeria monocyto-genes and NisR *Clostridium botulinum* are not resistant to common food preservatives. *Journal of Food Science* 65:888–890.

Mazzotta, A.S., Montville, T.J. 1997. Nisin induces changes in membrane fatty acid composi-tion of *Listeria monocytogenes* nisin-resistant strains at 10EC and 30EC. *Journal of Applied Microbiology* 82:32–38.

Mazzotta, A.S., Montville, T.J. 1999. Characterization of fatty acid composition, spore ger-mination, and thermal resistance in a nisin-resistant mutant of *Clostridium botulinum* 169B and in the wild-type strain. *Applied Environment Microbiology* 65:659–664.

McAuliffe, O., Ross, R.P., Hill, C. 2001. Lantibiotics: Structure, biosynthesis and mode of action. *FEMS Microbiology Reviews* 25:285–308.

Meyer, R.C., Winter, A.R., Weister, H.H. 1959. Edible protective coatings for extending the shelf life of poultry. *Food Technology* 13:146–148.

Millette, M., Tien, C.L., Smoragiewicz, W., Lacroix, M. 2007. Inhibition of *Staphylococcus aureus* on beef by nisin-containing modified alginate films and beads. *Food Control* 18:878–884.

Ming, X., Daeschel, M.A. 1993. Nisin resistance of foodborne bacteria and the specific resistance responses of *Listeria monocytogenes*. Scott A. *Journal of Food Protection* 56:944–948.

Ming, X., Weber, G.H., Ayres, J.W., Sandine, W.E. 1997. Bacteriocins applied to food packaging materials to inhibit *Listeria monocytogenes* on meats. *Journal of Food Sciences* 62:413–415.

Montville, T.J., Winkowski, K., Chikindas, M.L. 2001. Biologically based preservation systems. In: *Food Microbiology: Fundamentals and Frontiers*, M.P. Doyle, L.R. Beuchat, T.J. Montville (Eds.), 2nd ed. American Society for Microbiology, Washington, DC, pp. 629–647.

Mulders, J.W., Boerrigter, I.J., Rollema, H.S., Siezen, R.J., de Vos, W.M. 1991. Identification and characterization of the lantibiotic nisin Z, a natural nisin variant. *European Journal of Biochemistry* 201:581–584.

Mulet-Powell, N., Lacoste-Armynot, A.M., Vinas, M., Simeon de Buochberg, M. 1998. Interactions between pairs of bacteriocins from lactic bacteria. *Journal of Food Protection* 61:1210–1212.

Mulla, M., Ahmed, J., Al-Attar, H., Castro-Aguirre, E., Arfat, Y.A., Auras, R. 2017. Antimicrobial efficacy of clove essential oil infused into chemically modified LLDPE film for chicken meat packaging. *Food Control* 73:663–671.

Natrajan, N., Sheldon, B.W. 1995. Evaluation of bacteriocin-based packaging and edible film delivery systems to reduce *Salmonella* in fresh poultry. *Poultry Science* 74:31.

Natrajan, N., Sheldon, B.W. 2000a. Efficacy of nisin coated polymer films to inactivate *Salmonella typhimurium* on fresh broiler skin. *Journal of Food Protection* 63(9):1189–1196.

Natrajan, N., Sheldon, B.W. 2000b. Inhibition of *Salmonella* on poultry skin using protein- and polysaccharide-based films containing a nisin formulation. *Journal of Food Protection* 63:1268–1272.

Neetoo, H., Ye, M., Chen, H., Joerger, R.D., Hicks, D., Hoover, D.G. 2008. Use of nisin coated plastic films to control Listeria monocytogenes on vacuum packaged cold smoked salmon. *International Journal of Food Microbiology* 22:8–15.

Nguyen, V.T., Gidley, M.J., Dykes, G. 2008. A.Potential of nisin-containing bacterials cellulose film to inhibit *Listeria monocytogenes* on processed meats. *Food Microbiology* 25:471–478.

O'Callaghan, K.A.M., Kerry, J.P. 2014. Assessment of the antimicrobial activity of potentially active substances (nanoparticled and non-nanoparticled) against cheese-derived microorganisms. *International Journal of Dairy Technology* 67:483–489.

O'Sullivan, L., Ross, R.P., Hill, C. 2002. Potential of bacteriocin-producing lactic acid bacteria for improvements in food safety and quality. *Biochimie* 84:593–604.

Olasupo, N.A., Fitzgerald, D.J., Gasson, M.J., Narbad, A. 2003. Activity of natural antimicrobial compounds against *Escherichia coli* and *Salmonella enterica serovar Typhimurium*. *Letters in Applied Microbiology* 37:448–451.

Ooraikul, B. 1991. Further research in modified atmosphere packaging. In: *Modified Atmosphere Packaging of Food*, B. Ooraikul, M.E. Stiles (Eds.). Ellis Harwood Publishing, New York, NY, pp. 261–282.

Ooraikul, B., Stiles, M.E. 1991. Introduction: Review of the development of modified atmosphere packaging. In: *Modified Atmosphere Packaging of Food*, B. Ooraikul, M.E. Stiles (Eds.). Harwood Publishing, New York, NY, pp. 1–17.

Orr, R.V., Han, I.Y., Acton, J.C., Dawson, P.L. 1998. Effect of nisin in edible protein films on *Listeria monocytogenes* in milk. *Research and Development Activities Report for Military Food and Packaging Systems* 48:115–120.

Ouattara, B., Simard, R.E., Piette, G., Bégin, A., Holley, R.A. 2000. Inhibition of surface spoilage bacteria in processed meats by application of antimicrobial films prepared with chitosan. *International Journal of Food Microbiology* 62:139–148.

Oussalah, M., Caillet, S., Salmieri, S., Saucier, L., Lacroix, M. 2004. Antimicrobial and antioxidant effects of milk protein based film containig essential oils for the preservation of whole beef muscle. *Journal of Agriculture and Food Chemistry* 52:5598–5605.

Ozdemir, M., Floros, J.D. 2004. Active food packaging technologies. *Critical Reviews in Food Science and Nutrition* 44:185–193.

Padget, T., Han, I.Y., Dawson, P.L. 2000. Incorporation of food-grade antimicrobial compounds into biodegradable packaging films. *Journal of Food Protection* 61:1330–1335.

Padgett, T., Han, I.Y., Dawson, P.L. 2000. Effect of lauric acid addition on the antimicrobial efficacy and water permeability of protein films containing nisin. *Journal of Food Processing and Preservation* 24:423–432.

Papadokostaki, K.G., Amanratos, S.G., Petropoulos, J.H. 1997. Kinetics of release of particules solutes incorporated in cellulosic polymer matrices as a function of solute solubility and polymer swellability. I. Sparingly soluble solutes. *Journal of Applied Polymer Science* 67:277–287.

Pawar, D.D., Malik, S.V.S., Bhilegaonkar, K.N., Barbuddhe, S.B. 2000. Effect of nisin and its combination with sodium chloride on the survival of *Listeria monocytogenes* added to raw buffalo meat mince. *Meat Science* 56:215–219.

Pekcan, G., Köksal, E., Küçükerdönmez, O., Ozel, H. 2006. *Household Food Wastage in Turkey.* FAO, Rome, Italy.

Perez, L.M., Soazo, M.D., Balague, C.E., Rubiolo, A.C., Verdini, R.A. 2014. Effect of pH on the effectiveness of whey protein/glycerol edible films containing potassium sorbate to control non-O157 shiga toxin-producing *Escherichia coli* in ready-to-eat foods. *Food Control* 37:298–304.

Periago, P.M., Moezelaar, R. 2001. Combined effect of nisin and carvacrol at different pH and temperature levels on the viability of different strains of *Bacillus cereus*. *International Journal of Food Microbiology* 68:141–148.

Pintado, C.M.B.S., Ferreira, M.A.S.S., Sousa, I. 2010. Control of pathogenic and spoilage microorganisms from cheese surface by whey protein films containing malic acid, nisin and natamycin. *Food Control* 21:240–246.

Pongtharangkul, T., Demirci, A. 2004 Evaluation of agar bioassay for nisin quantification. *Applied Microbiology and Biotechnology* 65:268–272.

Pranoto, Y., Rakshit, S.K., Salokhe, V.M. 2005. Enhancing antimicrobial activity of chitosan films by incorporating garlic oil, potassium sorbate and nisin. *Lebensmittel-Wissenschaft Und-Technologie* 38:859–865.

Quintavalla, S., Vicini, L. 2002 Antimicrobial food packaging in meat industry. *Meat Science* 62:373–380.

Raheem, D. 2012. Application of plastics and paper as food packaging materials – An overview. *Emirates Journal of Food and Agriculture* 25:177–188.

Raju, C.V., Shamasundar, B.A., Udupa, K.S. 2003. The use of nisin as a preservative in fish sausage stored at ambient (28±2 °C) and refrigerated (6±2 °C) temperatures. *International Journal of Food Science and Technology* 38:171–185.

Ramos, O.L., Fernandes, J.C., Silva, S.I., Pintado, M.E., Malcata, F.X. 2012. Edible films and coatings from whey protein: A review on formulation, and on mechanical and bioactive peptides. *Critical Reviews in Food Science and Nutrition* 52:533–552.

Rasch, M., Knøchel, S. 1998. Variations in tolerance of *Listeria monocytogenes* to nisin, pediocin PA-1 and bavaricin. *Letters in Applied Microbiolology* 27:275–278.

Ray, B. 1992. Nisin of *Lactococcus lactis* spp. lactis as a food preservative. In: *Food Biopreservatives of Microbial Origin*, B. Ray, M. Daeschel (Eds.). CRC Press, Boca Raton, FL, pp. 207–264.

Reunanen, J., Saris, P.E.J. 2004. Bioassay for nisin in sausage; a shelf life study of nisin in cooked sausage. *Meat Science* 66:515–518.

Ripoche, A.C., Chollet, E., Peyrol, E., Sebti, I. 2006. Evaluation of nisin diffusion in a polysaccharide gel: Influence of agarose and fatty content. *Innovative Food Science & Emerging Technologies* 7:107–111.

Rodrigues, E.T., Han, J.H. 2000. Antimicrobial whey protein films against spoilage and-pathogenic bacteria. *Proceedings of the IFT* Annual *Meeting*, Dallas, TX, June 10–14. Institute of Food Technologists, Chicago, IL, p. 191.

Rodriguez-Martinez, A.V., Sendon, R., Abad, M.J., Gonzalez-Rodriguez, M.V., Barros-Velazquez, J., Aubourg, S.P., Paseiro-Losada, P., de Quiros, A.R.B. 2016. Migration kinetics of sorbic acid from polylactic acid and seaweed based films into food simulants. *LWT – Food Science and Technology* 65:630–636.

Rooney, M.L. 1995a. Active packaging in polymer films. In: *Active Food Packaging*, M.L. Rooney (Ed.). Blackie Academic and Professional, London, UK.

Rooney, M.L. 1995b. In: *Active Food Packaging*, M.L. Rooney (Ed.). Blackie Academic & Professional, New York, NY, pp. 74–110.

Rose, N., Palcic, M., Sporns, P., McMullen, L. 2002. A novel subtrate for glutathione S-transferase isolated from fresh beef. *Journal of Food Science* 67:2288–2293.

Rossi-Márquez, G., Han, J.H., García-Almendárez, B., Castano-Tostado, E., Regalado-González, C. 2009. Effect of temperature, pH and film thickness on nisin release from antimicrobial whey protein isolate edible films. *Journal of the Science of Food and Agriculture* 89:2492–2497.

Ruiz, A., Williams, K., Djeri, N., Hinton, A., Rodrick, G.E. 2009. Nisin, rosemary and ethylenediaminetetraacetic acid affect the growth of *Listeria monocytogenes* on ready-to-eat turkey ham stored at four degrees Celsius for sixty three days. *Poultry Science* 88:1765–1772.

Ruiz, A., Williams, K., Djeri, N., Hinton, A., Rodrick, G.E. 2010. Nisin affects the growth of *Listeria monocytogenes* on ready-to-eat turkey ham stored at four degrees Celsius for sixty-three days. *Poultry Science* 89:353–358.

Ryser, E.T., Maisnier-Patin, S., Gratadoux, J.J., Richard, J. 1994. Isolation and identification of cheese-smear bacteria inhibitory to *Listeria* spp. *International Journal of Food Microbiology* 21:237–246.

Sacharow, S., Griffin, R.C. 1970. The evolution of food packaging. In: *Food Packaging*, S. Sacharow, R.C. Griffin (Eds.). AVI Publishing Company, Inc., Westport, CT, pp. 1–62.

Salim, R., Nazir, F., Amin, F., Nissar, J. 2017. Antimicrobial food packaging. *International Journal of Engineering Technology Science and Research* 4:310–312.

Scannell, A.G.M., Hill, C., Ross, R.P., Marx, S., Hartmeier, W., Arendt, E.K. 2000. Development of bioactive food packaging materials using immobilised bacteriocins Lacticin 3147 and Nisaplin [R]. *International Journal of Food Microbiology* 60:241–249.

Schillinger, U., Chung, H.S., Keppler, K., Holzapfel, W.H. 1998. Use of bacteriocinogenic lactic acid bacteria to inhibit spontaneous nisin-resistant mutants of *Listeria monocytogenes* Scott A. *Journal of Applied Microbiology* 85:657–663.

Sebti, I., Carnet, A.R., Blanc, D., Saurel, R., Coma, V. 2003. Controlled diffusion of an antimicrobial peptide from a biopolymer film. *Chemical Engineering Research and Design* 81:1099–1104.

Sebti, I., Coma, V. 2002. Active edible polysaccharide coating and interactions between solution coating compounds. *Carbohydrate Polymers* 49:139–144.

Sebti, I., Delves-Broughton, J., Coma, V. 2003. Physicochemical properties and bioactivity of nisin-containing cross-linked hydroxypropylmethylcellulose films. *Journal of Agricultural and Food Chemistry* 51:6468–6474.

Silveira, M.F.A., Soares, N.F.F., Geraldine, R.M., Andrade, N.J., Goncalves, M.P.J. 2007. Antimicrobial efficiency and sorbic acid migration from active films into pastry dough. *Packaging Technology and Science* 20:287–292.

Sindt, R.H. 2001. *Agency Response Letter GRAS Notice*, April 2001. Available on http://www.cfsan.fda.gov/rdb/opa-g065.html.

Singh, B., Falahee, M.B., Adams, M.R. 2001. Synergistic inhibition of *L. monocytogenes* by nisin and garlic extract. *Food Microbiology* 18:133–139.

Sirgausa, G.R. 1992. Production of bacteriocin inhibitory to *Listeria* species by *Enterococcus hirae*. *Applied Environment Microbiology* 58:3508–3513.

Siragusa, G.R., Cutter, C.N. 1993. Brochocin-C, a new bacteriocin produced by *Brochothrix campestris*. *Applied Environment Microbiology* 59:2326–2328.

Siragusa, G.R., Cutter, C.N., Willett, J.L. 1999. Incorporation of bacteriocin in plastic retains activity and inhibits surface growth of bacteria on meat. *Food Microbiology* 16:229–235.

Siragusa, G.R., Dickson, J.S. 1992. Inhibition of *Listeria monocytogenes* on beef tissue by application of organic acids immobilized in a calcium alginate gel. *Journal of Food Science* 57:293–296.

Siragusa, G.R., Dickson, J.S. 1993. Inhibition of *Listeria monocytogenes*, *Salmonella typhimurium* and *Escherichia coli* O157:H7 on beef muscle tissue by lactic or acetic acid contained in calcium alginate gels. *Journal of Food Safety* 13:147–158.

Siripatrawan, U., Noipha, S. 2012. Active film from chitosan incorporating green tea extract for shelf life extension of pork sausages. *Food Hydrocolloids* 27:102–108.

Siripatrawan, U., Vitchayakitti, W. 2016. Improving functional properties of chitosan films as active food packaging by incorporating with propolis. *Food Hydrocolloids* 61:695–702.

Sivarooban, T., Hettiarachchy, N.S., Johnson, M.G. 2007. Inhibition of *Listeria monocytogenes* using nisin with grape seed extract on turkey frankfurters stored at 4 and 10°C. *Journal of Food Protection* 70:1017–1020.

Sivarooban, T., Hettiarachchy, N.S., Johnson, M.G. 2008. Physical and antimicrobial properties of grape seed extract, nisin, and EDTA incorporated soy protein edible films. *Food Research International* 41:781–785.

Sobrino-Lopez, A., Martin-Belloso, O. 2006. Enhancing inactivation of *Staphylococcus aureus* in skim milk by combining high-intensity pulsed electric fields and nisin. *Journal of Food Protection* 69:345–353.

Sofos, J.N., Beuchat, L.R., Davidson, P.M., Johnson, E.A. 1998. *Naturally Occurring Antimicrobials in Food*. Task Force Report # 132, *103*. Council for Agricultural Science and Technology, Ames, IA.

Sohaib, M., Anjum, F.M., Arshad, M.S., Rahman, U.U. 2016. Postharvest intervention technologies for safety enhancement of meat and meat based products: A critical review. *Journal of Food Science and Technology-Mysore* 53:19–30.

Solomakos, N., Govaris, A., Koidis, P., Botsoglou, N. 2008. The antimicrobial effect of thyme essential oil, nisin, and their combination against *Listeria monocytogenes* in minced beef during refrigerated storage. *Food Microbiology* 25:120–127.

Soriano, A., Ulmer, H.M., Scannell, A.G.M., Ross, R.P., Hill, C., Garcia-Ruiz, A., Arendt, E.K. 2004. Control of food spoiling bacteria in cooked meat products with nisin, lacticin 3147, and a lacticin 3147-producing starter culture. *European Food Research & Technology* 219:6–13.

Sorrentino, A., Gorrasi, G., Vittoria, V. 2007. Potential perspectives of bionanocomposites for food packaging applications. *Trends in Food Science and Technology* 18:84–95.

Soto, K.M., Hernández-Iturriaga, M., Loarca-Piña, G., Luna-Bárcenas, G., Gómez-Aldapa, C.A., Mendoza, S. 2016. Stable nisin food-grade electrospun fibers. *Journal of Food Science and Technology* 53:3787–3794.

Soysal, C., Bozkurt, H., Dirican, E., Guclu, M., Bozhuyuk, E.D., Uslu, A.E., Kaya, S. 2015. Effect of antimicrobial packaging on physicochemical and microbial quality of chicken drumsticks. *Food Control* 54:294–299.

Stevens, K.A., Sheldon, B.W., Klapes, N.A., Klaenhammer, T.R. 1991. Nisin treatment for inactivation of *Salmonella* species and other gram-negative bacteria. *Applied Environment Microbiology* 57:3613–3615.

Stöllman, U., Johansson, F., Leufvén, A. 1994. Packaging and food quality. In: *Shelf Life Evaluation of Foods*, C.M.C. Man, A.A. Jones (Eds.). Blackie Academic & Professional, New York, NY, pp. 52–71.

Sung, S.Y., Sin, L.T., Tee, T.T., Bee, S.T., Rahmat, A.R., Rahman, W.A.W.A., Tan, A.C., Vikhraman, M. 2013. Antimicrobial agents for food packaging applications. *Trends in Food Science and Technology* 33:110–123.

Suppakul, P., Miltz, J., Sonneveld, K., Bigger, S.W. 2003. Active Packaging Technologies with an emphasis on anti-microbial packaging and its applications. *Journal of Food Science: Concise Reviews and Hypotheses in Food Science* 68:2.

Szente, L., Szejtli, J. 2004. Cyclodextrins as food ingredients. *Trends in Food Science and Technology* 15:137–142.

Takala, P.N., Vu, K.D., Salmieri, S., Khan, R.A., Lacroix, M. 2013. Antibacterial effect of biodegradable active packaging on the growth of *Escherichia coli*, *Salmonella typhimurium* and *Listeria monocytogenes* in fresh broccoli stored at 4 degrees C. *LWT – Food Science and Technology* 53:499–506.

Teerakarn, A., Hirt, D.E., Acton, J.C., Rieck, J.R., Dawson, P.L. 2002. Nisin diffusion in rotein films: Effects of film type and temperature. *Journal of Food Science* 67:3019–3025.

Theivendran, S., Hettiarachchy, N.S., Johnson, M.G. 2006. Inhibition of *Listeria monocytogenes* by nisin combined with grape seed extract or green tea extract in soy protein film coated on turkey frankfurters. *Journal of Food Science* 71:39–44.

Tippayatum, P., Chonhenchob, V. 2007. Antibacterial activities of thymol, eugenol ad nisin against some food spoilage bacteria. *Kasetsart Journal (Natural Science)* 41:319–323.

Tramer, J., Fowler, G.G. 1964. Estimation of nisin in food. *Journal of Science and Food Agriculture* 15:522–528.

van Schaik, W., Gahan, C.G.M., Hill, C. 1999. Acid adapted *Listeria monocytogenes* displays enhanced tolerance against the lantibiotics nisin and lacticin 3147. *Journal of Food Protection* 62:536–539.

Vermereinen, L., Devlieghere, F., Debevere, J. 2002. Effectiveness of some recent antimicrobial packaging concepts. *Food Additives and Contaminants* 19:163–171.

Wang, H., She, Y., Chu, C., Liu, H., Jiang, S., Sun, M., et al. 2015. Preparation, antimicrobial and release behaviors of nisin-poly (vinyl alcohol)/wheat gluten/ZrO2 nanofibrous membranes. *Journal of Materials Science* 50:5068–5078.

Wang, H.L., Hao, L.L., Wang, P., Chen, M.M., Jiang, S.W., Jiang, S.T. 2017. Release kinetics and antibacterial activity of curcumin loaded zein fibers. *Food Hydrocolloids* 63:437–446.

Wilhoit, D.L. 1996a. Film and method for surface treatment of foodstuffs with antimicrobial compositions. U.S. Patent 5,573,797.

Wilhoit, D.L. 1996b. Surface treatment of foodstuffs with antimicrobial compositions. U.S. Patent 5573801.

Wilson, C. 2007. *Frontiers of Intelligent and Active Packaging for Fruits and Vegetables.* CRC Press, Boca Raton, FL, p. 360.

Winkowski, K., Ludescher, R.D., Monville, T.J. 1996. Physicochemical characterization of the nisin-membrane interaction with liposomes derived from *Listeria monocytogenes.* Applied Environment Microbiology 62:323–327.

Wong, D.W.S., Camirand, W.M., Pavlath, A.E. 1994. Development of edible coatings for minimally processed fruits and vegetables. In: *Edible Coatings and Films to Improve Food Quality*, J.M. Krochta, E.A. Baldwin, M.O. Nisperos-Carriedo (Eds.). Technomic Publishing Company, Lancaster, PA, pp. 65–88.

Yonema, H., Ando, T., Katsumata, R. 2004. Bacteriocins produced by lactic acid bacteria and their use for food preservation. *Tohoku Journal of Agricultural Research* 55:51–55.

Yoshida, C.M.P., Bastos, C.E.N., Franco, T.T. 2010. Modeling of potassium sorbate diffusion through chitosan films. LWT – Food Science and Technology 43:584–589.

Zendo, T., Nakayama, J., Fujita, K., Sonomoto, K. 2008. Bacteriocin detection by liquid chromatography/mass spectrometry for rapid identification. *Journal of Applied Microbiology* 104:499–507.

Zhou, F., Ji, B.P., Zhang, H., Jiang, H., Yang, Z.W., Li, J.J., Li, J.H., Ren, Y.L., Yan, W.J. 2007. Synergistic effect of thymol and carvacrol combined with chelators and organic acids against *Salmonella typhimurium. Journal of Food Protection* 70:1704–1709.

11 Application of Genetically Engineered Microbes for Sustainable Development of Agro-Ecosystem

Umesh Pankaj and Namo Dubey

CONTENTS

11.1 INTRODUCTION

Biotechnology is elucidated as the use of modern technology in the field of life science, which deals with the use and manipulation of genetic engineering of living microorganisms or their components to produce useful products for human welfare, genes being transferred to crops for making them resistant towards various abiotic and biotic stresses. As per the survey data, the world population is expected to double by 2033 (Aruna and Sreekanth 2015). This situation will pose a great challenge to agricultural systems. Increasing agricultural productivity in terms of yield through traditional farming systems will not be able to meet this challenge. The application of chemical fertilizers, pesticides, and other agricultural techniques has increased agricultural production; but in addition, they have adverse effects on the environment, soil productivity, and human health (Sharma and Singhvi 2017). These chemicals are harmful as well as expensive. Natural products are eco-friendly and will cause negligible or no adverse effects on animal and human health (Singh et al. 2017). Researchers are looking for eco-friendly and cost-effective alternatives for sustainable agriculture (Singh and Seneviratne 2017). The broad application of microbes in sustainable agriculture is due to the genetic dependency of plants on the beneficial functions provided by a symbiotic relationship (Aruna and Sreekanth 2015). Plants possess the necessary physiological, biochemical, and genetic characteristics to establish themselves as the ultimate choice for remediation of soil and water pollutants. Phytoremediation is the collection of diverse plant-based technologies that use either naturally occurring or genetically engineered plants to clean the environment (Salt et al. 1995; Flathman and Lanza 2010). However, the required time to clean up contaminants from soil inhibits its use on an industrial scale. It involves the cleaning up of contaminated soil and water by either plants or by the root-colonizing microbes and is best applied at sites with shallow contamination of inorganic and organic pollutants (Pilon-Smits 2005). Due to limitations, the utilization of biotechnological applications involving high biomass fast-growing vegetation for remediation purposes combined with biofuel manufacturing has gained interest in recent years (Oh et al. 2013; Pidlisnyuk et al. 2014; Singh et al. 2018, 2017a, b, c, 2019; Tiwari et al. 2018, 2019a, b; Kour et al. 2019).

The development of new genetic tools and a better knowledge of plant–microbe interactions have accelerated advancements in pathway-engineering techniques (referred to as designer plants and microbes) for improved toxic waste removal. This chapter's special focus is on the use of biotechnological applications and techniques for environmental protection, detoxification, and the removal of heavy metals and metalloids. This chapter also examines current developments and the future outlook for the bio/phytoremediation of toxic pollutants from infected water and soil.

Applications of microbial biotechnology in sustainable agriculture are increasing day by day (Mosa et al. 2016). The aim of the chapter is likewise to summarize and examine different aspects of beneficial microbes, their activities for the better management of agriculture, and also focused on the utility of "Omics" in the superior way for the development of sustainable agriculture.

11.2 MICROBIAL BIOTECHNOLOGY FOR AGRO-ECOSYSTEM

Microbial biotechnology plays an important role in maintaining sustainable agriculture by decreasing the various stresses on plants such as abiotic and biotic stresses. Many microorganisms such as *Pseudomonas spp.* control plant disease by providing induced systemic resistance. Plant growth and yield are mostly influenced by the plant–microbe interactions (Vimal et al. 2017; Tiwari et al. 2017). The plant rhizosphere also harbors microorganisms that may have a positive, negative, or neutral effect on the growth of the plant (Berg 2009). However, most rhizospheric microbes slow down the growth of plants and root (Tiwari et al. 2017). Microbes can be used for sustainable agriculture without harming the environment. It aids in sustainable agriculture (Figure 11.1) by various methods such as:

i) Surpass the technology for biomass-derived energy
ii) Enhance the productivity and quality of crops

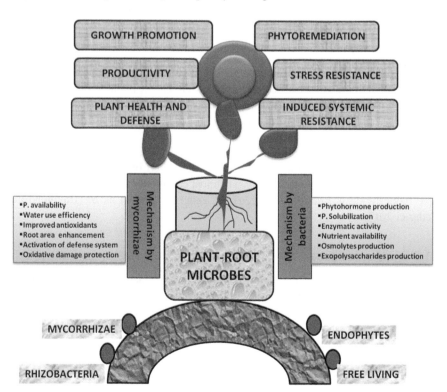

FIGURE 11.1 General plant growth-enhancing mechanism by soil microbes.

iii) Enhance the bioremediation process as compared to natural bioremediation
iv) Increase N2 fixation and nutrient uptake
v) Provide resistance against various plant pathogens
vi) Enhance the fertility of soil

Microbial biotechnology has increased the fertility of soil by generating the genetically modified organisms using the biotechnological techniques such as genetic
engineering for the transfer of genes among the different species to facilitate the
various functions such as resistance against pathogens, increasing the mobility and
uptake of the plant nutrients such as phosphorus, nitrogen. This reduces the dependence on agro-chemicals and increases the productivity.

11.2.1 BIOFERTILIZERS

Microbial biotechnology has provided various methods to facilitate the soil health
and introduced and modified several traditional roles of microbes in agriculture. A
microbe like Rhizobium is present in the nodules of leguminous plants, capable of
fixing atmospheric nitrogen to the organic form of nitrogen. Another example, a fungus *Penicillium bilaii*, releases an organic acid in its environment, which dissolves
the phosphate and makes it available to the plants (Mostafiz et al. 2012). Biofertilizers
are eco-friendly as they do not harm the ecosystem and are also economical in contrast to traditional chemical fertilizers (Mostafiz et al. 2012).

11.2.2 BIOPESTICIDES

Microorganisms are organisms present within the soil which are not always beneficial
for the plants. Some microbes cause disease or damage the plant. Beneficial microorganisms can be used as biological tools against weeds and pests. Biopesticides are the
pesticides which are derived from natural microbes, possessing harmful/lethal genes
against other living things, or their secretions of products may be lethal, causing the
death of various herbs and pests (Raymond and Federici 2017). Biopesticides do not
survive in the environment for a long time as their shelf lives are short and they do
not affect the environment negatively. The most commonly and commercially available biopesticide is *Bacillus thuringiensis*, generally known as *Bt* (Raymond and
Federici 2017).

11.2.3 BIOHERBICIDES

Bioherbicides comprise the microbes with invasive genes which can kill weeds by attack-
ing their defense genes. There are various benefits of using the bioherbicides instead of
chemical herbicides such as the fact that they are eco-friendly, non-toxic to the environment, and also persist to the next season when weeds reappear (Kremer 2005).

11.2.4 BIOREMEDIATION

Various microbes such as bacteria, fungi, viruses, and protozoa are being widely used
against various pathogens and weeds. The most common example is the bacterium

Bacillus thuringiensis (Bt), which is the most common pesticide used against pests of crops. Other fungal biopesticides include *Trichoderma harzianum*, or those used against various plant pathogens, e.g. Pythium, fusarium, and other soil phytopathogens (Lorito et al. 1993).

11.3   GENES IDENTIFIED FROM MICROBES TO OVERCOME STRESS IN PLANTS

Adverse conditions consisting of abiotic and biotic stresses are fundamental limiting factors for a decline in agricultural productivity. Although any accurate estimation of agricultural loss (reduction of crop production and soil health) in phrases of agro-ecological disturbances because of abiotic stresses cannot be made, it is obvious that such stresses affect big land areas and significantly impact qualitative and quantitative loss in crop production (Cramer et al. 2011).

Plants often cope with the rapid fluctuations and adversity of environmental conditions because of their intrinsic metabolic capabilities (Simontacchi et al. 2015). Variations in the outside environment could put the plant metabolism out of homeostasis (Foyer and Noctor 2005), and create a necessity for the plant to harbor some advanced genetic and metabolic mechanisms within its cellular system. Plants acquire an array of protective mechanisms during the course of evolution to combat adverse environmental conditions (Yolcu et al. 2016). Such processes cause metabolic re-programming in the cells to assist routine bio-physico-chemical processes irrespective of the external situations (Mickelbart et al. 2015). Many times, plants get facilitated in reducing the load of environmental stresses with the support of the microbiome they inhabit (Ngumbi and Kloepper 2014).

Microbial existence is the most fundamental live system on the earth. Being an important living part of the soils, they naturally become an integral part of the crop production system as soon as a seed comes into the soil to begin its life cycle. Microorganisms are important inhabitants of seeds also, and proliferate as the seeds grow in the soils to form a symbiotic relationship at the surface or endophytic interactions inside the roots, stems, and leaves. Microbial intrinsic genetic and metabolic capabilities make them suitable organisms to mitigate extreme conditions of the environment (Singh et al. 2014).

Various mechanisms highlighting the role of microorganisms in stress alleviation have been proposed. Soil-inhabiting microbes belonging to genera *Azospirillum, Achromobacter, Bacillus, Variovorax, Azotobacter, Enterobacter, Aeromonas, Klebsiella,* and *Pseudomonas* have been shown to increase plant growth even under adverse environmental conditions (Hamdia et al. 2004; Ortiz et al. 2015; Kaushal and Wani 2016; Sorty et al. 2016). Literature relating to the involvement of microbes for the alleviation of stressors signifies the role of microbes in this field (Table 11.1).

11.4   EXPRESSION OF MICROBIAL GENES IN TRANSGENIC PLANTS TO IMPROVE DEFENSE MECHANISMS

One of the major challenges facing modern agriculture is to achieve a satisfactory, but environmentally friendly, control of plant diseases. Although the extensive use

TABLE 11.1

Microbe-Associated Abiotic Stress Resistance in Plants

Crop	Gene Donor Organism	Gene Engineered	Resistance To	Reference
Arabidopsis and rice	*Arthrobacter globiformis*	COD gene	Salt and cold stress	Sakamoto and Murata 1998
Rice plants	Spinach (*Spinacia oleracea*)	CMO gene	Salt and temperature stress	Shirasawa et al. 2006
Tobacco (*Nicotiana tabacum*)	*E. coli*	Genes for Glycine betaine biosynthesis	Salt stress	Yang et al. 2005
Cotton	*Atriplex hortensis*,	AhCMO	Resistance to salinity stress	Yao et al. 2011
Transgenic potato (*Solanum tuberosum* L.)	*Arthrobacter globiformis*	choline oxidase (codA) gene	Resistance to NaCl and drought stress	Ahmad et al. 2008
Tobacco plant	Bacterial	Mothbean (*Vigna aconitifolia*) Δ1-pyrroline-5-carboxylate synthase (p5cs) gene	Improved freezing tolerance and increased salt tolerance	Kishor et al. 1995
Tansgenic Arabidopsis	Sheepgrass (*Leymus chinensis*)	LcSAIN2 (*Leymus chinensis* salt-induced 2) gene	Salt	Li et al. 2013
Tobacco, Arabidopsis, *Lotus tenuis* and sweet potato plants	Bacterial	Arginine decarboxylase	Environmental stress	Rangel et al. 2016; Espasandin et al. 2017
Chinese cabbage (*Brassica campestris* ssp. pekinensis) and lettuce (*Lactuca sativa*	*B. napus*	*B. napus* group 3 LEA proteins	Increased growth under salt and drought conditions	Park et al. 2005
Rice	Barley	HVA1	Drought condition	Babu et al. 2004
				(Continued)

TABLE 11.1 (CONTINUED)
Microbe-Associated Abiotic Stress Resistance in Plants

Crop	Gene Donor Organism	Gene Engineered	Resistance To	Reference
Creeping bentgrass (*Agrostis stolonifera* var. palustris),	Barley	HVA1 gene	Drought condition	Fu et al. 2007
Tobacco	*Tamarix androssowii*	LEA gene (DQ663481)	Drought stress	Wang et al. 2006
Transgenic rice	Wheat	PMA80 and PMA1959	Drought	Cheng et al. 2002
Tobacco plant	*Boea hygrometrica*	BhLEA1 and BhLEA2	Drought	Liu et al. 2009
Transgenic tobacco	*Escherichia coli*	mannitol 1-phosphate dehydrogenase (mtlD)	Salinity-tolerant phenotype	Thomas et al. 1995
Wheat	*Escherichia coli*	mtlD	Tolerance to water stress and salinity	Abebe et al. 2003
Arabidopsis	*Glycine max*	IMT	Tolerance to dehydration stress treatment and high salinity stress treatment	Ahn et al. 2011
Transgenic *Arabidopsis thaliana*	*Mesembryanthemum crystallinum*	myo-inositol-O-methyltransferase (Imt1)	Elevated cold tolerance	Zhu et al. 2012
Transgenic *Nicotiana tabaccum*	halophilic bacteria	ectB, ectA, and ectC	Increased tolerance to hyperosmotic shock	Nakayama et al. 2000
Tomato	*Halomonas elongate*	ectA, ectB, and ectC	Salt stress	Moghaieb et al. 2011
Transgenic tomatoes	Yeast	trehalose-6-phosphate synthase (TPS1) gene	Salt, drought and oxidative stresses	Cortina and Culiáñez-Maciá 2005
Rice	*E. coli*	otsA and otsB)	Salt and drought stress	Garg et al. 2002
Tomato	Tobacco	osmotin gene	Salt and drought stress	Goel et al. 2010
Soyabean	*Glomus intraradices*	–	Salt	Porcel and Ruiz-Lozano 2004

of chemicals remains the main strategy of disease control, a variety of alternative approaches have been taken into consideration which include the use of biocontrol agents and pathogen-resistant crop cultivars. Classical and conventional
plant breeding has been very useful in producing disease-resistant varieties, but
this has no access to resistance available in sexually incompatible species. Since
the first plant expressing a transgene was obtained by Agrobacterium-mediated
transformation more than 10 years ago, a variety of desirable traits have been
transferred to model and crop plants by using a few but powerful techniques
(Mourgues et al. 1998). In many situations, endogenous plant genes have been
reengineered to improve crop quality or resistance, such as the introduction of
sense and antisense configurations to alter ripening and senescence processes
(Fray and Grierson 1993). Two major strategies have been followed to express
microbial genes in plants to improve resistance against pathogens. Disease resistant plants have been obtained that (i) synthesize transgenic compounds which
directly improve the antimicrobial and insecticidal activity of the plant, or (ii)
express transgenes, often pathogen-derived, capable of activating plant defense
response or providing tolerance to pathogen toxins and enzymes. The capacity and usage of microbial genomes to genetically engineer new crop varieties
with enhanced resistance traits are considered, and some of the most promising
research and gene sources are presented. In addition, most of the microbial genes
already expressed in different plants for a variety of purposes other than disease
control are listed and briefly discussed.

11.4.1 BACTERIAL AND FUNGAL TRANSGENES WITH ANTIMICROBIAL AND INSECTICIDAL EFFECT IN PLANT

Antifungal and antibacterial proteins are key components of defense and offense
mechanisms of many groups of fungi and bacteria. They are often effective
on a broad range of targets and function synergistically in combinations, also
with other biologically active compounds (i.e. antibiotics) (Lorito et al. 1996).
Although a number of such genes have been identified, not many have been tested
transgenically in plants to improve disease tolerance. Chitinase encoding genes
have been among the most used to increase plant defense against fungal pathogens (Table 11.2).

11.5 ACTIVATION OF PLANT DEFENSE RESPONSE VIA MICROBIAL TRANSGENES

Genes that may protect plants by activating defense responses or inhibiting pathogen
virulence factors have also been exploited in engineering resistance mechanisms.
Avirulence genes (Avr) in pathogens and their matching resistance (R) genes in host
plants have received considerable attention (Lauge and De Wit 1998). In plant-patho-
gen systems following the gene for gene relationship, the interactions between the
product of an Avr gene (elicitor) and the product of the corresponding R gene (receptor) activate a signal transduction pathway which leads to resistance, often through

TABLE 11.2

Expression of Microbial Genes in Plants That Increase Defense against Pathogens

Source	Genes Encoded	Activity	Plant	Reference
Serratia marcescens	Chitinase (chiA)	Antifungal chitinase	Tobacco	Howie et al. 1994
Chitosanase Streptomyces sp.	Chitosanase	Antifungal chitosanase	Tobacco	El Quackfaoui et al. 1995
Aspergillus niger	Glucose-oxidase	Generate reactive oxygen and HR.	Potato	Wu et al. 1995
Bacillus amyloliquefaciens	Bacteria ribonuclease (barnase) with inducible promoter (prp1- 1) and barstar	Ribonuclease localized at infection site.	Potato	Strittmatter et al. 1995
B. thuringensis	B.t. toxin	Insecticidal	Potato, Corn, poplar, cotton, cranberry, tomato, rice, Arabidopsis, carrot, etc.	Diehn et al. 1996
Vibrio cholera	Cholera toxin, subunit A	Antibacterial	Tobacco	Beffa et al. 1995
Ustilago maydis (mycovirus)	Killer toxin	Toxic to closely related Ustilago species	Tobacco	Park et al. 1996

TABLE 11.3

Expression of Microbial Genes in Plants That Trigger Plant Defense Responses against Pathogen Toxins

Source	Genes Encoded	Activity	Plant	Reference
P. infestans, Cladosporium fulvum	Avirulence (avr9, inf1)	Elicit HR	Tomato, tobacco	Kamoun et al. 1999
P. syringae pv. tomato	Avirulence (avrPto)	Elicit defense response	Tomato	Tobias et al. 1999
C. fulvum	Virulence (Ecp2)	Elicit HR	Tomato	Lauge et al. 1998
P. cryptogea	β-cryptogein	Elicit defense response	Potato, tobacco	Tepfer et al. 1998; Keller et al. 1999
Halobacterium halobium	Bacterio-opsin proton pump (bO)	Activate local and systemic response	Tobacco, potato	Abad et al. 1977
P. syringae pv. phaseolicola	Ornithine carbamoyltransferase (argK)	Resistance to bacteria phaseolotoxin	Bean, tobacco	De la Fuente Martinez et al. 1992

a hypersensitive response (Lauge and De Wit 1998). To neglect unwanted expression of Avr/R, which leads to generalized hypersensitive response and leads to plant death, one of the genes must be controlled by a pathogen inducible promoter. A few examples have been listed (Table 11.3).

The work performed thus far with microbial genes indicates that a significant improvement of plant resistance can be acquired in a variety of ways, but attained mainly through producing enzymes that degrade pathogen structures, enzymes that synthesize antimicrobial compounds or selective toxins. Clearly, each of these approaches will need a tailor-made strategy to minimize any negative effects on the acceptor plant and the ecosystem.

11.6   NEW VISTAS TO REGULATE OR OVERCOME BIOTIC STRESS IN PLANTS VIA MICROBIAL INTERACTIONS

With the burgeoning global population and the erosion of agricultural land, it has become important to look for ways to improve food production. Bacteria, fungi, oomycetes, viruses, and insect pests pose threats to food growth and transport. Globally, pre- and post-harvest losses in crops are a considerable 30% (Bebber and Gurr 2015). Pathogens have an arsenal of effectors to induce effector-triggered susceptibility (ETS), while plants have evolved new resistance (R) proteins to recognize the new effectors. This interplay of defense and counter defense between pathogen and host has resulted in different types of pathogen effectors and resistance genes (Huang et al. 2016).

11.7 THE ROLE OF PLANT GROWTH-PROMOTING BACTERIA (PGPBS)

Plants are able to acquire induced systemic resistance (ISR) to pathogens after inoculation with PGPBs. PGPBs, in association with plant roots, can prime the innate immune system of and confer resistance to a wide array of pathogens with a minimal impact on yield and growth (Kannojia et al. 2017). The plant hormones salicylic acid (SA), jasmonic acid (JA), and ethylene (ET) mediate a signaling network resulting in molecular recognition between the plant and microbe. Several PGPBs, including *Acinetobacter lwoffii, Azospirillum brasilense, Bacillus pumilus, Chryseobacterium balustinum, Paenibacillus alvei, Pseudomonas putida, Pseudomonas fluorescens*, and *Serratia marcescens* colonize roots and provide protection to different plant species of crops including vegetables and trees against foliar diseases field and greenhouse trials (Kannojia et al. 2017).

11.8 THE ROLE OF BENEFICIAL FUNGUS IN PLANT DEFENSE

Besides the classic mycorrhizal fungi, many other fungi such as Trichoderma spp. and *Piriformospora indica* suppress plant diseases leading to plant growth stimulation (Van Wees et al. 2008). These microorganisms are able to form endophytic associations and interact with other microbes in the rhizosphere, thereby influencing disease protection, plant growth, and yield. For example, Trichoderma genomes have revealed mycotrophy and mycoparasitism as ancestral lifestyles of species of this genus. Some Trichoderma strains have become established in the plant rhizosphere and evolved as intercellular root colonizers. As a result, they stimulate plant growth and defenses against pathogens. This depends on the Trichoderma strains, their concentrations, the plant material, the stage of the plant, and the timing of the interaction. The phytohormones ET and IAA produced by Trichoderma play roles in interconnecting plant development and defense responses. The expression of Trichoderma genes in plants has beneficial results, mainly in the control of plant diseases and resistance to adverse environmental conditions (Kannojia et al. 2017). *Piriformospora indica* confers disease resistance systemically. *P. indica* colonizes the roots of many plant species and stimulates growth, biomass, and seed production of the hosts. The host colonization by the fungus stimulates it to produce phosphatidic acid, which triggers the OXI1 pathway (Kannojia et al. 2017). This pathway is activated when there is a pathogen attack and triggers host defense (Rentel et al. 2004).

11.9 RNAI-MEDIATED DEFENSE MECHANISM IN PLANT

Gene expression in eukaryotes is regulated by 20 to 30 nucleotide (nt)-long noncoding RNA molecules called small RNAs (sRNAs) (Kannojia et al. 2017). They are differentiated by their biogenesis pathway and precursor structure, and in plants, RNAs are of two types: microRNA (miRNA) and small interfering RNA (siRNA). miRNAs are derived from RNAs with imperfectly base-paired hairpin structures and are usually 21–24 nt long (Chen 2009). Several RNAi strategies have shown success in plant improvement against biotic stresses. The first miRNA identified to be involved in PTI is Arabidopsis miR393. miR393 was induced in response to *Pseudomonas syringae* attacked by a flagellin-derived peptide, flg22. SA provides defense against biotrophic pathogens, while glucosinolates are antimicrobial molecules that contribute to plant defense against pests and disease.

11.10   CROP DESIGNING USING TRANSGENIC APPROACH

Transgenic plants have been produced with genes involved in different pathways to enhance disease resistance against fungal pathogens. For controlling disease, an apt approach would be the expression of pathogenesis-related genes and defensins. Enhanced resistance in tobacco plants against *Rhizoctonia solani* has been shown by the chit1 gene from the entomopathogenic fungus *Metarhizium anisopliae*, encoding the endochitinase Chit42 (Kern et al. 2010). Three genes, ech42, nag70, and gluc78, encoding hydrolytic enzymes from a biocontrol fungus *Trichoderma atroviride* were introduced in rice. Gluc78-overexpressing transgenic plants showed enhanced resistance to *Magnaporthe grisea* (Sanghera et al. 2011). A few instances have been listed below (Table 11.4).

11.11   EFFICIENT BIOTECHNOLOGICAL APPLICATIONS FOR THE CLEAN-UP OF METALLOIDS AND HEAVY METALS

Plants possess the necessary genetic, biochemical, and physiological characteristics to establish themselves as the ultimate choice for soil and water pollutant remediation.

TABLE 11.4

Transgenic-Engineered Crops to Enhance Defense against Fungal Pathogens

Gene Donor	Crop	Target Fungal Pathogen	Gene/Gene Product Inserted in Plant
Phytophthora cryptogea	Tobacco	*Phytophthora parasitica*	β-Cryptogein elicitor
Phytolacca americana	Tobacco	Broad-spectrum resistance to viral and fungal pathogens	PAPII
Trichoderma harzianum	Potato	Foliar and soilborne fungal pathogen	Endochitinase
Pseudomonas fluorescence	Carrot	*Alternaria dauci*, *Alternaria radicina*, and *Botrytis cinerea*	Microbial factor 3 (MF3)
Erwinia amylovora	Tobacco	*Botrytis cinerea*	hrp N
W. japonica	Tobacco	*B. cinerea*	PR1
Trichoderma harzianum	Grape	*Botrytis cinerea*	Endochitinase
Trichoderma harzianum	Apple	*Venturia inaequalis*	Endochitinase, exochitinase
Aspergillus giganteus	Rice	*Magnaporthe grisea*	AFP
Fungi	Rice	Multiple pathogen	Glucose oxidase gene
Aspergillus giganteus	Pearl millet	Rust and downy mildew	Afp

Phytoremediation is a cost-effective, green-clean technology with long-term applicability for the cleaning up of contaminated sites. However, the required time frame to clean up contaminants from soil prevents its use on an industrial scale. It involves the cleaning up of contaminated soil and water by either root-colonizing microbes or by the plants themselves, and is best applied at sites with shallow contamination of organic and inorganic pollutants (Mosa et al. 2016). Due to this shortcoming, the utilization of biotechnological approaches involving high biomass fast-growing crops for remediation purposes combined with biofuel production has gained momentum in recent years (Mosa et al. 2016). The development of new genetic tools and a better understanding of microbe and plant gene structures and functions have accelerated advancements in pathway-engineering techniques (referred to as designer microbes and plants) for improved hazardous waste removal.

Heavy metals are a threat to the environment as they are toxic, non-degradable, and can bio-accumulate in the ecosystem and persistent in the environment for a long time (Nriagu 1989). Thus, the native form of microbes is not so effective in degrading the heavy metal pollutants. Soil heavy metal pollution is caused by various metals such as Ni, Cd, Cu, Zn, Cr, and Pb. Due to the toxic effects of heavy metals, there is a change in microbe diversity, along with the population size of the soil microbial community. Heavy metal contamination may result in stunted plant growth, and a decrease in yield and nutrient uptake. Heavy metal uptake by plants affects the physiological processes, and thus decreases the productivity of plant (Mosa et al. 2016). The heavy metals Hg, Cr, As, Zn, Cd, Ur, Se, Ag, Au, and Ni are hazardous heavy metals that contaminate the environment and adversely affect the quality of the soil and crop production, as well as public health (Turpeinen et al. 2002) Several strains of yeast such as *Hansenula polymorpha*, *S. cerevisiae*, *Yarrowia lipolytica*, *Rhodotorula pilimanae*, *Pichia guilliermondii*, and *Rhodotorula mucilage* have been used to bio-convert Cr (VI) to Cr (III) (Mosa et al. 2016).

11.11.1 Efficient Strategies for Bioremediation

Microorganisms are mainly used in bioremediation to eliminate heavy metals (elements with densities above 5 g/cm^3) from the polluted environment (Mosa et al. 2016). In addition to the natural occurrence of heavy metals, they are widely used in industry, agriculture, and military operations. These processes have led to the continuous accumulation of heavy metals in the environment, which raises threats to public health and ecosystems. The high concentrations of heavy metals in the environment were also attributed to several life-threatening diseases, including cancer and cardiovascular ailments (Mosa et al. 2016). The elimination of heavy metals requires their concentration and containment as they cannot be degraded by any biological, physical, or chemical processes (Mosa et al. 2016). Microbes perform the remediation of heavy metals through three different processes (Figure 11.2)

11.11.2 Biosorption and Bioaccumulation

Biosorption and bioaccumulation are processes by which the microorganisms, or biomass, bind to and concentrate heavy metals and contaminants from the environment

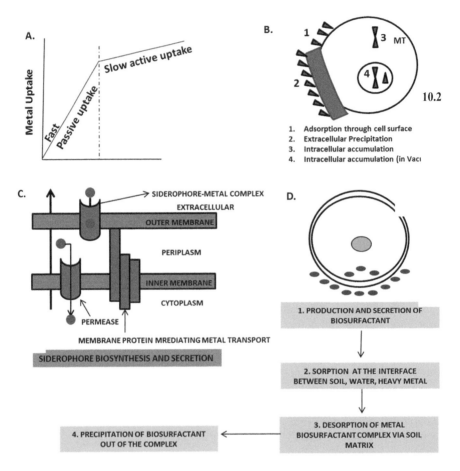

FIGURE 11.2 Mechanism of microbial remediation. (**A**) Heavy metal uptake by biological materials, (**B**) Biosorption of heavy metal by bacterial cells, (**C**) Siderophore formation for heavy metal remediation, (**D**) Mechanism of bacterial heavy metal remediation through biosurfactant production.

(Joutey et al. 2015). However, both biosorption and bioaccumulation work in distinct ways. During biosorption, contaminants are adsorbed onto the sorbent's cellular surface in amounts that depend on the composition and kinetic equilibrium of the cellular surface. Thus, it is a passive metabolic process (Figure 11.2A, 11.2B) that does not require energy/respiration (Mosa et al. 2016). Bioaccumulation, on the other hand, is an active metabolic process that needs energy and requires respiration (Velásquez and Dussan 2009). Since contaminants (such as heavy metals) bind to the cellular surface of microorganisms during biosorption, it is a revisable process. In contrast, bioaccumulation is only partially reversible. Biosorption is also shown to be faster and to produce a greater number of concentrations (Velásquez and Dussan 2009). The biosorption capacity of chromium ions in the dead Bacillus sphaericus was increased by 13–20% in comparison with living cells of the same strain (Velásquez and Dussan 2009)

A study reported improved heavy metal biosorption capacities for *E. coli* cells engineered with mice MT1, demonstrating the potential of the genetic engineering approach in developing organisms with tailored and improved biosorption capacities (Mosa et al. 2016). The introduction of modern research technologies in genomics such as next generation sequencing (Mosa et al. 2016) and high throughput genome editing techniques (Bao et al. 2016) allowed the study of organisms with potential biosorption capacities that are expected to have several bioremediation applications in future. On the other hand, several different organisms are used for the study of bioaccumulation and as indicators for increased levels of pollutants, including plants, fungi, fish, algae, mussels, oysters, and bacteria (Mosa et al. 2016). In *E. coli*, for instance, the expression of metal-binding peptides with the repetitive metal-binding motif $(Cys-Gly-Cys-Cys-Gly)_3$, as well as their high affinity and selectivity for target metals, was investigated for potential use in bioremediation (Mosa et al. 2016). Furthermore, bioinformatics and mathematical modeling have been utilized to investigate the properties and potentials of candidate organisms in order to predict the concentration of chemicals that can be tolerated by them (Mosa et al. 2016).

11.11.3 SIDEROPHORE FORMATION

Siderophores are selective and specific iron-chelating agents secreted by living organisms such as bacteria, yeasts, fungi (Figure 11.2C), and plants. Siderophores have a relatively low molecular weight and an extremely high binding affinity to trivalent metal ions (Fe^{3+}), which is poorly soluble and predominantly found in oxygenated environments (Mosa et al. 2016). There are three different types of siderophores; namely, hydroxamate siderophores, catecholates (phenolates) siderophores, and carboxylate siderophores. Systems biology approaches have been recently used in the investigation of siderophore formation on an unprecedented scale. Genomics and metagenomics were used to evaluate the environmental consequences of the devastating 2011 earthquake and tsunami in Japan that drastically altered the soil environment (Hiraoka et al. 2016). Bioinformatics analysis, coupled with high throughput experimental techniques, demonstrated a remarkable approach in addressing challenges such as identifying the secondary metabolites produced by cryptic genes in bacteria (Mosa et al. 2016), as well as creating online resources for the field (Flissi et al. 2016). Systems biology is a very promising approach for the study of genes, mechanisms, and signaling pathways during siderophores formation.

11.11.4 BIOSURFACTANT PRODUCTION

Biosurfactants are surfactants produced or secreted by living organisms such as microbes (Figure 11.2D). Although biosurfactants have been commonly used for organic pollutants remediation, several studies have also reported that biosurfactants are able to complex and remediate heavy metals such as Cd, Pb, and Zn (Mosa et al. 2016). Rhamnolipids are a major class of biosurfactants that are produced by *P. aeruginosa* and several other organisms (Maier and Soberón-Chávez 2000). They are glycolipids with rhamnose moiety comprising of a glycosyl head group and a 3-(hydroxyalkanoyloxy) alkanoic acid (HAA) fatty acid as the tail (Desai and Banat

1997). These complexations lead to an increase in the apparent solubility of metals (Mosa et al. 2016).

11.11.5 EFFICIENT STRATEGIES FOR PHYTOREMEDIATION

Biotechnological approaches are currently being used for the phytoremediation of heavy metals and metalloids such as mercury (Hg), cadmium (Cd), lead (Pb), selenium (Se), copper (Cu), and arsenic (As). Three main biotechnological approaches are being used to engineer plants for the phytoremediation of heavy metals and metalloids (Figure 11.3): (1) manipulating metal/metalloid transporter genes and uptake systems; (2) enhancing metal and metalloid ligand production; (3) conversion of metals and metalloids to less toxic and volatile forms (Mosa et al. 2016).

11.11.6 MECHANISM OF METAL/METALLOID TRANSPORTER GENES AND UPTAKE SYSTEM

Improved metal tolerance and accumulation has been achieved in different plant species by manipulating metal transporters (Figure 11.3). For instance, *Arabidopsis thaliana* overexpressing yeast YCF (Yeast Cadmium Factor 1) resulted in enhanced tolerance to Pb(II) and Cd(II) and accumulated higher amounts of these metals in plants (Mosa et al. 2016). YCF1 is involved in Cd transport into vacuoles by

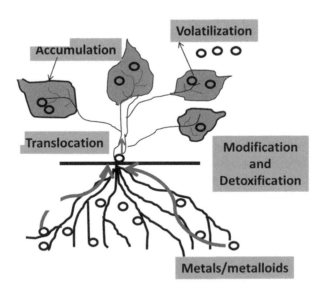

FIGURE 11.3 Effective biotechnological strategies for phytoremediation. Hazardous elements can be mobilized and transported (influx) into roots through plasma membrane transporters. At this stage, plant resistance capacity to toxic elements may be enhanced through manipulation of influx/efflux transporters or by enhancing the levels of ligands/chelators. Volatilization of the toxic elements through enzymes is possible that modify these hazardous elements. Efflux transporters or chelators can also be used to export these elements out of the cytosol.

conjugation with glutathione (GSH). Overexpression of full-length *Nicotiana tabacum* plasma membrane channel protein (*NtCBP4*) showed Pb^{2+} hypersensitivity and enhanced Pb^{2+} accumulation in the transgenic plants. Arsenic (As) is a highly toxic metalloid which is classified as a group I carcinogen for humans by the International Agency for Research on Cancer (IARC) (IARC Monographs 2004). Furthermore, Cu-tolerant genes have been identified in the *Paeonia ostii* plant using the *de novo* transcriptome sequencing approach (Wang et al. 2016).

11.11.7 ENHANCEMENT OF METALS AND METALLOIDS LIGAND PRODUCTION

There are several reports of using Cys-rich peptides such as MTs, PCs, and GSH as metal-binding ligands for the detoxification or accumulation of heavy metals (Figure 11.3). For instance, *A. thaliana* overexpressing an MT gene, PsMTA from pea (*Pisum sativum*), showed increased Cu^{2+} accumulation in roots (Mosa et al. 2016). Shrub tobacco overexpressing the wheat TaPCS1 gene encoding PC synthase increased its tolerance to Pb and Cd significantly; transgenic seedlings grown in soil containing 1572 ppm Pb accumulated double the amount of Pb than WT plants (Mosa et al. 2016). Arsenic (As) tolerance in plants can also be increased by modifying GSH and PCs. These researches showed that the manipulation of genes for increasing the production of metal chelation agents hold great potential for improving heavy metal and metalloid resistance and accumulation in plants.

Recently, the de novo sequencing approach for radish (*Raphanus sativus* L.) root under Cd stress has been used to identify differentially expressed genes and microRNAs (miRNAs) involved in Cd-responsive regulatory pathways. Furthermore, the degradome sequencing approach has been used to identify miRNAs and their target genes under Pb stress in *Platanus acerifolia* tree plants (Mosa et al. 2016).

11.11.8 CONVERSION OF METALS AND METALLOIDS TO LESS HAZARDOUS AND VOLATILE FORMS

Several research groups have focused their efforts on developing phytoremediation strategies for Se and Hg using biotechnological approaches employing the conversion of these metals to less toxic and volatile forms (Figure 11.3). Selenium (Se) is an essential micronutrient for many organisms. However, in excess concentrations, it is very toxic and is a worldwide environmental pollutant (Zwolak and Zaporowska 2012). Se occurs naturally in soil, and is chemically similar to sulfur (S). Therefore, plants uptake the inorganic and organic forms of Se via S transporters and metabolize them to volatile forms through S assimilation pathways to relatively non-toxic forms, such as dimethylselenide (DMSe). A constitutive overexpression of *A. thaliana* ATP sulfurylase (APS), converting selenate to selenite, in *Brassica juncea* showed enhanced reduction of selenate to organic Se forms in the APS overexpressed plants, whereas WT plants accumulated mainly selenate. Arabidopsis expressing a mouse selenocysteine lyase (Scly) gene, showed enhanced shoot Se concentrations (up to 1.5-fold), compared to the WT (Mosa et al. 2016). The overexpression of the SMT gene from the Se hyperaccumulator *Astragalus bisulcatus* in *A. thaliana* and *B. juncea* improved the tolerance of transgenic plants to selenium and enhanced Se accumulation in shoots. The transgenic plants also increased Se volatilization rates. Recently, small RNA and degradome sequencing analyses were used to identify several miRNAs induced by Se treatment in the Se hyperaccumulator *Astragalus chyrsochlorus* plant callus (Mosa et al. 2016). Furthermore, differentially expressed genes in *A. chyrsochlorus* under selenate treatment were identified using de novo transcriptome analysis. Mercury (Hg) is a global pollutant threatening human and environmental health cycling between air, water, sediment, soil, and organisms (Mosa et al. 2016). Various plant species such as *A. thaliana* (Rugh et al. 1996), rice (Heaton et al. 2003), and tobacco (Heaton et al. 2005) constitutively expressing modified merA were resistant to levels up to 25–250 μM HgCl2 and exhibited significant levels of Hg(0) volatilization as compared to control plants. *Arabidopsis* expressing both genes, merA, and merB, grow on 50-fold higher methylmercury concentrations than WT plants and up to 10-fold higher concentrations than plants that express merB alone. Transgenic plants were also seen to detoxify organic mercury by converting it to volatile and significantly less toxic elemental mercury (Mosa et al. 2016).

11.12   RHIZOREMEDIATION: THE COMBINED EFFECTS OF BIO/PHYTOREMEDIATION

Rhizoremediation is the combination of two approaches, i.e. phytoremediation and bioaugmentation, for cleaning contaminated substrates. Rhizoremediation refers to the exploitation of microbes present in the rhizosphere of plants utilized for phytoremediation purposes. Bacteria, mainly plant growth-promoting rhizobacteria (PGPR), and fungi, mainly arbuscular mycorrhizal fungi (AMF), are used as pure cultures or co-cultures for bioaugmentation. PGPR such as *Agrobacterium, Alcaligenes, Arthrobacter, Azospirillum, Bacillus, Burkholderia, Serretia, Pseudomonas*, and *Rhizobium* are generally used for metal extraction with plants (Glick 2003). The presence of rhizobacteria in soil can improve the efficiency of arsenic phytoextraction in hyperaccumulator plant species as well (Mosa et al. 2016). Abou-Shanab et al. (2006) reported that the bacterial strain *Microbacterium oxydans* AY509223 plays a role in nickel (Ni) mobilization and increased Ni uptake of *Alyssum murale* grown in low, medium, and high Ni soils by 36.1, 39.3, and 27.7%, respectively. The results provided potential development of inoculum for enhanced uptake during commercial phytoremediation or phytomining of Ni. The rhizo- and endophytic bacterial communities of *Prosopis juliflora* also harbor some novel heavy metal-resistant bacteria. The three different endophytic bacterial strains, viz., *Pantoea stewartii* strain ASI11, *Microbacterium arborescens* strain HU33, and *Enterobacter* sp. strain HU38 improved plant growth and heavy metal removal from tannery effluent contaminated soils, and showed that these bacteria play a role in improving the phytoremediation efficiency of heavy metal-degraded soils (Mosa et al. 2016). Genomic analysis of three legume growth-promoting rhizobia (*Mesorhizobium amorphae* CCNWGS0123, *Sinorhizobium meliloti* CCNWSX0020 and *Agrobacterium tumefaciens* CCNWGS0286) that survive at regions with high levels of heavy metals in China confirmed the existence of different transporters involved in nickel, copper, zinc, and chromate resistance (Xie et al. 2015). In another report, Zhao et al. (2015)

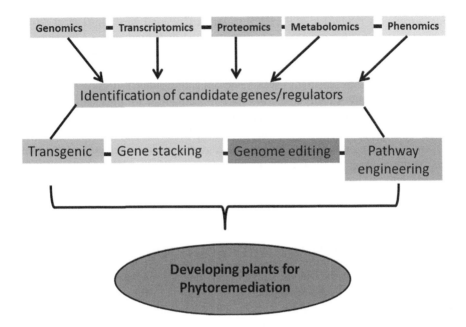

FIGURE 11.4 Integration of "Omics" for developing plants for phytoremediation. Genomics, transcriptomics, proteomics, metabolomics, and phenomics could help on identifying the candidate genes which can be used for developing plants for phytoremediation through different approaches including transgenic, gene- stacking, genome editing, etc.

used an RNA-seq approach to investigate the genes associated with Cd stress in the Dark Septate Endophytic (DSE) fungal Cd-tolerant strain.

Recently, a novel metal transporter homology to natural resistance-associated macrophage protein (Nramp) isolated from *Exophiala pisciphila* has been reported to increase Cd^{2+} sensitivity and accumulation when heterologously expressed in yeast (Wei et al. 2016). The combinatorial effects of bioaugmentation and phytoremediation leading to rhizoremediation may solve the problems encountered during the application of both techniques individually. Moreover, phytoextraction could be enhanced through the application of genetically engineered plant-associated microorganisms. Collectively, these efforts promise an upcoming generation of tailored organisms with higher bio/phytoremediation efficiencies and lower costs (Figure 11.4).

11.13   MICROBIAL GENETIC ENGINEERING

With the advances in genetic engineering, microbes are engineered with desired characteristics such as the ability to tolerate metal stress, overexpression of metal-chelating proteins and peptides, and ability of metal accumulation. Frederick and Taylor (2013) engineered microorganisms to produce trehalose and establish that reduce 1 mM Cr (VI) to Cr (III). Engineered *Chlamydomonas reinhardtii*-generated significant increase in tolerance to Cd toxicity and its accumulation (Igiri et al. 2018). Genetically engineered microbes for heavy metal remediation involve the use of *Escherichia coli* (*E. coli* ArsR (ELP153AR)) to target As(III) (Igiri et al. 2018) and

Saccharomyces cerevisiae (CP2 HP3) to target Cd^{2+} and Zn^{2+} (Igiri et al. 2018). *Corynebacterium glutamicum* was genetically modified using overexpression of *ars* operons (*ars*1 and *ars*2) to decontaminate As-contaminated sites (Igiri et al. 2018).

Bioremediation of heavy metals has been extensively studied and the performance of several bioremediators were reviewed and summarized. Bioremediation is an environmentally friendly and cost-effective technology for the clean-up of complex industrial tannery effluent containing heavy metals. Many natural biosorbents of microbial origins have been identified with efficient biosorption characteristics. Recent surface modifications on these bioremediators have helped to ameliorate their metal-binding properties and increase the overall cost of the process. In spite of such shortcomings, both native and modified biosorbents have demonstrated their compatibility when tested with tannery effluent. These biosorbents showed effective metal removal over a wide range of temperature, pH, and solution conditions.

11.14   CONCLUSION AND FUTURE OUTLOOK

Plant responses toward various stresses and microbe-mediated stress mitigation approaches in plants have been studied on sound grounds of physiological, molecular, biochemical, and ultrastructural parameters. Such studies have been carried out encompassing different omics approaches (genomics, metagenomics, metatranscriptomics, proteomics, metaproteomics etc.) that have strengthened our understanding behind the mechanisms of microbial interactions, gene cascades and metabolic pathways, accumulation and enhancement of various metabolites, proteins, enzymes, and the up- and down-regulation of different genes. Such studies could yield dynamic data related to combined responses of plants to multiple adverse conditions, and the same is also pertinent with the naturally associated or artificially inoculated microorganisms. This chapter provides a new direction for improvising the existing protocols in the field of plant–microbe interactions under stress, and use of microorganisms and microbial metabolite molecules for the alleviation of diverse adverse conditions encountered by plants.

Tannery effluent and biosorbent characteristics need to be assessed prior to application. Keeping in focus the inhibitions of bioremediation technology, the future prospect looks promising on microbial genetic technologies and the development of increased specificity using biofilms which could be achieved by optimization process and immobilization techniques. Hence, more effort should be made in biofilm-mediated bioremediation, genetically modified microbes, and microbial fuel cell in the bioremediation of heavy metals in the ecosystem.

In future, efforts should be made to develop strategies to improve the tolerance, uptake, and hyperaccumulation of heavy metals/metalloids using genomic and metabolic engineering approaches. Pathways that control the uptake, detoxification, transport from root to shoot tissues, and translocation and hyperaccumulation in the aboveground storage tissues can be engineered using gene stacking approaches. Additionally, genome editing strategies can be designed using TALENs (transcription activator-like effector nucleases) technology or the powerful CRISPR-Cas9 (clustered regularly interspaced short palindromic repeats) system to produce microbes/plants for bio/phytoremediation purposes. Additionally, efforts should be made to develop breeding programs to improve the biomass and growth habits of natural hyperaccumulators and breed those traits into non-food, high biomass, fast-growing plants for the commercial phytoremediation of heavy metals and metalloids.

REFERENCES

Abad, M.S., Hakimi, S.M., Kaniewski, W.K. et al. 1997. Characterization of acquired resistance in lesion mimic transgenic potato expressing bacterio-opsin. *Molecular Plant–Microbe Interactions* 10:635–645.

Abebe, T., Guenzi, A., Martin, B., Cushman, J. 2003. Tolerance of mannitol-accumulating transgenic wheat to water stress and salinity. *Plant Physiology* 131:1748–1755.

Abou-Shanab, R.A.I., Angle, J.S., Chaney, R.L. 2006. Bacterial inoculants affecting nickel uptake by *Alyssum murale* from low, moderate and high Ni soils. *Soil Biology and Biochemistry* 38:2882–2889.

Ahmad, R., Kim, M.D., Back, K.H. et al. 2008. Stress-induced expression of choline oxidase in potato plant chloroplasts confers enhanced tolerance to oxidative, salt, and drought stresses. *Plant Cell Reports* 27:687–698.

Ahn, C., Park, U., Park, P. 2011. Increased salt and drought tolerance by D-ononitol production in transgenic *Arabidopsis thaliana*. *Biochemical and Biophysical Research Communications* 415:669–674.

Aruna, K.K., Sreekanth, R.M. 2015. Microbial biotechnology: Role of microbes in sustainable agriculture. In: *New Horizons in Biotechnology* (Eds. Viswanath, B., Indravathi, G.). Paramount Publishing House, India, pp. 51–52.

Babu, R.C., Zhang, J., Blum, A. et al. 2004. HVA1, a LEA gene from barley confers dehydration tolerance in transgenic rice (*Oryza sativa* L.) via cell membrane protection. *Plant Science* 166:855–862.

Bao, Z., Cobb, R.E., Zhao, H. 2016. Accelerated genome engineering through multiplexing. *Wiley Interdisciplinary Reviews: Systems Biology and Medicine* 8:5–21.

Bebber, D.P., Gurr, S.J. 2015. Crop-destroying fungal and oomycete pathogens challenge food security. *Fungal Genetics Biology* 74:62–64.

Beffa, R., Szell, M., Meuwly, P. et al. 1995. Cholera toxin elevates pathogen resistance and induces pathogenesis related expression in tobacco. *European Molecular Biology Organization Journal* 14:5753–5761.

Berg, G. 2009. Plant–microbe interactions promoting plant growth and health: Perspectives for controlled use of microorganisms in agriculture. *Applied Microbiology and Biotechnology* 84:11–18.

Chen, Z., Zheng, Z., Huang, J., Lai, Z., Fan, B. 2009. Biosynthesis of salicylic acid in plants. *Plant Signal and Behavior* 4:493–496.

Cheng, Z., Targoli, J., Huang, X., Wu, R. 2002. Wheat LEA genes, PMA80 and PMA1959, enhance dehydration tolerance of transgenic rice *Oryza sativa* L. *Molecular Breeding* 10:71–82.

Cortina, C., Culiáñez-Maciá, F.A. 2005. Tomato abiotic stress enhanced tolerance by trehalose biosynthesis. *Plant Science* 169:75–82.

Cramer, G.R., Urano, K., Delrot, S., Pezzotti, M., Shinozaki, K. 2011. Effects of abiotic stress on plants: A systems biology perspective. *BMC Plant Biology* 11:163.

De la Fuente-Martinez, J.M., Mosqueda-Cano, G., AlvarezMorales, A., Herrera-Estrella, L. 1992. Expression of a bacterial phaseolotoxin resistant ornithyl transcarbamylase in transgenic tobacco confers resistance to *Pseudomonas syringae* pv. phaseolicola. *Biotechnology* 10:905–909.

Desai, J.D., Banat, I.M. 1997. Microbial production of surfactants and their commercial potential. *Microbiology and Molecular Biology Review* 61:47–64.

Diehn, S.H., De Rocher, E.J., Green, P.J. 1996. Problems that can limit the expression of foreign genes in plants: Lessons to be learned from B.t. toxin genes. *Genetic Engineering* 18:83–99.

El Quackfaoui, S., Potvin, C., Brzezinski, R., Asselin, A. 1995. A streptomyces chitosanase is active in transgenic tobacco. *Plant Cell Reports* 15:222–226.

Espasandin, F.D., Calzadilla, P.I., Maiale, S., Ruiz, O.A., Sansberro, P. 2017. Overexpression of the arginine decarboxylase gene improves tolerance to salt stress in *Lotus tenuis* plants. *Journal of Plant Growth Regulation* 37:1–10.

Flathman, P.E., Lanza, G.R. 2010. Phytoremediation: Current views on an emerging green technology. *Journal of Soil Contamination* 7:415–432.

Flissi, A., Dufresne, Y., Michalik, J. et al. 2016. Norine, the knowledgebase dedicated to non-ribosomal peptides, is now open to crowdsourcing. *Nucleic Acids Research* 44:1113–1118.

Foyer, C.H., Noctor, G. 2005. Oxidant and antioxidant signalling in plants: A re-evaluation of the concept of oxidative stress in a physiological context. *Plant, Cell & Environment* 28:1056–1071.

Fray, R.G., Grierson, D. 1993. Molecular genetics of tomato fruit ripening. *Trends in Genetics* 9:438–443.

Frederick, T.M., Taylor, E.A. 2013. Chromate reduction is expedited by bacteria engineered to produce the compatible solute trehalose. *Biotechnology Letters* 35(8):1291–1296.

Fu, B.Y., Xiong, J.H., Zhu, L.H. et al. 2007. Identification of functional candidate genes for drought tolerance in rice. *Molecular Genetics and Genomics* 278:599–609.

Garg, A.K., Kim, J.K., Owens, T.G. et al. 2002. Trehalose accumulation in rice plants confers high tolerance levels to different abiotic stresses. *Proceedings of National Academy of Science USA* 99:15898–15903.

Glick, B.R. 2003. Phytoremediation: Synergistic use of plants and bacteria to clean up the environment. *Biotechnology Advances* 21:383–393.

Goel, D., Singh, A.K., Yadav, V., Babbar, S.B., Bansal, K.C. 2010. Overexpression of Osmotin gene confers tolerance to salt and drought stresses in transgenic tomato (*Solanum lycopersicum* L.). *Protoplasma* 245:133–141.

Hamdia, A.B.E., Shaddad, M.A.K., Doaa, M.M. 2004. Mechanisms of salt tolerance and interactive effects of *Azospirillum brasilense* inoculation on maize cultivars grown under salt stress conditions. *Plant Growth Regulation* 44:165–174.

Heaton, A.C P., Rugh, C.L., Kim, T., Wang, N.J., Meagher, R.B. 2003. Toward detoxifying mercury-polluted aquatic sediments with rice genetically engineered for mercury resistance. *Environment Toxicology and Chemistry* 22:2940–2947.

Heaton, A.C.P., Rugh, C.L., Wang, N.J., Meagher, R.B. 2005. Physiological responses of transgenic merA-TOBACCO (*Nicotiana tabacum*) to foliar and root mercury exposure. *Water, Air and Soil Pollution* 161:137–155.

Hiraoka, S., Machiyama, A., Ijichi, M. et al. 2016. Genomic and metagenomic analysis of microbes in a soil environment affected by the 2011 Great East Japan Earthquake tsunami. *BMC Genomics* 17:53.

Howie, W., Joe, L., Newbigin, E., Suslow, T., Dunsmuir, P. 1994. Transgenic tobacco plants which express the chiA gene from *Serratia marcescens* have enhanced tolerance to *Rhizoctonia solani*. *Transgenic Research* 3:90–98.

Huang, J., Yang, M., Zhang, X. 2016. The function of small RNAs in plant biotic stress response. *Journal of Integrated Plant Biology* 58:312–327.

IARC Monographs. 2004. *IARC Monographs on the Evaluation of Carcinogenic Risks to Humans.* Available at: http://monographs.iarc.fr/ENG/Monographs/vo l83/mono83.pdf.

Igiri, B.E., Okoduwa, S., Idoko, G.O., Akabuogu, E.P., Adeyi, A.O., Ejiogu, I.K. 2018. Toxicity and bioremediation of heavy metals contaminated ecosystem from tannery wastewater: A review. *Journal of Toxicology* 2018:16.

Joutey, N.T., Sayel, H., Bahafid, W., El Ghachtouli, N. 2015. Mechanisms of hexavalent chromium resistance and removal by microorganisms. *Review of Environmental Contamination and Toxicology* 233:45–69.

Kamoun, S., Honee, G., Weide, R. et al. 1999. The fungal gene Avr9 and the oomycete gene inf1 confer avirulence to potato virus X on tobacco. *Molecular Plant–Microbe Interactions* 12:459–462.

Kannojia, P., Sharma, P., Kashyap, A. et al. 2017. Microbe-mediated biotic stress management in plants. doi:10.1007/978-981-10-6593-4_26.

Kaushal, M., Wani, S.P. 2016. Plant-growth-promoting rhizobacteria: Drought stress alleviators to ameliorate crop production in drylands. *Annals of Microbiology* 66:35–42.

Keller, H., Pamboukdjian, N., Ponchet, M. et al. 1999. Pathogen induced elicitin production in transgenic tobacco generates a hypersensitive response and non-specific disease resistance. *Plant Cell* 11:223–235.

Kern, M.F., Maraschin, S.D.F., Endt, D.V. et al. 2010. Expression of a chitinase gene from *Metarhizium anisopliae* in tobacco plants confers resistance against *Rhizoctonia solani*. *Applied Biochemistry and Biotechnology* 160:1933–1946.

Kishor, P., Hong, Z., Miao, G.H., Hu, C., Verma, D. 1995. Overexpression of δ-pyrroline-5-carboxylate synthetase increases proline production and confers osmotolerance in transgenic plants. *Plant Physiology* 108:1387–1394.

Kour, D., Rana, K.L., Yadav, N., Yadav, A.N., Rastegari, A.A., Singh, C., Negi, P., Singh, K., Saxena, A.K. 2019. Technologies for biofuel production: Current development, challenges, and future prospects. In: *Prospects of Renewable Bioprocessing in Future Energy Systems, Biofuel and Biorefinery Technologies* (Eds. Rastegari, A.A. et al.), Vol. 10. Springer, Berlin, pp. 1–50.

Kremer, R. 2005. The role of bioherbicides in weed management. *Biopesticides International* 1:127–141.

Lauge, R., De Wit, P. 1998. Fungal avirulence genes: Structure and possible functions. *Fungal Genetics and Biology* 24:285–297.

Lauge, R., Joosten, M.H., Haanstra, J.P., Goodwin, P.H., Lindhout, P., de Wit, P.J. 1998. Successful search for a resistance gene in tomato targeted against a virulence factor of a fungal pathogen. *Proceedings of the National Academy of Sciences USA* 95:9014–9018.

Li, X., Gao, Q., Liang, Y. et al. 2013. A novel salt-induced gene from sheep grass, LcSAIN2, enhances salt tolerance in transgenic Arabidopsis. *Plant Physiology and Biochemistry* 64C:52–59.

Liu, X., Wang, Z., Wang, L., Wu, R., Phillips, J., Deng, X. 2009. LEA 4 group genes from the resurrection plant *Boea hygrometrica* confer dehydration tolerance in transgenic tobacco. *Plant Science* 176:90–98.

Lorito, M., Harman, G.E., Hayes, C.K. et al. 1993. Chitinolytic enzymes produced by *Trichoderma harzianum*: Antifungal activity of purified endochitinase and chitobiosidase. *Phytopathology* 83:302–307.

Lorito, M., Peterbauer, C., Sposato, P., Mach, R.L., Strauss, J., Kubicek, C.P. 1996. Mycoparasitic interaction relieves binding of the Cre1 carbon catabolite repressor protein to promoter sequences of the ech-42 (endochitinase encoding) gene in *Trichoderma harzianum*. *Proceedings of the National Academy of Sciences USA* 93:14868–14872.

Maier, R.M., Soberón-Chávez, G. 2000. *Pseudomonas aeruginosa* rhamnolipids: Biosynthesis and potential applications. *Applied Microbiology and Biotechnology* 54:625–633.

Mickelbart, M.V., Paul, M., Hasegawa, P.M., Bailey-Serres, J. 2015. Genetic mechanisms of abiotic stress tolerance that translate to crop yield stability. *Nature Reviews Genetics* 16:237–251.

Moghaieb, A.E.R., Nakamura, A., Saneoka, H., Fujita, K. 2011. Evaluation of salt tolerance in ectoine-transgenic tomato plants (*Lycopersicon esculentum*) in terms of photosynthesis, osmotic adjustment, and carbon partitioning. *GM Crops* 2:58–65.

Mosa, A.K., Saadoun, I., Kumar, K., Helmy, M., Dhankher, P.O. 2016. Potential biotechnological strategies for the cleanup of heavy metals and metalloids. *Frontiers in Plant Science* 7:303.

Mostafiz, B., Suraiya, R., Mizanur, M., Rahman, M. 2012. Biotechnology: Role of microbes in sustainable agriculture and environmental health. *Internet Journal of Microbiology* 10:1937–8289.

Mourgues, F., Brisset, M.N., Chevreau, E. 1998. Strategies to improve plant resistance to bacterial diseases through genetic engineering. *Trends in Biotechnology* 16:203–210.

Nakayama, H., Yoshida, K., Ono, H., Murooka, Y., Shinmyo, A. 2000. Ectoine, the compatible solute of *Halomonas elongata*, confers hyperosmotic tolerance in cultured tobacco cells. *Plant Physiology* 122:1239–1247.

Ngumbi, E., Kloepper, J. 2014. Bacterial-mediated drought tolerance: Current and future prospects. *Applied Soil Ecology* 105:109–125.

Nriagu, J.O. 1989. A global assessment of natural sources of atmospheric trace metals. *Nature* 338:47–49.

Oh, K., Li, T., Cheng, H., Hu, X., He, C., Yan, L., Shinich, Y. 2013. Development of profitable phytoremediation of contaminated soils with biofuel crops. *Journal of Environmental Protection* 4:58–64.

Ortiz, N., Armadaa, E., Duque, E., Roldánc, A., Azcóna, R. 2015. Contribution of arbuscular mycorrhizal fungi and/or bacteria to enhancing plant drought tolerance under natural soil conditions: Effectiveness of autochthonous or allochthonous strains. *Journal of Plant Physiology* 174:87–96.

Park, B.J., Liu, Z., Kanno, A., Kameya, T. 2005. Genetic improvement of Chinese cabbage for salt and drought tolerance by constitutive expression of a B. napus LEA gene. *Plant Science* 169:553–558.

Park, C.M., Berry, J.O., Bruenn, J.A. 1996. High level secretion of a virally encoded antifungal toxin in transgenic tobacco plants. *Plant Molecular Biology* 30:359–366.

Pidlisnyuk, V., Stefanovska, T., Lewis, E.E., Erickson, L.E., Davis, L.C. 2014. Miscanthus as a productive biofuel crop for phytoremediation. *Critical Reviews in Plant Sciences* 33:1–19.

Pilon-Smits, E. 2005. Phytoremediation. *Annual Review of Plant Biology* 56:15–39.

Porcel, R., Ruiz-Lozano, J. 2004. Arbuscular mycorrhizal influence on leaf water potential, solute accumulation, and oxidative stress in soybean plants subjected to drought stress. *Journal of Experimental Botany* 55:1743–1750.

Rangel, D.S., Ana, I.C.M., Aída, A.R.H. et al. 2016. Simultaneous silencing of two arginine decarboxylase genes alters development in *Arabidopsis. Frontiers in Plant Science* 7:300.

Raymond, B., Federici, B.A. 2017. In defence of *Bacillus thuringiensis*, the safest and most successful microbial insecticide available to humanity—A response to EFSA. *FEMS Microbiology and Ecology* 93(7):fix084.

Rentel, M.C., Lecourieux, D., Ouaked, F. et al. 2004. OXI1 kinase is necessary for oxidative burst-mediated signaling in Arabidopsis. *Nature* 427:858–861.

Rugh, C.L., Wilde, H.D., Stack, N.M., Thompson, D.M., Summers, A.O., Meagher, R.B. 1996. Mercuric ion reduction and resistance in transgenic *Arabidopsis thaliana* plants expressing a modified bacterial merA gene. *Proceedings of the National Academy of Sciences USA* 93:3182–3187.

Sakamoto, A., Murata, A.N. 1998. Metabolic engineering of rice leading to biosynthesis of glycinebetaine and tolerance to salt and cold. *Plant Molecular Biology* 38:1011–1019.

Salt, D., Blaylock, J.M., Kumar, N., Dushenkov, V., Ensley, B., Chet, I., Raskin, I. 1995. Phytoremediation: A novel strategy for the removal of toxic metals from the environment using plants. *Biotechnology* (Nature Publishing Company) 13:468–474.

Sanghera, G.S., Wani, S.H., Singh, G., Kashyap, P.L., Singh, N.B. 2011. Designing crop plants for biotic stresses using transgenic approach. *International Journal of Plant Research* 24:1–25.

Sharma, N., Singhvi, R. 2017. Effects of chemical fertilizers and pesticides on human health and environment: A review. *International Journal of Agriculture, Environment and Biotechnology* 10:675–679.

Shirasawa, K., Takabe, T., Takabe, T., Kishitani, S. 2006. Accumulation of glycinebetaine in rice plants that overexpress choline monooxygenase from spinach and evaluation of their tolerance to abiotic stress. *Annals of Botany* 98:565–571.

Simontacchi, M., Galatro, A., Ramos-Artuso, F., Santa-Maria, G.E. 2015. Plant survival in a changing environment: The role of nitric oxide in plant responses to abiotic stress. *Frontier of Plant Science* 6:977.

Singh, C., Tiwari, S., Boudh, S., Singh, J.S. 2017a. Biochar application in management of paddy crop production and methane mitigation. In: *Agro-Environmental Sustainability: Managing Environmental Pollution* (Eds. Singh, J.S., Seneviratne, G.), 2nd ed. Springer, Switzerland, pp. 123–146.

Singh, C., Tiwari, S., Gupta, V.K., Singh, J.S. 2018. The effect of rice husk biochar on soil nutrient status, microbial biomass and paddy productivity of nutrient poor agriculture soils. *Catena* 171:485–493.

Singh, C., Tiwari, S., Singh, J.S. 2017b. Impact of rice husk biochar on nitrogen mineralization and methanotrophs community dynamics in paddy soil. *International Journal of Pure and Applied Bioscience* 5:428–435.

Singh, C., Tiwari, S., Singh, J.S. 2017c. Application of biochar in soil fertility and environmental management: A review. *Bulletin of Environment, Pharmacology and Life Sciences* 6:07–14.

Singh, C., Tiwari, S., Singh, J.S. 2019. Biochar: A sustainable tool in soil 2 pollutant bioremediation. In: *Bioremediation of Industrial Waste for Environmental Safety* (Eds. Bharagava, R.N., Saxena, G.). Springer, Berlin, pp. 475–494.

Singh, J., Seneviratne, G. 2017. *Agro-Environmental Sustainability Volume-2: Managing Environmental Pollution*, Vol. 2. Springer, Dodrecht, The Netherlands, p. 229.

Singh, K.D., Faridi, S.H., Lodhi, M. 2017. Possible mechanisms of development of carcinoma of breast in patients with early-onset cataract. *International Surgery Journal* 4:1394–1397.

Singh, R., Singh, P., Sharma, R. 2014. Microorganism as a tool of bioremediation technology for cleaning environment: A review. *Proceedings of the International Academy of Ecology and Environmental Sciences* 4:1–6.

Sorty, A.M., Meena, K.K., Choudhary, K., Bitla, U.M., Minhas, P.S., Krishnani, K.K. 2016. Effect of plant growth promoting bacteria associated with halophytic weed (*Psoralea corylifolia* L.) on germination and seedling growth of wheat under saline conditions. *Applied Biochemistry and Biotechnology* 180:872–882.

Strittmatter, G., Janssens, J., Opsomer, C., Botterman, J. 1995. Inhibition of fungal disease development in plants by engineering controlled cell death. *Biotechnology* 13:1085–1089.

Tepfer, D., Boutteaux, C., Vigon, C. et al. 1998. Phytophthora resistance through production of a fungal protein elicitor (β-cryptogein) in tobacco. *Molecular Plant–Microbe Interactions* 11:64–67.

Thomas, J.C., Sepahi, M., Arendall, B., Bohnert, H.J. 1995. Enhancement of seed germination in high salinity by engineering mannitol expression in *Arabidopsis thaliana*. *Plant Cell Environment* 18:801–806.

Tiwari, S., Prasad, V., Chauhan, P.S., Lata, C. 2017. *Bacillus amyloliquefaciens* confers tolerance to various abiotic stresses and modulates plant response to phytohormones through osmoprotection and gene expression regulation in rice. *Frontiers in Plant Science* 8:1510.

Tiwari, S., Singh, C., Boudh, S., Rai, P.K., Gupta, V.K., Singh, J.S. 2019a. Land use change: A key ecological disturbance declines soil microbial biomass in dry tropical uplands. *Journal of Environmental Management* 242:1–10.

Tiwari, S., Singh, C., Singh, J.S. 2018. Land use changes: A key ecological driver regulating methanotrophs abundance in upland soils. *Energy, Ecology, and the Environment* 3:355–371.

Tiwari, S., Singh, C., Singh, J.S. 2019b. Wetlands: A major natural source responsible for methane emission. In: *Restoration of Wetland Ecosystem: A Trajectory Towards a Sustainable Environment* (Eds. Upadhyay, A.K. et al.). Springer, Berlin, pp. 59–74.

Tobias, C.M., Oldroyd, G.E., Chang, J.H., Staskawicz, B.J. 1999. Plants expressing the Pto disease resistance gene confer resistance to recombinant PVX containing the avirulence gene AvrPto. *Plant Journal* 17:41–50.

Turpeinen, R., Kairesalo, T., Haggblom, M. 2002. Microbial activity community structure in arsenic, chromium and copper contaminated soils. *Journal of Environmental Microbiology* 35:998–1002.

Van Wees, S.C.M., van der Ent, S., Pieterse, C.M.J. 2008. Plant immune responses triggered by beneficial microbes. *Current Opinion in Plant Biology* 11:443–448.

Velásquez, L., Dussan, J. 2009. Biosorption and bioaccumulation of heavy metals on dead and living biomass of *Bacillus sphaericus*. *Journal of Hazardous Material* 167:713–716.

Vimal, S., Singh, J., Arora, N., Singh, S. 2017. Soil–plant–microbe interactions in stressed agriculture management: A review. *Pedosphere* 27:177–192.

Wang, Y., Dong, C., Xue, Z., Jin, Q., Xu, Y. 2016. De novo transcriptome sequencing and discovery of genes related to copper tolerance in *Paeonia ostii*. *Gene* 576:126–135.

Wang, Y., Jiang, J., Zhao, X. et al. 2006. A novel lea gene from *Tamarix androssowii* confers drought tolerance in transgenic tobacco. *Plant Science* 171:655–662.

Wei, Y.F., Li, T., Li, L.F., Wang, J.L., Cao, G.H., Zhao, Z.W. 2016. Functional and transcript analysis of a novel metal transporter gene EpNramp from a dark septate endophyte (*Exophiala pisciphila*). *Ecotoxicology and Environment Safety* 124:363–368.

Wu, G., Shortt, B.J., Lawrence, E.B., Levine, E.B., Fitzsimmons, K.C., Shah, D.M. 1995. Disease resistance conferred by expression of a gene encoding H2O2-generating glucose oxidase in transgenic potato plants. *Plant Cell* 9:1357–1368.

Xie, P., Hao, X., Herzberg, M., Luo, Y., Nies, D.H., Wei, G. 2015. Genomic analyses of metal resistance genes in three plant growth promoting bacteria of legume plants in Northwest mine tailings, China. Journal Environment Science 27:179–187.

Yang, X., Liang, Z., Lu, C. 2005. Genetic engineering of the biosynthesis of glycinebetaine enhances photosynthesis against high temperature stress in transgenic tobacco plants. *Plant Physiology* 138:2299–2309.

Yao, D., Zhang, X., Zhao, X. et al. 2011. Transcriptome analysis reveals salt-stress regulated biological processes and key pathways in roots of cotton (*Gossypium hirsutum* L.). *Genomics* 98:47–55.

Yolcu, S., Ozdemir, F., Güler, A., Bor, M. 2016. Histone acetylation influences the transcriptional activation of POX in *Beta vulgaris* L. and *Beta maritima* L. under salt stress. *Plant Physiology and Biochemistry* 100:37–46.

Zhao, D., Li, T., Shen, M., Wang, J., Zhao, Z. 2015. Diverse strategies conferring extreme cadmium (Cd) tolerance in the dark septate endophyte (DSE), *Exophiala pisciphila*: Evidence from RNA-seq data. *Microbiological Research* 170:27–35.

Zhu, B., Peng, R., Xiong, A. et al. 2012. Transformation with a gene for myo-inositol O-methyltransferase enhances the cold tolerance of *Arabidopsis thaliana*. *Biologia Plantarum* 56:135–139.

Zwolak, I., Zaporowska, H. 2012. Selenium interactions and toxicity: A review. Selenium interactions and toxicity. *Cell Biology and Toxicology* 28:31–46.

12 Interaction Scenario of Insects, Plants, and Mycorrhizal Fungi

V.K. Mishra, Usha, Umesh Pankaj, and M. Soniya

CONTENTS

12.1 INTRODUCTION

Insects are programmed to recognize and rapidly respond to patterns of host cues. Particularly, specialist insect species have to find specific plant species on which they can feed and reproduce (host plants). Some plant species do not support feeding and/ or reproduction of insects (non-host plants). Thus, in an environment with a changing availability and quality of host plants, phytophagous insects are under selection pressure to find quality hosts. To maximize their fitness, they need to locate suitable plants and avoid unsuitable hosts. Thus, they have evolved a finely tuned sensory system for the detection of host cues, and a nervous system capable of integrating inputs from sensory neurons with a high level of spatio-temporal resolution time and space, which also influence plant responses to insects. The time dimensions are of major significance, since whether or not odors arrive simultaneously at the antenna can change the type of behavioral response elicited in the insect. The huge number of species of flowering plants on our planet (approximately 275,000) is thought to be the result of adaptive radiation driven by the coevolution between plants and their beneficial animal pollinators (Yuan et al. 2013). The fossil record shows that pollination originated 250 million years ago (Labandeira 2013). Some plants have evolved with their pollinators and produce olfactory messages which make them unique for their specific pollinators (Grajales-Conesa et al. 2011). For example, certain orchid flowers mimic aphid alarm pheromones to attract hoverflies for pollination (Stoekl et al. 2011). Furthermore, insect herbivores can drive real-time ecological and evolutionary change in plant populations. Recent studies provide evidence for the rapid evolution of plant traits that confer resistance to herbivores when herbivores are present but the evolution of traits confer increased competitive ability when herbivores are absent (Agrawal et al. 2012; Züst et al. 2012). While phytophagous insects have been adapting to exploit their hosts, the plants have simultaneously been evolving defensive systems to counteract herbivore attack (Anderson and Mitchell-Olds 2011; Johnson 2011). The evolutionary development of the tremendous variety of insects that we are able to see today is from the fossil record. From this, we can speculate that the earliest insects were wingless, Thysanura-like forms that abounded in the Silurian and Devonian periods. The major advance made by their descendants was the evolution of wings, facilitating dispersal and, therefore, colonization of new habitats. During the Carboniferous and Permian periods, there was a massive adaptive radiation of winged forms, and it was at this time that most modern orders had their beginnings. Although members of many of these orders retained a life history similar to that of their wingless ancestors, in which the change from juvenile to adult form was gradual (hemimetabolous or exopterygote), in other orders a life history evolved in which the juvenile and adult phases were separated by a pupal stage (holometabolous or endopterygote). The great advantage of having a pupal stage is that the juvenile and adult stages can become very different from each other in their habits, thereby avoiding competition for the same resources. The insects and mycorrhizal fungi interact with one another in complex ways that are beneficial to fungi, insects, and host plants (Gehring and Whitham 2002; Gange and Bower 1997; Gange 2007). There is a need to place insects, mycorrhizal fungi, and their host plants into a broader community context that includes other interactions; and potential for the mediation of interactions between insects and mycorrhizal fungi by fungus species or host plant.

12.2   INTERACTION BETWEEN PLANTS AND INSECTS

It is a well-known fact that plants and insects have coexisted together for about 4 million years. This has resulted in the development of certain offensive and defensive strategies by both plants and insects for their existence. In the natural ecosystem, the survival of plants through the evolutionary time is mainly due to their own defense mechanism which deters the feeding of herbivores. Moreover, plants also have the capacity to either tolerate or escape attack. On the other hand, with regards to insects, their selective ability, such as mechanical adaptations, detoxification, sequestration, host manipulation, association with microorganisms, and avoidance enables them to overcome these plant defense mechanisms and continue feeding.

In natural ecosystems, phytophagous insects have coexisted in a complex relationship with plant communities. Different species of plant-feeding insects had to seek out their host plants from the mixed wild vegetation. For this purpose, insects made use of various types of olfactory stimuli which work from long distances and taste stimuli which work in the vicinity of the host plants. In this search, they

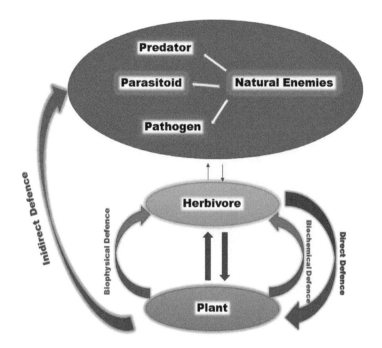

FIGURE 12.1   Tri-trophic interaction between plant, herbivores, and natural enemies.

had to face the dangers of annihilation from various climatic and biotic agencies. Therefore, the damage caused by insects was probably quite limited. Pest problems originated with the origin of agriculture. As soon as the land was cleared of natural vegetation and replaced by a single species of food plant, humans came into conflict with phytophagous insects. These insects and other organisms feeding on the valuable crops which humans planted were called "pests". Not only did the insect pests originate with agriculture, but the intensity of pests continued to increase with the intensification of agriculture.

The evolution of land plants (especially flowering plants) are a major force driving the diversity of insects. As the diversity of land plants has increased, the diversity of insects has also risen. Interaction between plants and insects is an example of co-evolution (Figure 12.1). Co-evolution can occur between: a single plant and a single insect, or a single plant and a group of insects, or a group of insects, and a group of closely related plants. Interactions are often examined from the plant's perspective. Insect–plant interaction can occur in one of two ways, depending on whether the interaction is beneficial to both parties (mutualism) or is beneficial to insects but harmful to plants (herbivory). Phytophagous insects have coexisted in a complex relationship with plant communities.

12.3   INTERACTION BETWEEN FUNGI, INSECTS, AND PLANTS

The majority of mycorrhizal fungi, plants, and insect studies have focused on insect herbivores and observed many ways in which mycorrhizal fungi can influence the

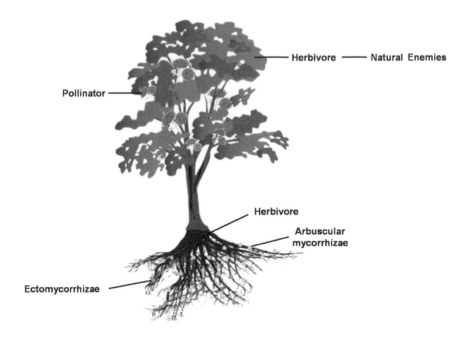

FIGURE 12.2 Interaction among herbivores/pollinator, ecto-mycorrhiza, and arbuscular-mycorrhizal fungi on common host plant.

interaction between plants and their herbivores (Figure 12.2). Both ecto-mycorrhizal (EM) and arbuscular mycorrhizal (AM) fungi have been shown to increase plant size and alter plant quality through changes in nutrient content (Smith and Read 1997). Mycorrhizal fungi may also alter plant–herbivore interactions through changes in constitutive and inducible defenses as well as tolerance to herbivory (Bennett et al. 2006). There are a growing number of studies focused on the indirect effects of EM and AM fungi on plant herbivores with the following basic protocol: plants are grown with a single mycorrhizal fungal species (for AM fungal studies, species are most commonly from the genus *Glomus*) and subjected to herbivory (often by a single herbivore species), and various traits (herbivore survival or growth, plant chemical content, or plant tolerance to herbivory) are measured and compared with control plants not inoculated with mycorrhizal fungi. However, this narrow glimpse into the role of mycorrhizal fungi in plant–herbivore interactions has largely ignored the variation in the ecology of mycorrhizal fungal species (Hart and Reader 2002, 2005; Klironomos 2003; Karst et al. 2008; Singh et al. 2018, 2017a, b, c; 2019; Tiwari et al. 2018, 2019a, b; Kour et al. 2019a). Species of mycorrhizal fungi can vary greatly (from parasitic to mutualistic) in the benefit they provide to hosts (Klironomos 2003; Karst et al. 2008), and in AM fungi, variation in host growth benefit is thought to derive from variation in colonization or competitive ability in host roots (Hart and Reader 2002).

12.3.1 MUTUALISTIC CO-EVOLUTION

- In mutualistic co-evolution, the two parties evolve in such a way to enhance the effectiveness of the interaction.

- This type of co-evolution is common in interactions between plants and their insect pollinators, as well as in some other specialized cases.
- The relationship between plant and their insect pollinators are mutualistic in which the plants provide sugar or amino acid-rich nectar and pollen to insects and insects, in turn assisting the plants by transmitting the male gametes to the female flowers (cross-pollination).
- If the co-evolution is "extreme" enough, the two partners may become completely dependent on each other (obligate mutualism like ants and *Acacia*). Many species of plants in the genus *Acacia* have mutualistic ants of the genus *Pseudomyrmex* associated with them. Ants prevent herbivores from feeding on plants by killing them or chasing them off. Ants also remove any other plants growing nearby or on "their" *Acacia* (which decreases competition). On the other hand, plants provide a safe home to ants (ants live in the trunk, and enter by chewing through the large hollow thorns). It also provides two food sources: (a) extra floral nectarines, and (b) Beltian bodies which are high in proteins (Magallon and Sanderson 2001).
- *Effects of obligate mutualism*: Effects of ant removal on *Acacia*. When ants were removed and kept off *Acacia* plants, the growth rate of ants decreased dramatically due to the unavailability of food resources. The number of herbivorous insects increased dramatically; as a result, survival of the plant over the course of a year declined drastically.

12.3.2 A Detailed Study of Figure and Figure Wasp, *Blastophaga Psenes* (Agaonitae: Hymenoptera) Mutualism

Fig is pollinated by fig wasp only. There is no other mode of pollination. There are two types of fig, caprifig and Symrnafig. (i) *Caprifig*: (a) is a wild type of fig, i.e. not edible; (b) it has both male and female flowers; (c) pollen is produced in plenty; and (d) it is the natural host of the fig wasp. (ii) *Symrnafig*: (a) is the cultivated type of fig, i.e. edible; (b) it has only female flowers; (c) pollen is not produced; and (d) it is not the natural host of the fig wasp.

- Fig wasp: male – wingless, present in caprifig; and female – winged.
- Female wasp lays eggs in caprifig, larvae develop in galls in the base of the flowers.
- Male mates with female even when the female is inside the gall.
- Mated female wasp emerges out of flower (caprifig) with a lot of pollen dusted around its body.
- The female fig wasp enters Symrnafig with lot of pollen and deposits it on the stigma.
- It cannot oviposit in the ovary of Symrnafig which is deep-seated.
- It again moves to caprifig for egg-laying. In this process, Symrnafig is pollinated.
- Caprifig will be planted next to Symrnafig to aid in pollination.

12.3.3   Herbivores and Herbivory: A Negative Interaction

Herbivores are organisms that are anatomically and physiologically adapted to eat plant-based food, principally autotrophs such as plants, algae, and photosynthesizing bacteria. Herbivory is the act of feeding by herbivores. Examples of herbivores are: mammals, birds (parakeet, parrot, scarlet, goose, etc.), reptiles like tortoise, invertebrates like insects (grass hoppers, butterflies, leaf hoppers, tree hoppers, aphids, caterpillars, etc.), or other invertebrates like garden snails, earthworms, etc. Herbivores form an important link in the food chain. Insect-eating plants are referred to as phytophagous species. Phytophagous insects may be monophagous, oligophagous or polyphagous.

12.3.3.1   Categories of Phytophagous Insects
- Leaf chewers (caterpillars, coleopterans, and orthopterans)
- Plant miners and plant borers (larvae of dipterans and lepidopterans)
- Sap suckers (hemipterans)
- Seed feeders (harvester ants, coleopterans, and dipterans)
- Nectar or pollen feeders (hymenopterans, lepidopterans, coleopterans, dipterans, and thysanopterans)
- Gall formers (hemipterans, dipterans, or hymenopterans)

12.3.3.2   Plant Defense against Herbivory
- Plant defense is a trait which increases survival and/or reproduction (fitness) of plants under pressure of consumption by herbivores.
- Defense can be divided into two main categories, i.e. tolerance and resistance.
- Tolerance is the ability of a plant to withstand damage without a reduction in fitness.
- Resistance is the ability of a plant to reduce the amount of damage caused by herbivores and can occur by avoiding space and time (Milchunas and Noy-Meir 2002).
- Defenses can either be constitutive, i.e. always present in the plant or induced i.e. produced or translocated by the plant following damage or stress (Edwards and Wratten 1985; Nishida 2002).
- Mechanisms of plant defenses are given below:
    1. *Physical defenses (mechanical defenses)* – These are the physical structures that act as barriers against herbivores. For example, thorns found on roses, acacia trees, spines on cactuses, small hair-like structure known as trichomes may cover leaves or stems and are especially effective against invertebrate herbivores.
    2. *Chemical defenses* – These are due to secondary metabolites produced by the plant that deter herbivores (Tilmon 2008).
        - Plants normally maintain a baseline level of metabolites. Plants respond to insect damage by releasing a range of volatiles from the damaged site. These volatiles attract parasitoids and predators to the plant under attack. Parasitoids and predators pick up these biochemical cues with the help of specific receptors present

in their antenna. Each host plant-insect pest pair is believed to release a range of different and combination specific volatiles (Rashid et al. 2012).

– *The adaptation dance* – The back and forth relationship of plant defense and herbivore offense can be seen as a sort of "adaptation dance" in which one partner makes a move and the other counters it (Karban and Agrawal 2002). This reciprocal change drives co-evolution between many plants and herbivores, resulting in a "co-evolutionary arms race" (Mead et al. 1985).

12.3.4   Molecular Basis of Insect–Plant Interactions

- The ultimate goal of ecology is to understand how the traits of an individual contribute to its fitness in terms of reproductive success.
- This was investigated by comparing the performance of individuals or populations that differed in certain traits without (much) information on the underlying mechanisms and genes.
- Recent developments in molecular genetics have opened stimulating new avenues for ecologists through a molecular genetic approach.
- With the sequencing of plant genomes and the availability of well-characterized mutants and genetically modified traits that mediate interactions between plants and their biotic community members, ecologists can now address the ecological function of individual traits in very precise ways (Dicke 2004).

12.4   CONCLUSION

Thus it can be concluded that there exists an interaction between plants and insects. Insects play one of the most important roles in their ecosystems, which include many roles, such as soil turning and aeration, dung burial, pest control, pollination, and wildlife nutrition. This has resulted in the development of certain offensive and defensive strategies by both plants and insects for their existence. The evolution of land plants (especially flowering plants) is a major force driving the diversity of insects. With the increase in diversity of land plants, the diversity of insects has also increased. Interaction between plants and insects is an example of coevolution. Insect–plant interaction can occur in one of two ways, depending on whether the interaction is beneficial to both parties (mutualism) or is beneficial to insects but harmful to plants (herbivory). Insects play one of the most important roles in their ecosystems, which include many roles, such as soil turning and aeration, dung burial, pest control, pollination, and wildlife nutrition.

Interactions among insect-mycorrhizal fungus-plant have shown great community perspective in both above and below ground in a natural habitat with a complex community of organisms. However, research conducted under an isolated environment may not represent the whole story of these interactions (insects, plants, and mycorrhizal fungi). Further research is needed on the mechanisms of insect-plant-mycorrhizal fungal interactions, thereby enabling predictions as to their place in an evolutionary context.

REFERENCES

Agrawal, A.A., Hastings, A.P., Johnson, M.T., Maron, J.L., Salminen, J.P. 2012. Insect herbi-
   vores drive real-time ecological and evolutionary change in plant populations. *Science*
   338:113–116.
Anderson, J.T., Mitchell-Olds, T. 2011. Ecological genetics and genomics of plant defences:
   Evidence and approaches. *Functional Ecology* 25:312–324.
Bennett, A.E., Alers-Garcia, J., Bever, J.D. 2006. Three way interactions among mutualistic
   mycorrhizal fungi, plants, and plant enemies: Hypotheses and synthesis. *The American
   Naturalist* 167:141–152.
Dicke, A.E. 2004. Nutritional interactions in insect-microbial symbioses: Aphids and their
   symbiotic bacteria *Buchnera*. *Annual Review of Entomology* 43:17–37.
Edwards, P.J., Wratten, S.D. 1985. Induced plant defences against grazing: Fact or artefact?
   Oikos 44:70–74.
Gange, A.C. 2007. Insect-mycorrhizal interactions: Patterns, processes, and consequences.
   In: T. Ohgushi, T.P. Craig, P.W. Price (Eds.), *Ecological Communities: Plant Mediation
   in Indirect Interaction Webs*. London, United Kingdom: Cambridge University Press,
   pp. 124–143.
Gange, A.C., Bower, E. 1997. Interactions between insects and mycorrhizal fungi. In: A.C.
   Gange, V.K. Brown (Eds.), *Multitrophic Interactions in Terrestrial Systems*. Oxford,
   United Kingdom: Blackwell, pp. 115–132.
Gehring, C.A., Whitham, T.G. 2002. Mycorrhizae herbivore interactions: Population and
   community consequences. In: M.G.A. van der Heijden, I.R. Sanders (Eds.), *Mycorrhizal
   Ecology*. Berlin, Germany: Springer, pp. 295–320.
Grajales-Conesa, J., Melendez-Ramirez, V., Cruz-Lopez, L. 2011. Floral scents their interac-
   tion with insect pollinators. *Revista Mexicana De Biodiversidad* 82:1356–1367.
Hart, M.M., Reader, R.J. 2002. Taxonomic basis for variation in the colonization strategy of
   arbuscular mycorrhizal fungi. *New Phytologist* 153:335–344.
Hart, M.M., Reader, R.J. 2005. The role of the external mycelium in early colonization for
   three arbuscular mycorrhizal fungal species with different colonization strategies.
   Pedobiologia 49:269–279.
Johnson, M.T.J. 2011. Evolutionary ecology of plant defences against herbivores. *Functional
   Ecology* 25:305–311.
Karban, R., Agrawal, A.A. 2002. Herbivore offense. *Annual Review of Ecology and
   Systematics* 33:641–664.
Karst, J., Marczak, L., Jones, M.D., Turkington, R. 2008. The mutualism-parasitism con-
   tinuum in ectomycorrhizas: A quantitative assessment using meta-analysis. *Ecology*
   89:1032–1042.
Klironomos, J.N. 2003. Variation in plant response to native and exotic arbuscular mycor-
   rhizal fungi. *Ecology* 84:2292–2301.
Kour, D., Rana, K.L., Yadav, N., Yadav, A.N., Rastegari, A.A., Singh, C., Negi, P., Singh, K.,
   Saxena, A.K. 2019a. Technologies for biofuel production: Current development, chal-
   lenges, and future prospects. In: A.A. Rastegari et al. (Eds.), *Prospects of Renewable
   Bioprocessing in Future Energy Systems, Biofuel and Biorefinery Technologies*, Vol.
   10. Berlin: Springer, pp. 1–50.
Labandeira, C.C. 2013. A paleobiologic perspective on plant–insect interactions. *Current
   Opinion in Plant Biology* 16:414–421.
Magallón, S., Sanderson, M.J. 2001. Absolute diversification rates in angiosperm clades.
   Evolution 55:1762–1780.
Mead, R.J., Oliver, A.J., King, D.R., Hubach, P.H. 1985. The co-evolutionary role of fluoroac-
   etate in plant–animal interactions in Australia. *Oikos* 44:55–60.

Milchunas, D.G., Noy-Meir, I. 2002. Grazing refuge, external avoidance of herbivory and plant diversity. *Oikos* 99:113–130.

Nishida, R. 2002. Sequestration of defensive substances from plants by Lepidoptera. *Annual Review of Entomology* 47:57–92.

Rashid, A., Dicke, M., Van Poecke, RM.P. 2012. Mechanisms of plant defense against insect herbivores. *Plant Signaling and Behavior* 7:1306–1320.

Singh, C., Tiwari, S., Boudh, S., Singh, J.S. 2017a. Biochar application in management of paddy crop production and methane mitigation. In: J.S. Singh, G. Seneviratne (Eds.), *Agro-Environmental Sustainability: Managing Environmental Pollution*, 2nd ed. Switzerland: Springer, pp. 123–146.

Singh, C., Tiwari, S., Gupta, V.K., Singh, J.S. 2018. The effect of rice husk biochar on soil nutrient status, microbial biomass and paddy productivity of nutrient poor agriculture soils. *Catena* 171:485–493.

Singh, C., Tiwari, S., Singh, J.S. 2017b. Impact of rice husk biochar on nitrogen mineralization and methanotrophs community dynamics in paddy soil. *International Journal of Pure and Applied Bioscience* 5:428–435.

Singh, C., Tiwari, S., Singh, J.S. 2017c. Application of biochar in soil fertility and environmental management: A review. *Bulletin of Environment, Pharmacology and Life Sciences* 6:07–14.

Singh, C., Tiwari, S., Singh, J.S. 2019. Biochar: A sustainable tool in soil 2 pollutant bioremediation. In: R.N. Bharagava, G. Saxena (Eds.), *Bioremediation of Industrial Waste for Environmental Safety*. Berlin: Springer, pp. 475–494.

Smith, S.E., Read, D.J. 1997. *Mycorrhizal Symbiosis*, 2nd ed. London, United Kingdom: Academic.

Stoekl, J., Brodmann, J., Dafni, A., Ayasse, M., Hansson, B.S. 2011. Smells like aphids: Orchid flowers mimic aphid alarm pheromones to attract hoverflies for pollination. *Proceedings of the Royal Society B: Biological Sciences* 278:1216–1222.

Tilmon, K.J. (Ed). 2008. *Specialization, Speciation, and Radiation: The Evolutionary Biology of Herbivorous Insects*. USA: Univ. California.

Tiwari, S., Singh, C., Boudh, S., Rai, P.K., Gupta, V.K., Singh, J.S. 2019a. Land use change: A key ecological disturbance declines soil microbial biomass in dry tropical uplands. *Journal of Environmental Management* 242:1–10.

Tiwari, S., Singh, C., Singh, J.S. 2018. Land use changes: A key ecological driver regulating methanotrophs abundance in upland soils. *Energy, Ecology, and the Environment* 3(6):355–371.

Tiwari, S., Singh, C., Singh, J.S. 2019b. Wetlands: A major natural source responsible for methane emission. In: A.K. Upadhyay et al. (Eds.), *Restoration of Wetland Ecosystem: A Trajectory Towards a Sustainable Environment*. Berlin: Springer, pp. 59–74.

Yuan, Y.W., Byers, K.J.R.P., Bradshaw, J.R.H.D. 2013. The genetic control of flower–pollinator specificity. *Current Opinion in Plant Biology* 16:422–428.

Züst, T., Heichinger, C., Grossniklaus, U., Harrington, R., Kliebenstein, D.J., Turnbull, L.A. 2012. Natural enemies drive geographic variation in plant defenses. *Science* 338:116–119.

13 Tannery Wastewater
A Major Source of Residual Organic Pollutants and Pathogenic Microbes and Their Treatment Strategies

Ashutosh Yadav, Pooja Yadav, Abhay Raj, Luiz Fernando R. Ferreira, Ganesh Dattatraya Saratale, and Ram Naresh Bharagava

CONTENTS

13.1   INTRODUCTION

Global trade was becoming liberalized through the 1970s and 80s. From a global turnover of $4 billion, the leather and leather products industry grew to an estimated $70 billion in 2000 (UNIDO 2010). Leather is an intermediate industrial product, with numerous applications in downstream sectors of the consumer industry. For the latter, leather is often the major material input, and is cut and assembled into shoes, clothing, leather goods, furniture and many other items of daily use (Ballard 2001; Lofrano et al. 2013; Yadav et al. 2016a). Different applications require different types of leather for other assorted goods flown from industrialized countries in the West to the developing countries of the East. Such applications are in heavy demand, primarily motivated by cost considerations.

Unfortunately, rapid industrialization and modernization around the world have resulted in the release of various types of wastes containing different types of toxic pollutants into the environment (Mishra and Bharagava 2016). Water is the most important and essential element on the earth for sustainable life. Industrialization and the extraction of natural resources have resulted in large-scale environmental contamination and severe health hazards to animals and human beings. The contamination of soils, groundwater, sediments and surface water with toxic metals and chemicals is one of the major problems facing the world today (Alam et al. 2009).

Alarming growth in population, urbanization, industrialization, agricultural activities, climate change, socio-economic growth, along with high living standards, have generated an unrelenting demand for water resources. However, due to this heavy population load and human activity, the quality of our water resources is continually deteriorating. Every year, millions of tons of tannery wastes containing thousands of organic (phthalates, biphenyls, phenols, tannin oils, greases), inorganic (heavy metals, compounds containing, fluoride, phosphate, sulfate, nitrate) and biological pollutants or contaminants (virus, bacteria, fungi, algae, amoebas, planktons) have been released into our water resources (Haydar and Aziz 2009; Chowdhary et al. 2018). Nowadays, the tannery industry occupies pride of place, due to its massive potential for employment growth, and export, and thus plays an important role in the economy of many developing countries (Yadav et al. 2019; Lofrano et al. 2013; Lefebvre et al. 2006).

India is the third largest leather producer in the world (Ramteke et al. 2010). There are more than 3,000 tanneries in India, and most of them (nearly 80%) are engaged in the chrome tanning process (Shukla et al. 2009). The tannery industry is a major industry on an international scale and is of significant economic importance as an agro-based sector, producing a host of products in one of the world's finest natural materials. The conversion of animal hides/skin into useful leather artifacts may be man's oldest technology, hence tanning is claimed to be the second oldest profession in the world. India is the third largest producer of leather and its valued products around the world (Ramteke et al. 2010). The production of leather and its products by the use of raw skins/hides is one of the oldest technologies of human civilization. The processing of leather is a high water consuming process, utilizing large volumes of fresh water. Only 30% of this fresh water is used in leather manufacturing; the remaining 70% water is discharged into the environment as wastewater containing various toxic pollutants and thus becomes a great challenge for the environment (Chowdhury et al. 2015, 2013).

Our earth is getting progressively polluted with different types of inorganic and organic compounds, primarily as a result of anthropogenic activities (Mishra et al. 2019; Yadav et al. 2017). Natural resources (soil, air and water) have faced a tremendous amount of pressure due to the rising human population and their associated activity. The uncontrolled discharge of effluent from industries in water has led to a rapid increase in effluent concentration which alters the nature of the ecosystem and adversely affects the health of human beings, plants and animals (Yadav et al. 2019; Chowdhary et al. 2018; Matsumoto et al. 2006). There are several ways by which a huge number of toxic compounds enter the environment. The tannery industry is a major industry on an international scale and is of significant economic importance as an agro-based sector, producing a host of products in one of the world's finest natural materials. Rapid industrialization and modernization in developing countries, different types of industrial wastes (solid and liquid) containing a large number of organic pollutants and toxic metals in high concentration are directly or indirectly discharged into the environment without adequate treatment (Saxena et al. 2016; Dixit et al. 2015; Chandra et al. 2011).

In India, the tannery industry plays a prominent role, contributing 15% of the total production capacity of the world. There are more than 3000 tanneries, producing about 500,000 tons of hide and 314 of skin (Verma et al. 2008; Kumari et al. 2014). In Unnao, Kanpur CEPT for tannery, received a credit line at a total of 19.3 million USD from the World Bank (the company operated under Unnao Pollution Controls Co. Ltd.). This includes primary and secondary treatments but various organic pollutants do not degrade during secondary and primary treatment, enter the major water source, and consequently, contaminate both water and soil (Yadav et al. 2016a).

For the past few years, anthropogenic activities have contributed to ecological contamination, causing an increase in the concentration of various heavy metals; for example, chromium, lead, cadmium and nickel, etc. Industrial waste is disposed of in the nearby water bodies and ultimately absorbed in the surroundings. These heavy metals are utilized in several industries such as tanneries, electroplating, mining, textiles, pesticide industries, etc. (Mishra et al. 2019; Yadav et al. 2017; Dixit et al. 2015). Environmental pollution has caused various illnesses because the recommended threshold limit value given by the WHO (World Health Organization) has not been respected. These industries are releasing their effluents continuously in their surroundings, which is a leading threat to environmental safety. Because chromium is a non-degradable pollutant, it persists in the environment (Bharagava and Mishra 2017; Yadav et al. 2016b).

The residual organic pollutants present in tannery wastewater (TWW) may allow a variety of pathogenic microbes to flourish and contaminate the receiving water bodies as well as surrounding environments because they may act as a source of nutrients for a diversity of microbes (Yadav et al. 2016a; Verma et al. 2008). However, various authors have reported that numerous residual organic pollutants and pathogenic microbes also remain in wastewater discharged from different industries even after the secondary treatment process, enter the environment and persist in nature for long periods (Chandra et al. 2011; Naraian et al. 2012; Kapley et al. 2001).

Waterborne pathogens infect around 250 million people each year, resulting in 10–20 million deaths (Saxena and Bharagava 2015). Many of these infections occur in developing nations that suffer from lower levels of sanitation, problems associated with low socio-economic conditions and less public health awareness than in more developed nations. The risk from microbial pathogens in water necessitates the proper monitoring of wastewaters and contaminated aquatic resources for various types of organic pollutants and microbial pathogens (Yadav et al. 2016a).

Therefore, this chapter provides detailed information on the various types of organic and inorganic pollutants, as well as different types of pathogenic and non-pathogenic microbes present in tannery wastewater, along with various approaches to treatment, which can be used for the removal of the aforementioned pollutants for the safety of the environment as well as human and animal health.

13.2   VARIOUS STEPS AND CHEMICALS USED IN THE TANNING PROCESS

The tanning process aims to transform skins into stable and rot-resistant products, namely leather (Figure 13.1). There are four major groups of sub-processes such as pre-tanning, tanning, wet finishing and finishing Operations which are applied to obtain the finished leather (U.S. EPA 1986; Tunay et al. 1995; Cooman et al. 2003).

13.2.1   PRE-TANNING (BEAM HOUSE OPERATIONS)

The pre-tanning process is also known as a "beam house operation" as it is carried out in a beam house for the cleaning and conditioning of raw hides/skins leading to the generation of the major part of effluent.

13.2.1.1   Soaking

Raw skins are treated with water and small quantities of imbibing substances in order to hydrate the skin proteins, to solubilize the denatured proteins, to eliminate

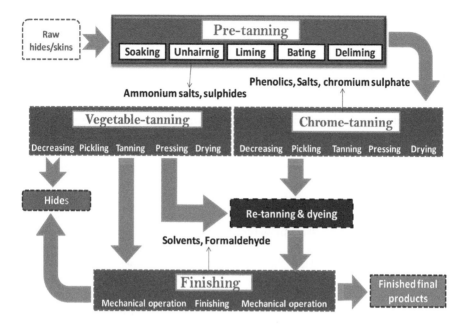

FIGURE 13.1 Different steps and chemicals used in a typical leather industry to obtain a finished leather product from raw hides/skins.

the salt used in the preservation step, to hydrate and to open the contract fibers of the dried skins, and eliminate the residuals of blood, excrement and earth attached to the skin.

13.2.1.2 Fleshing and Trimming

Extra tissues are removed. Dehairing is accomplished by the chemical dissolution of hair and epidermis with an alkaline medium of sulfide and lime. After skinning at the slaughterhouse, the hide appears to contain excessive meat, and fleshing usually precedes the dehairing and liming. This process produces wastewater with a very high COD value.

13.2.1.3 Deliming and Bating

The unhaired, fleshed and alkaline hide are neutralized with acid ammonium salts and treated with enzymes, similar to those found in the digestive system to remove the hair remnants and to degrade proteins. During this process, hair roots and pigments are removed to counteract a major part of the ammonium salt load in the generated wastewater.

13.2.1.4 Pickling

Pickling increases the acidity of the hides to up pH 3 by the addition of acid liquor and salts, enabling chromium tannins to penetrate the hide. However, the value salts are added to prevent the hide from swelling. For preservation purposes, 0.03–2% by weight fungicides and bactericides are used.

13.2.1.5   Degreasing

This process is used to remove the excess of material fatty substances from skins by using the organic solvents /surfactant.

13.2.2   TANNING (TANYARD OPERATION)

Tanning allows the stabilization of collagen fibers through a cross-linking action. The tanned hides and skins are tradable intermediate products (wet- blue). Tanning agents can be categorized into three main groups, namely: mineral (chrome) tanning agent; vegetable tanning agents; and alternative tanning agents (e. g. syntans, alde-hydes, and oil tanning agents).

13.2.2.1   Chrome Tanning

Chrome tanning is the most common type of tanning process used in most of the tan-ning industry around the world. After pickling, when the pH value is low, chromium (III) salts are added. To fixate the chromium, the pH is slowly increased through the addition of a base. The chrome tanning process is based on the cross-linkage of chromium ions with free carboxyl groups in the collagen fibers. It makes the hides resistant to bacteria and high temperature. Chrome tanned leather is characterized by top handling quality, high hydrothermal stability, user-specific properties and versatile applicability. Waste chrome from leather manufacturing, however, poses a significant disposal problem.

13.2.2.2   Vegetable Tanning

Vegetable tanning is usually accomplished in a series of vats with increasing concen-trations of tanning liquor, vegetable tanning produces relatively dense, pale brown leather that tends to darken on exposure to natural light. Vegetable tanning is fre-quently used to produce sole leather, belts and other leather goods. Unless specifi-cally treated, the vegetable-tanned leathers have low hydrothermal stability, limited water resistance, and are hydrophilic in mature.

13.2.2.3   Alternative Tanning

Tanning with organic tanning agents, using polymers or condensed plant polyphe-nols with aldehydic cross-linkers, can produce mineral-free leather with high hydro-thermal stability similar to chrome-tanned leather. However, organic all-tanned leather usually is more filled (e.g. leather with interstices filled with a filler material) and hydrophilic than chrome-free leather, with an equally high hydrothermal stabil-ity. This tanning process is carried out with a combination of metal salts, preferably but not exclusively aluminum (III), and a plant polyphenol containing pyrogallol groups, often in the form of hydrolysable tannins.

13.2.3   WET FINISHING (POST TANNING)

Post tanning operations involved neutralization and bleaching, followed by re-tan-ning, dyeing and fat liquoring. These processes are mostly carried out in a single processing vessel. Specialized operations may also be performed to add certain properties to the leather product (e.g. water repellence or resistance, oleophobicity, gas permeability, flame retardancy, abrasion resistance and anti- electrostatic properties).

13.2.4 FINISHING

The crust resulting from the re-tanning and drying process is subjected to a number of finishing operations. The purpose of these operations is to make the hide softer and to mask any small faults. The hide is treated with an organic solvent on water-based dye and varnish. Environmental aspects are mainly related to the finishing chemicals, which can also reach the generated wastewater. The quantity of the pollution load in tannery wastewater as reported by many authors is shown in Table 13.1.

13.3 RESIDUAL ORGANIC AND INORGANIC POLLUTANTS PRESENT IN TANNERY WASTEWATER

The environmental pollution due to wastewater discharged from the tannery industry causes serious problems in the environment, as well as health threats to both humans and animals. Since it contains a complex mixture of both organic and inorganic pollutants, which do not degrade much during the secondary treatment process in industry, it enters the environment, causing significant problems (Chandra et al. 2011).

13.3.1 ORGANIC POLLUTANTS PRESENT IN TANNERY WASTEWATER

During the end of the 20th century, the global environment became polluted with a number of industrial pollutants. The pollution of the global environment with a complex mixture of organic pollutants has resulted from industrial discharge from anthropogenic activities, as well as the inadvertent formation of by-products of various industrial processes. Organic pollutants are a group of chemical compounds that originated from different anthropogenic activities but have some common characteristics such as semi-volatility, lipophilicity, bioaccumulation, which make them resistant to biological photolysis, as well as chemical degradation, and persist in the environment for a long period of time (Yadav et al. 2017; Samarandaand and Gavrilescu 2008).

Most of the organic pollutants have three common characteristics: (i) one or more cyclical ring structures of either aromatic or aliphatic nature, (ii) a lack of polar functional groups and (iii) a variable amount of halogen atoms, usually chlorine. If some key properties of organic pollutants are known, then the environmental chemists can make predictions about their fate and behavior in the natural environment. These properties include aqueous solubility, vapor pressure, partition coefficients between water:solid and air:solid or liquid, and half-live in air, water and soil (Chandra and Chaudhary 2013). A large amount of waste discharged from tannery industries includes a variety of gaseous, liquid and solid wastes, which persist for a long period of time in the environment and cause serious threats to the environment, humans animals, in addition to plants (Chandra and Chaudhary 2013).

TABLE 13.1

Quantity of Pollution Load in Tannery Wastewater Contributed by Individual Operations of Leather Tanning Process

Pollution load		Pollution Load (kg/T of raw hide/skins processed operations)					
	Soaking	Unhairing/Liming	Deliming and Bating	Chrome Tanning	Post Tanning	Finishing	Total Load
Wastewater generated (m3/T)	9–12	4–6	1.5–2	1–2	1–1.5	1–2	17.5–25.5
Suspended solids	11–17	53–97	8–12	5–10	6–11	0–2	83–149
COD	22–33	79–122	13–20	7–11	24–40	0–5	145–231
BOD	7–11	28–45	5–9	2–4	8–15	0–2	50–86
Chromium	–	–	–	2–5	1–2	–	3–7
Sulfides	–	3.9–8.7	0.1–0.3	–	–	–	49
NH$_3$-N	0.1–0.2	0.4–0.5	2.6–3.9	0.6–0.9	0.3–0.5	–	4–5.8
Total Nitrogen	1–2	6–8	3–5	0.6–0.9	1–2	–	11.6–17.9
Chlorides	85–113	5–15	2–4	40–60	5–10	–	137–202
Sulfates	1–2	1–2	10–26	30–55	10–25	–	52–110

Source: Modified from Sabumon (2016), Dixit et al. (2015), Lofrano et al. (2013)

In the tannery industry during the tanning process, a number of chemical compounds (mainly chromium salts) are used to convert the raw hides/skins into leather and generate a large volume of wastewater containing huge amounts of organic matter, phenolics, phthalates, tannins, salts, a variety of organic pollutants and toxic heavy metals, mainly chromium-making wastewater unsuitable for irrigation and aquatic life (Chandra et al. 2011). These chemicals are genotoxic in nature and compounds can act at various levels in the cell (causing gene, chromosome or genome mutations), necessitating the use of a range of genotoxicity assays designed to detect these different types of mutations (Chowdhary et al. 2017; Bartling et al. 2005; Taylor et al. 2004).

Most of the compounds detected (Table 13.2) in the present study have also been reported by various authors by using different solvent systems (Kumari et al. 2016; Chandra et al. 2011; Alam et al. 2010, 2009).

13.3.2 INORGANIC POLLUTANT IN TANNERY WASTEWATER

Rapid industrialization and urbanization have given rise to the problem of heavy metal contamination in the environment. Environmental damage caused by tannery discharge has created a critical problem in India and signifies a technical challenge for an efficient and safe cleaning process. The environment has been severely contaminated with inorganic pollutants because different types of pollutants are

TABLE 13.2

Previously Reported Organic Pollutants Identified in Tannery Wastewater Using Different Extraction Solvents through GC-MS/MS Analysis

Solvents	Identified Organic Pollutants
Acetonitrile + acetone	Benzene, 2,2,3-Trimethyl oxepane
Methanol	3-Nitropthalic acid, 1,2-Benzenedicarboxylic acid, di-isooctyl ester, 1,3-Hexadien-5-yn
Diethyl ether + Chloroform	2-phenylethanol, Nonadec-1-ene, bis (2-methoxyethyl)phthalate, Hexatriacontane, Heneicosane, 2,4-Di-tert-butylphenol, Tricosane, 2,3-Epoxypinane
Choloform + Hexane	Dibutyl phthalate, Hexatriacontane, Bis(2-ethylhexyl) phthalate, bis (2-methoxyethyl) phthalate, 1,2-Benzenedicarboxylic acid, diisooctyl ester (diisooctyl phthalate)
Dichloromethane	Trimethyl1(2,6 ditert-butylphenoxy) silane, Dibutyl phthalate, Phenyl N-methylcarbamate, Di-n-octyl phthalate, 1,2-Benzenedicarboxylic acid, diisooctyl, Oleic acid, 2,4-bis(1,1-dimethyl) phenol, aminobiphenyl,4-nonylphenol, Hexadecanoic acid, Phenol, 2,4-bis(1,1-dimethylethyl), hexachlorobenzene, 4-Trimethylsiloxyphenylphenoxysulfone
Ethyl acetate	1,1-dimethylethyl-2-phenylethiazole, Acetic acid, Benzene,3-methoxy-4-benzaldehyde, Benzoic acid, Decanoic acid, Benzene propanoic acid, 2-hydroxy-3-methyl-butanoic acid

Source: Modified from Yadav et al. (2019), Kumari et al. (2016), Chandra et al. (2011), Alam et al. (2009)

released into the environment through the activity of the tannery industry. A number of toxic metals such as chromium (Cr), copper (Cu), cadmium (Cd), zinc (Zn), nickel (Ni), etc. are known as serious environmental pollutants and act as potentially toxic or carcinogenic agents in nature in very low concentrations and cause serious health hazards in human and animals, if entering the food chain (Montalvão et al. 2017; Khan et al. 2008; Jarup 2003; Hutchinson and Meema 1987).

13.3.2.1 Cr (VI) as a Major Pollutant in Tannery Wastewater

Environmental damage caused by tannery discharge has created a critical problem in India and signifies a technical challenge for an efficient and safe cleaning process. Chromium, a brittle, hard, steel gray, shiny metal present in the environment in a combined form of around 0.1–0.3 mg/kg of the earth's surface. It exists in several oxidation states (–2 to +6) and the most stable are Cr (III) and Cr (VI) (Mishra and Bharagava 2016; Molokwane et al. 2008). Cr (III) solubility is affected by the formation of oxides and hydroxides. Chromium is mainly employed in the metallurgy industry, particularly stainless steel production. Other Cr salts are used for the manufacturing of pigments, leather tanning, metal finishing, etc. Tanned hide is approximately 80–90% made from chromium compounds (Yadav et al. 2016a; Papp 2004).

Discharged effluents from tanneries contain about 40% of Cr as Cr (III) and Cr (VI). For each 200 kg of hide, more than 600 kg of waste is produced by a tannery (Khan 2001). Cr chemicals have also been used for the production of metal castings and mortars, refractory bricks and as wood preservative. Conversely, the U.S. Environmental Protection Agency (USEPA) has prohibited the use of Cr (VI) compounds as a wood preservative due to surrounding health issues.

Because of such diverse applications, a huge amount of Cr waste is released into the environment each year. In 2003, USEPA declared that approximately 32,589.6 metric tons of Cr compounds were disposed of; half of the quantity was in surrounding landfill. Potable water guidelines by the WHO state 0.05 mg/l as the maximum permissible limit for total chromium. Cr is hazardous, but also spreads fast over aquatic systems and underground waterways. Consequently, Cr has been recognized as a toxic environmental pollutant by US EPA (Narayani and Shetty 2013).

Chromate is present naturally but anthropogenic activities give rise to Cr (VI) pollution in the environment. Natural sources contribute 54,000 tons of chromium. Studies showed that atmospheric Cr reenters the soil and water bodies by rain. The Cr-estimated time in the atmosphere is less than ten days (Agency for Toxic Substance and Diseases Registry (ATSDR) 2015). Chromate present in soils can seep into surface water because of its highly soluble and mobile nature (Coetzee et al. 2018). It is a common practice to irrigate agricultural land by wastewater. Tannery effluent has a large content of valuable nutrients, however, it also contains toxins such as Cr that might damage the soil quality and crop production (Alvarez-Bernal et al. 2006). A high percentage of Cr in soils can prevent the germination of seeds and the production of seedlings. The toxic effects of Cr are less apparent on seed development than on the growth of seedling. Barley seeds were able to germinate in soil under a chromate stress of 100 mg/kg. Nevertheless, it showed slow growth due to the Cr (VI) inhibition of diastase that is necessary for mobilizing the starch reserved for early growth (Zayed and Terry 2003). In plants, the toxicity of Cr greatly depends on the ionic species of elements. Cr (VI) and Cr (III) were supplied in the range of 0 to 100 mg/kg. When chromate (100 mg/l) stress was applied to plants, up to 3000–5000 mg/kg of Cr (VI) was accumulated and was up to 300-400 mg/kg when Cr(III) (100 mg/l) stress was applied to hydroponic culture. These high levels of chromium caused leaf chlorosis, reduced root and shoot growth, stimulation of chitinase activity and low levels of water content in leaves (Chowdhary et al. 2018; Zayed and Terry 2003).

13.3.2.1.1 Health Hazards

Cr (VI) can enter the body when people breathe air, eat food or drink water contaminated with it. Cr (VI) is also found in house dust and soil, which can be ingested or inhaled from the various forms of chromium; Cr (VI) is the most common form and toxic in nature. Many Cr (VI) compounds have been found to be carcinogenic in nature, but the evidence to date indicates that the carcinogenicity is site-specific and limited to lung and sinonasal cavities, and dependent on the exposure intensity (Mishra et al. 2019). Inhaling a relatively high concentration of Cr (VI) can cause a runny nose, sneezing, itching, nosebleeds, ulcers and holes in the nasal septum. Short-term high-level inhalational exposure can cause adverse effects at the contact site, including ulcers, irritation of the nasal mucosa and holes in the nasal septum. Ingestion of very high Cr (VI) doses can cause kidney and liver damage, nausea, irritation of the gastrointestinal tract, stomach ulcers, convulsions and death. At the same time, dermal exposure may cause skin ulcers or allergic reactions. Cr (VI) is one of the most highly allergenic metals, second to nickel. Studies on mice given high doses of Cr (VI) have shown reproductive abnormalities including a reduced litter size and decreased fetal weight (ATSDR 2000; Mishra and Bharagava 2016).

13.4 PATHOGENIC MICROBES PRESENT IN TANNERY WASTEWATER

Waterborne pathogens infect around 250 million people each year, resulting in 10 20 million deaths (El-Lathy et al. 2009; Toze 1999). Many of these infections occur in developing countries, which suffer from lower levels of sanitation, problems associated with low socio-economic conditions and less public health awareness than in developed countries. A number of chromium-resistant microorganisms have been reported, such as *Pseudomonas* spp., *Proteus* spp., *Enterobacter* spp., *Enterobacter* spp., *Escherichia coli* and *Bacillus* spp. (Naraian et al. 2012).

The microbial pathogens in water necessitate the proper monitoring of wastewater presence in contaminated aquatic resources for various types of organic pollutants and microbial pathogens. For an appropriate risk assessment, the type of organic pollutants and microbial pathogen present in wastewater and its relative numbers need to be determined. This is particularly important for the effective treatment of industrial wastewaters and its safe reuse/recycle or disposal into the environment.

In this literature study, the current and emerging conventional and molecular approaches for characterizing the bacterial community, composition and structure in water and wastewater processes are shown in Figure 13.2 (Gilbride et al. 2006).

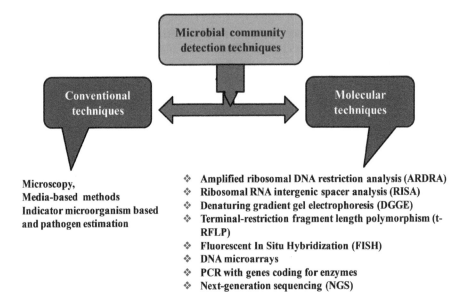

FIGURE 13.2 Current and emerging conventional and molecular techniques used to detect wastewater microorganisms.

In recent years, molecular techniques have been used for examining microbial diversity and detecting specific microorganisms. The wastewater released from the tannery industry may contain a variety of organic pollutants and millions of pathogenic and non-pathogenic bacteria per milliliter, including coliform, *Streptococci*, *Staphylococci*, anaerobic spore forming bacilli and many other types of health hazards organisms (Naraian et al. 2012).

Previous studies have detected and characterized various antibiotic resistant bacteria such as *Citrobacter, Enterobacter, Salmonella typhi, Klebsiella pneumoniae, Escherichia cloacae, Pseudomonas aeruginosa, Citrobacter freundii, Serratia marcescens, Shigella flexneri, Shigella sonnei* and *proteus mirabilis* from industrial wastewater systems (Verma et al. 2004; Chandra et al. 2011; Naraian et al. 2012; Ramteke et al. 2010).

13.5 TREATMENT APPROACHES FOR THE REMOVAL OF ORGANIC AND INORGANIC POLLUTANTS AND PATHOGENIC MICROBES FROM TANNERY WASTEWATER

13.5.1 Physico-Chemical Treatment

13.5.1.1 Coagulation and Flocculation

Various authors have investigated the coagulation and flocculation of TWW by using various inorganic coagulants such as aluminum sulfate ($AlSO_4$), ferric chloride ($FeCl_3$), ferrous sulfate ($FeSO_4$) to reduce the total organic load (BOD and COD), total solids (TDS and TSS), as well as to remove the toxic metals such as chromium before the biological treatment of TWW (Song et al. 2005; Lofrano et al. 2006). However, each coagulant operates most effectively on the specific pH and the extent of the pH range largely depends on the nature of coagulants, as well as on the dosage of the coagulant characteristics of the wastewater to be treated (Song and Burns 2005).

There are several studies that have been conducted to investigate the effectiveness of different coagulants used for the treatment of TWW in terms of COD and chromium removal (Song et al. 2005). Despite various physicochemical methods being found to be effective, their application is limited due to the large amount of chemicals used, the huge quantity of sludge generation, and disposal problems in the environment, high installation, as well as operating costs. Thus, for the biological treatment method to be viable, a cost-effective and environment-friendly alternative to the physico-chemical method for the treatment of industrial wastewater is necessary.

13.5.2 Biological Treatment

13.5.2.1 Aerobic Processes

Biological treatment processes are generally used for the treatment of industrial wastewaters to reduce the organic content as these processes have many economic advantages over the physicochemical treatment methods. But the high concentration of tannins and other poorly degradable compounds, as well as toxic metals present in tannery wastewater, have negative effects on the biological treatment processes (Lofrano et al. 2013).

A typical sequencing batch reactor (SBR) has been proved to be more capable of carrying out the biological processes such as nitrification and denitrification in the presence of inhibitors, due to the selection and enrichment of a particular microbial species (Murat et al. 2006). Studies of the performance of SBR for nitrogen removal in TWW, with a wide range of temperatures (7–30°C), were studied and achieved full nitrification and denitrification. This was achieved by the adjustment of sludge age for each temperature range. The biodegradation of naphthalene-2-sulfonic acid, which is a main component of the naphthalene sulfonate, by *Arthrobacter* sp. 2AC and *Comamonas* sp. 4BC, was reported and these two bacterial strains were isolated from tannery-activated sludge (Song et al. 2005). Song and Burns (2005) described the degradation of all components of the condensation product of 2-naphthalene-sulfonic acid and formaldehyde (CNSF) by fungus *Cunninghamella Polymorpha*, and suggested that the combination of *C. polymorpha* and *Arthrobacter* sp. 2 AC or *Comamonas* sp. 4BC was effective for the treatment of tannery wastewater. Nevertheless, conventional cultures could not treat saline wastewaters of values higher than 3%–5% (weight/volume, w/v) and a shift in salt concentration causes significant reactor failures in the system performance. Senthilkumar et al. (2008) have isolated *Pseudomonas aeruginosa, Bacillus flexus,* and *Staphylococcus aureus* from soak liquor, marine soil, salt lake saline liquor and seawater, respectively, and studied the biodegradation of tannery soak liquor by these halotolerant bacterial consortia. An appreciable COD removal (80%) was observed at 8% (w/v) salinity for mixed salt tolerant consortia, but an increase in salt concentration to 10% (w/v) resulted in a decrease in COD-removal efficiency.

The presence of sulfide, chromium, chloride and fluctuations in pH and temperature has adverse effects on the nitrification process. The impact of temperature on the organic carbon and on the efficiency of a full-scale industrial-activated sludge plant nitrogen removal has been studied during the treatment of TWW (Gorgun et al. 2007). It was observed that temperature changes had a minor influence on COD-removal efficiency (4%–5%), while the total nitrogen removal was affected significantly by the temperature. Insel et al. (2009) also investigated the performance of the intermittent aeration type of operation when the temperature fluctuated between 21°C and 35°C. It was found that an increase in the aeration intensity improved the nitrification performance and the application of intermittent aeration also improved the total nitrogen removal up to 60%.

13.5.2.1.1 *Activated Sludge Process (ASP)*

The ASP is the most generally applied biological (aerobic) wastewater treatment method that primarily removes the dissolved organic solids as well as settleable and non-settleable suspended solids (SSs). In ASP, a suspension of bacterial biomass (the activated sludge) is mainly used for the removal of organics. These organisms are cultivated in aeration tanks, where they are provided with dissolved oxygen (DO) and food from the wastewater. Depending on the design and specific applications, an activated sludge treatment plant can remove organic nitrogen (N) removal and phosphorus (P), as well as organic carbon substances.

Currently numerous processes are used for the biological treatments of TWW in CETPs. In India, the ASP and up flow anaerobic sludge blanket (UASB) process is the most common. However, the biological treatment of tannery wastewater using ASP has been reported by many workers (Ramteke et al. 2010; Tare et al. 2003).

13.5.3 Emerging Treatment Technologies

13.5.3.1 Membrane Processes

In recent years, membrane technologies have been a focus of attention; their cost is continuing to decrease, while their application possibilities are ever-expanding. The use of membrane technologies applied to the leather industry represents an economic advantage, especially in the recovery of chromium from the residual waters from the leather tanning process. Several studies have shown that cross flow microfiltration, ultrafiltration, nanofiltration, reverse osmosis (RO), and supported liquid membranes can be applied. This may be in the leather industry for the recovery of chromium from spent liquors (Ashraf et al. 1997; Cassano et al. 2001; Labanda et al. 2009); the reuse of wastewater and chemicals of deliming/bating liquor (Gallego-Molina et al. 2013); the reduction in the polluting load of unhairing and degreasing removal of salts; and in the biological treatment of TWW in the light of their reuse. Reverse osmosis with a plane membrane has been used in a post-treatment process to remove refractory organic compounds (chloride and sulfate) by De Gisi et al. (2009). The high quality of permeate produced by the RO system with a plane membrane allowed the reuse of TWW within the production cycle, and thus reduced the groundwater consumption.

13.5.3.2 Membrane Bioreactors

The membrane Bio-Reactor (MBR) has been attracting much attention from scientists and engineers for TWW treatment due to the numerous advantages over CASP, such as the elimination of settling basins, independence of the process performance from filamentous bulking, or other phenomena affecting settleability (Suganthi et al. 2013). An MBR system essentially consists of a combination of membranes and biological reactor systems. The separation of biomass from wastewater by membranes also allows the concentration of MLSS in bioreactor to be increased significantly. However, from the studies of Munz et al. (2009) it is possible to infer how the kinetics of nitrification are effectively reduced by the presence of tannins, without large differences between biomass selected with either the CASP or the MBR. One of the main drawbacks of membrane application is significant fouling due to the clogging, adsorption and cake layer formation by the pollutants onto the membrane. In recent years, extensive work is in progress to reduce such a befouling phenomenon.

13.5.3.3 Advanced Oxidation Processes

There has been increasing interest in the study of advanced oxidation processes (AOPs) to treat TWW for the removal of organic pollutants and pathogenic microbes. AOPs refer to the set of chemical treatment processes that use strong oxidizing agents O_3, hydrogen peroxide (H_2O_2)] and/or catalysts (Fe, Mn, TiO_2), and are also sometimes supported in activity by high-energy radiation; for example, UV light (Schrank et al. 2004). All these processes are based on the production and utilization of hydroxyl radicals, which are very powerful oxidants that quickly and unselectively oxidize a broad range of organic compounds. The scientific interest toward the application of AOPs to high strength wastewater has increased remarkably in the last 20 years. AOPs can reduce the concentration of pollutants several hundred ppm to less than 5 ppm and therefore significantly reduce the level of COD and TOC, which have earned it the title of "wastewater treatment processes of the 21st century".

13.6   FUTURE CONSIDERATIONS

The major problem with tannery wastewater is its complex nature, due to the organic pollutants such as tannins and other poorly degradable compounds that can inhibit the biological treatment process. Besides organic pollutants, tannery wastewater also contains many toxic metals ions such as Cu^{2+}, Cr^{6+}, Cd^{2+}, Zn^{2+} and Ni^{2+}; all these metals ions also have high inhibitory and antimicrobial activity reducing the anaerobic digestion of tannery wastewater.

A bioremediation approach such as in situ remediation is applied to reduce or eliminate residual organic pollutants and toxic heavy metals that have led to environmental hazards and risks. In situ remediation involves the direct inoculation of microbes and reagents into the polluted aquifer and is becoming a progressively common technique. The cost-effectiveness, simplicity of procedure and minimal interference in the site render it more applicable to this technique. The removal processes that utilize permeable reactive barriers are also gaining acceptance. No single technique is adequate for the removal of the majority of the pollutants that might exist at a site or to accomplish compliance with cleanup standards. To accomplish such goals, the use of a treatment train strategy is frequently required. For instance, inorganic reductants might be applied for mass removal of chromate contaminants, followed by the use of anaerobic bioremediation and/or to additionally check natural reduction.

Recently, the sequential applications of bacteria and wetland plants have been also reported to be very promising for the degradation and detoxification of tannery wastewater, but this has to be optimized yet with the detailed microbiology of wetland plants, plants rhizosphere, interaction and detoxification mechanisms. Moreover, the nature of residual organic pollutants in tannery wastewater and the extent of their toxicity need to be explored in detail for their complete degradation and detoxification during the treatment process of CETPs.

13.7 CONCLUSION

This chapter has highlighted the current and future pollution details measures in tannery wastewater. It is, therefore, concluded that since the treated tannery wastewater is destined to be discharged into the environment, and thus the organic pollutants, pathogenic microbes, heavy metals and other chemicals reported will enhance the pollution load and other health hazards. Further, the disposal of treated wastewater through drains may also contaminate shallow ground water often used for irrigation and potable purposes in the agricultural field on both sides of a drain before entering the Ganga River.

Thus, an urgent need to address the cost-effectiveness is necessary. The simplicity of procedure treatment technologies which has been deemed essential has been developed and applied to tannery wastewater treatment including biological, precipitation, co-precipitation, solvent extraction, adsorption, coagulation, flocculation, ion-exchange and membrane technology. These techniques are exploited to reduce pollutants or contaminants from tannery wastewater and to remove toxic chemicals from tannery wastewater and to recover the quality of raw wastewater for irrigation purposes. It is therefore hoped that our investigation will prove to be of most relevance to the regulatory agencies and public health departments for water quality, safety management, in addition to the clinical management of disease among workers caused due to the contamination of aquatic resources by tannery wastewater.

ACKNOWLEDGMENTS

The author Abhay Raj acknowledges financial support from the DST-SERB (Grant No. EEQ/2017/000571) and Director of the CSIR-Indian Institute of Toxicology Research (IITR), Lucknow (India) for the encouragement of this work. The financial support received by Dr. R N Bharagava from the "Science and Engineering Research Board" (SERB), Department of Science and Technology (DST), Government of India (GOI), New Delhi, India, as "Major Research Project" (Grant No. EEQ/2017/000407).

REFERENCES

Agency for Toxic Substance and Disease Registry (ATSDR). 2015. *Toxicological Profile for Chromium*. U.S Department of Health and Human Services, Public Health Services, ATSDR, Atlanta.

Alam, M.Z., Ahmad, S., Malik, A. 2009. Genotoxic and mutagenic potential of agricultural soil irrigated with tannery effluents at Jajmau (Kanpur), India. *Archives of Environmental Contamination and Toxicology* 57:463–476.

Alam, M.Z., Ahmad, S., Malik, A., Ahmad, M. 2010 Mutagenicity and genotoxicity of tannery effluents used for irrigation at Kanpur, India. *Ecotoxicology and Environmental Safety* 57:463–476.

Alvarez-Bernal, D., Contreras-Ramos, S.M., Trujillo-Tapia, N., Olalde-Portugal, V., Frías-Hernández, J.T., Dendooven, L. 2006. Effects of tanneries wastewater on chemical and biological soil characteristics. *Applied Soil Ecology* 33:269–277.

Ashraf, C.M., Ahmad, S., Malik, M.T. 1997. Supported liquid membrane technique applicability for removal of chromium from tannery wastes. *Waste Management* 17:211–218.

Ballard, R. 2001. *A Preliminary Study on the Bovine Leather Value Chain in South Africa*, Industrial Restructuring Project School of Development Studies, University of Natal, Durban, CSDS, Research Report No. 40, May 2001.

Bartling, B., Rehbein, G., Somoza, V., Silber, R.E., Simm, A. 2005. Maillard reaction product-rich foods impair cell proliferation and induce cell death in vitro. *Signal Transduction* 5:303–313.

Cassano, A., Molinari, R., Romano, M., Drioli, E. 2001. Treatment of aqueous effluent of the leather industry by membrane processes: A review. *Journal of Membrane Science* 181:111–126.

Chandra, R., Bharagava, R.N., Kapley, A., Purohit, H.J. 2011. Bacterial diversity, organic pollutants and their metabolites in two aeration lagoons of common wastewater treatment plant (CETP) during the degradation and detoxification of tannery wastewater. *Bioresource Technology* 102:2333–2334.

Chandra, R., Chaudhary, S. 2013. Persistent organic pollutants in environment and health hazards. *International Journal Bioassays* 2:1232–1238.

Chowdhary, P., Yadav, A., Kaithwas, G., Bharagava, R.N. 2017. Distillery wastewater: A major source of environmental pollution and its biological treatment for environmental safety. In: Singh, R., Kumar, S. (Eds.), *Green Technologies and Environmental Sustainability*. Springer, Berlin, pp. 409–435.

Chowdhary, P., Yadav, A., Singh, R., Chandra, R., Singh, D.P., Raj, A., Bharagava, R.N. 2018. Stress response of *Triticum aestivum* L. and *Brassica juncea* L. against heavy metals growing at distillery and tannery wastewater contaminated site. *Chemosphere* 206:122–131.

Chowdhury, M., Mostafa, M.G., Biswas, T.K., Mandal, A. 2015. Characterization of the effluents from leather processing industries. *Environmental Process* 2:173–187.

Chowdhury, M., Mostafa, M.G., Biswas, T.K., Saha, A.K. 2013. Treatment of leather industrial effluents by filtration and coagulation processes. *Water Resource Industry* 3:11–22.

Coetzee, J.J., Bansal, N., Chirwa, E.M. 2018. Chromium in environment, its toxic effect from chromite-mining and ferrochrome industries, and its possible bioremediation. *Exposure and Health* 1–12:51–62.

Cooman, K., Gajardo, M., Nieto, J., Bornhardt, C., Vidal, G. 2003. Tannery wastewater characterization and toxicity effects on *Daphnia spp. Environmental Toxicology* 18:45–51.

De Gisi, S., Galasso, M., De Feo, G. 2009. Treatment of tannery wastewater through the combination of a conventional activated sludge process and reverse osmosis with a plane membrane. *Desalination* 249:337–342.

Dixit, S., Yadav, A., Dwivedi, P.D., Das, M. 2015. Toxic hazards of leather industry and technologies to combat threat: A review. *Journal of Cleaner Production* 87:39–49.

El-Lathy, M.A., El-Taweel, G.E., El-Sonosy, W.M., Samhan, F.A., Moussa, T.A.A. 2009.
Determination of pathogenic bacteria in wastewater using conventional and PCR techniques. *Environmental Biotechnology* 5:73–80.

Gallego-Molina, A., Mendoza-Roca, J.A., Aguado, D., Galiana-Aleixandre, M.V. 2013.
Reducing pollution from the deliming-bating operation in a tannery. Wastewater
reuse by microfiltration membranes. *Chemical Engineering Research and Design*
91:369–376.

Gilbride, K.A., Le, D.Y., Beaudette, L.A. 2006. Molecular techniques in wastewater:
Understanding microbial communities, detecting pathogens, and real-time process
control. *Journal of Microbiology Methods* 66:1–20.

Gorgun, E., Insel, G., Artan, N., Orhon, D. 2007. Model evaluation of temperature dependency for carbon and nitrogen removal in a full-scale activated sludge plant treating
leather-tanning wastewater. *Journal of Environmental Science and Health Part A:
Toxic Hazardous Substance Environmental Engineering* 42:747–756.

Haydar, S., Aziz, J.A. 2009. Characterization and treatability studies of tannery wastewater using chemically enhanced primary treatment (CEPT) – A case study of Saddiq
Leather Works. *Journal of Hazardous Materials* 163:1076–1083.

Hutchinson, T.C., Meema, K.M. (Eds.) 1987. *Lead, Mercury, Cadmium and Arsenic in the
Environment*. John Wiley & Sons, Chichester, pp. 279–303.

Insel, G.H., Gorgun, E., Artan, N., Orhon, D. 2009. Model based optimization of nitrogen
removal in a full scale activated sludge plant. *Environmental Engineering Science*
26:471–480.

Jarup, L. 2003. Hazards of heavy metal contamination. *British Medical Bulletin* 68:167–182.

Kapley, A., Lampel, K., Purohit, H.J. 2001. Development of duplex PCR for *Salmonella* and
Vibrio. *World Journal of Microbiology and Biotechnology* 16:457–458.

Khan, A.G. 2001. Relationships between chromium biomagnifications ratio, accumulation factor, and mycorrhizae in plants growing on tannery effluent-polluted soil.
Environmental International 26:417–423.

Khan, S., Cao, Q., Zheng, Y.M., Huang, Y.Z., Zhu, Y.G. 2008. Health risks of heavy metals in contaminated soils and food crops irrigated with wastewater in Beijing, China.
Environmental Pollution 152:686–692.

Kumari, V., Kumar, S., Haq, I., Yadav, A., Singh, V.K., Ali, Z., Raj, A. 2014. Effect of tannery
effluent toxicity on seed germination á-amylase activity and early seeding growth of
Mung Bean (*Vigna Radiata*) seeds. *International Journal of Latest Research Science
and Technology* 3:165–170.

Kumari, V., Yadav, A., Haq, I., Kumar, S., Bharagava, R.N., Singh, S.K., Raj, A. 2016.
Genotoxicity evaluation of tannery effluent treated with newly isolated hexavalent chromium reducing *Bacillus cereus*. *Journal of Environment Management* 183:204–211.

Labanda, J., Khaidar, M.S., Llorens, J. 2009. Feasibility study on the recovery of chromium
(III) by polymer enhanced ultrafiltration. *Desalination* 249:577–581.

Lefebvre, O., Vasudevan, N., Torrijosa, M., Thanasekaran, K., Moletta, R. 2006. Anaerobic
digestion of tannery soak liquor with an aerobic post-treatment. *Water Research*
40:1492–1500.

Lofrano, G., Belgiorno, V., Gallo, M., Raimo, A., Meric, S. 2006. Toxicity reduction in
leather tanning wastewater by improved coagulation flocculation process. *Global Nest
Journal* 8:151–158.

Lofrano, G., Meric, S., Zengi, G.E., Orhon, D. 2013. Chemical and biological treatment technologies for leather tannery chemicals and wastewaters: A review. *Science of the Total
Environment* 461–462:265–281.

Matsumoto, S.T., Mnlovani, S.M., Malaguttii, M.I.A., Dias, A.L., Fonseca, I.C., Morales, M.A.M. 2006. Genotoxicity and mutagenicity of water contaminated with tannery effluent, as evaluated by the micronucleus test and comet assay using the fish *Oreochromis niloticus* and chromosome aberrations in onion root tips. *Genetic Molecular Biology* 29:148–158.

Mishra, S., Bharagava, R.N. 2016. Toxic and genotoxic effects of hexavalent chromium in environment and its bioremediation strategies. *Journal Environment Science and Health Part C* 34:1–32.

Mishra, S., Bharagava, R.N., More, N., Yadav, A., Zainith, S., Mani, S., Chowdhary, P. 2019. Heavy metal contamination: An alarming threat to environment and human health. In: Sobti, R.C., Arora, N.K., Kothari, R. (Eds.), *Environmental Biotechnology: For Sustainable Future*. Springer, Singapore, pp. 103–125.

Molokwane, P.E., Meli, K.C., Nkhalambayausi-Chirwa, E.M. 2008. Chromium (VI) reduction in activated sludge bacteria exposed to high chromium loading: Brits culture (South Africa). *Water Research* 42:4538–4548.

Montalvão, F.J.M., de Souza, A.T.B., Guimarães, I.P.P., et al. 2017. The genotoxicity and cytotoxicity of tannery effluent in bullfrog (*Lithobates catesbeianus*). *Chemosphere* 183:491–502.

Munz, G., De Angelis, D., Gori, R., Mori, G. 2009. The role of tannins in conventional ango- gated membrane treatment of tannery wastewater. *Journal of Hazardous Materials* 164:733–739.

Murat, S., Insel, G., Artan, N., Orhon, D. 2006. Performance evaluation of SBR treatment for nitrogen removal from tannery wastewater. *Water Science and Technology* 53:275–284.

Naraian, R., Ram, S., Kaistha, S.D., Srivastava, J. 2012. Occurrence of plasmid linked multiple drug resistance in bacterial isolates of tannery effluent. *Cell & Molecular Biology* 58:134–141.

Narayani, M., Shetty, K.V. 2013. Chromium-resistant bacteria and their environmental condition for hexavalent chromium removal: A review. *Critical Review in Environmental Science and Technology* 43:955–1009.

Papp, J.F. 2004. Chromium use by market in the United States. Paper Presented at 10th International Ferroalloys Congress, Cape Town, South Africa, 1–4.

Ramteke, P.W., Awasthi, S., Srinath, T., Joseph, B. 2010. Efficiency assessment of Common Effluent Treatment Plant (CETP) treating tannery effluents. *Environmental Monitoring and Assessment* 169:125–131.

Sabumon, P.C. 2016. Perspectives on biological treatment of tannery effluent. *Advances in Recycling & Waste Management* 1:1. doi:10.4172/arwm.1000104.

Samaranda, C., Gavrilescu, M. 2008. Migration and fate of persistent organic pollutants in the atmosphere – A modeling approach. *Environmental Engineering and Management Journal* 7:743–761.

Saxena, G., Bharagava, R.N. 2015. Persistent organic pollutants and bacterial communities present during the treatment of tannery wastewater. In: Chandra, R. (Ed.), *Environmental Waste Management*, 1st ed. CRC Press, Taylor & Francis Group, Boca Raton, FL, pp. 217–247.

Saxena, G., Chandra, R., Bharagava, R.N. 2016. Environmental pollution, toxicity profile and treatment approaches for tannery wastewater and its chemical pollutants. *Review of Environment Contamination and Toxicology* 240:31–69.

Schrank, S.G., Bieling, U., Jose, H.J., Moreira, R.F.P.M., Schroder, H.F. 2004. Generation of endocrine disruptor compounds during ozone treatment of tannery wastewater confirmed by biological effect analysis and substance specific analysis. *Water Science Technology* 59:31–38.

Senthilkumar, S., Surianarayanan, M., Sudharshan, S., Susheela, R. 2008. Biological treatment of tannery wastewater by using salt-tolerant bacterial strains. *Microbial Cell
Factories* 7:15.

Shukla, O.P., Rai, U.N., Dubey, S. 2009. Involvement and interaction of microbial communities in the transformation and stabilization of chromium during the composting of
tannery effluent treated biomass of *Vallisneria spiralis* L. *Bioresource Technology*
100:2198–2203.

Song, Z., Burns, R.G. 2005. Depolymerisation and biodegradation of a synthetic tanning
agent by activated sludges, the bacteria *Arthrobacter globiformis* and *Comamonas tes-
tosterone* and the fungus *Cunninghamella polymorpha*. *Biodegradation* 16:305–318.

Song, Z., Edwards, S.R., Burns, R.G. 2005. Biodegradation of naphthalene-2-sulfonic
acid present in tannery wastewater by bacterial isolates *Arthrobacter* sp. 2AC and
Comamonas sp. 4BC. *Biodegradation* 16:237–252.

Suganthi, K.V., Mahalaksmi, M., Balasubramanian, N. 2013. Development of hybrid membrane bioreactor for tannery effluent treatment. *Desalination* 309:231–236.

Tare, V., Gupta, S., Bose, P. 2003. Case studies on biological treatment of tannery effluents in
India. *Journal of Air Waste and Management Association* 53:976–982.

Taylor, J.L., Demyttenaere, J.C., Abbaspour, T.K., et al. 2004. Genotoxicity of melanoidin
fractions derived from a standard glucose/glycine model. *Journal of Agriculture and
Food Chemistry* 28:18–23.

Toze, S. 1999. PCR and the detection of microbial pathogens in water and wastewater. *Water
Research* 33:3545–3556.

Tunay, O., Kabdasli, I., Orhon, D., Ates, E. 1995. Characterization and pollution profile of
leather tanning industry in Turkey. *Water Science and Technology* 32:1–9.

United Nations Industrial Development Organization (UNIDO). 2010. *Future Trends in the
World Leather and Leather Products Industry and Trade*. UNIDO, Vienna.

United State Environmental Protection Agency (USEPA). 1986. *Guidelines for the
Health Risk Assessment of Chemical Mixtures*, (PDF) EPA/630/R-98/002, USEPA,
Washington, DC.

Verma, T., Ramteke, P.W., Garg, S.K. 2004. Occurrence of chromium resistant thermotolerant
coliforms in tannery effluent. *Indian Journal of Experimental Botany* 42:1112–1116.

Verma, T., Ramteke, P.W., Garg, S.K. 2008. Quality assessment of treated tannery wastewater with special emphasis on pathogenic *E. coli* detection through serotyping.
Environmental Monitoring and Assessment 145:243–249.

Yadav, A., Chowdhary, P., Kaithwas, G., Bharagava, R.N. 2017. Toxic metals in environment,
threats on ecosystem and bioremediation approaches. In: Das, S., Singh, H.R. (Eds.),
Handbook of Metal–Microbe Interactions and Bioremediation. CRC Press, Taylor &
Francis Group, Boca Raton, FL, USA, pp. 128–141.

Yadav, A., Mishra, S., Kaithwas, G., Raj, A., Bharagava, R.N. 2016a. Organic pollutants
and pathogenic bacteria in tannery wastewater and their removal strategies. In: Singh,
J.S., Singh, D.P. (Eds.), *Microbes and Environmental Management*. Studium Press Pvt.
Ltd., India, pp. 101–127.

Yadav, A., Raj, A., Bharagava, R.N. 2016b. Detection and characterization of a multi-drug
and multi-metal resistant enterobacterium *pantoea* sp. from tannery wastewater after
secondary treatment process. *International Journal of Environment and Botany*
1(2):37–42.

Zayed, A.M., Terry, N. 2003. Chromium in the environment: Factors affecting biological
remediation. *Plant Soil* 249:139–156.

14 Role of Microbes in Bioremediation of Pollutants (Hydrocarbon) in Contaminated Soil

Snigdha Singh and R. Y. Hiranmai

CONTENTS

14.1　INTRODUCTION

Throughout the world, environmental pollution has become a matter of grave concern. It has involved adulteration to numerous media, namely water, soil, and air. Soil is a vital environmental medium to endorse the life of all organisms indirectly or directly. Contamination of soil has been declared to be of the utmost threats among the three afore-mentioned media, as the contaminants have the ability to pose adverse effects on the inherent organism in the soil and obliterate the food chain. It is the eventual "sink" for all pollutants, which has undesirably affected the quality of the soil. Despite this, it has been ignored for a long time. The dumping of pollutants has led to the introduction of toxicity into the soil, as well as adulteration in its properties (Sakshi et al. 2019).

14.2   HYDROCARBONS AND TYPES

Environmental contaminants can be of two types: organic or inorganic. Quantitatively, hydrocarbons in their various forms are organic pollutants of utmost concern. Hydrocarbon pollution has become one of the major environmental issues caused by various uses practiced in the petrochemical industry. Generally, petroleum and its products enter the environment through natural seepages, transportation, accidental spills, deliberate disposal, offshore production, and the breakage of pipelines (Rodrigues et al. 2010; Tang et al. 2012; McAlexander 2014). The presence of petroleum hydrocarbon compounds in the environment can affect both human health and the environment.

There are various types of hydrocarbons. On the basis of carbon-carbon bonds present, they can be categorized into three main classes: (i) saturated, (ii) unsaturated, and (iii) aromatic hydrocarbons. Carbon-carbon and carbon-hydrogen single bonds constitute the saturated hydrocarbons. Different atoms of carbon are arranged together to formulate an open chain of single-bonded carbon and are denoted as "alkanes". On the other hand, atoms of carbon are arranged in a ring or closed chain, they are named "cycloalkanes". The unsaturated hydrocarbons are comprised of carbon-carbon multiple bonds. An exceptional form of cyclic compounds is known as "aromatic hydrocarbons" (Figure 14.1).

These organic pollutants include simple aliphatic as well as complex aromatic compounds such as benzene, toluene, naphthalene, methylbenzene, polychlorinated biphenyls, polycyclic aromatic compounds, nitroaromatics, or straight chain-halogenated hydrocarbons.

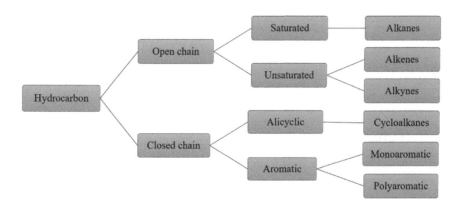

FIGURE 14.1   Classification of hydrocarbons.

Due to the deficiency of functional groups, hydrocarbons are mostly polar, hence allowing very weak chemical reactions in ambient temperature. The chemical reactivity of hydrocarbons is mainly specified via type, existence, and the order of the aromatic bonds (π bonds) in these compounds. Thus, these compounds are classified regarding their bonding types into the following categories: aliphatic and aromatic hydrocarbons.

The aromatic compounds present in the environment are from natural sources and anthropogenic activities. The chemical characteristic of these compounds is the presence of one benzene ring (monoaromatic hydrocarbon—MAHs) or more than one fused ring (polyaromatic hydrocarbon—PAHs) (Favre and Powell 2013). The ring provides structural and chemical stabilities due to a symmetric π-electron system and therefore recalcitrance of these compounds (Vogt et al. 2011). The physic-chemical properties of aromatic hydrocarbons have environmental significance because they determine fate in soil, water, and the atmosphere. For instance, adsorption in soils or sediments, due to hydrophobicity, is a major factor in their transportation and eventual degradation (Karickhoff 1981).

The polycyclic aromatic hydrocarbon (PAH) compounds are mostly stable, toxic, and carcinogenic (Abdel-Shafy and Mansour 2016). The occurrence of polycyclic aromatic hydrocarbons (PAHs), polychlorinated biphenyls, pesticides, and heavy metals disturbs all forms of life since these chemicals have accompanying mutagenicity, carcinogenicity, and toxicity.

Hydrocarbons can cause deleterious effects to the flora, aquatic fauna, and humans as they are bioaccumulative, as well as biopersistent in nature (Benson et al. 2007).

A distinctive group of pollutants of the soil are PAHs which cause variation in porosity, water-holding capacity, and the grain size of soil affecting the diversity/population of microbes adversely (Sakshi et al. 2019) (Table 14.1).

One of the major concerns of the world today is the contamination of the environment by petroleum hydrocarbons. The consequences of these contaminants are devastating and catastrophic on human beings, along with other biotic components of the ecosystem (Spierings and Mulder 2017).

Petroleum hydrocarbons are readily adsorbed in the soil and strongly hydrophobic. Due to their low water solubility and biological availability, these compounds do not degrade naturally by bionts in soils.

14.3 SOURCES OF HYDROCARBON IN THE ENVIRONMENT

The foremost cause of pollution of soil is the discharge of petroleum into the environment, either naturally or because of anthropogenic activities.

Equipment maintenance, inappropriate operations, along with other reasons like refining, storage and sales, accidental spilling during oil and gas exploration processes, results in overspill, in addition to the release of petroleum hydrocarbons (Shi 2013). Much of this pollution has resulted from the increased activities associated with petroleum exploration, transport, and processing. In addition, further increases in the quantity of adulterated sites are due to the lack of dumping of hazardous oil wastes into landfills without sufficient management, as well as the lack of waste oil recycling (Koshlaf and Ball 2017). Refining of oil, unfinished incineration,

TABLE 14.1

Molecular Formula and Chemical Structure of Some Hydrocarbons

S. No.	Hydrocarbon Classification	Mol. Formula	Chemical Structure
1.	Alkane	n-butane	n-butane
2.	Alkene	butadiene	
3.	Alkyne	1-Butyne	
4.	Cycloalkanes	CnH2n	
5.	Naphthalene	$C_{10}H_8$	
6.	Fluorene	$C_{13}H_{10}$	
7.	Anthracene	$C_{14}H_{10}$	
8.	Phenanthrene	$C_{14}H_{10}$	

(*Continued*)

TABLE 14.1 (CONTINUED)
Molecular Formula and Chemical Structure of Some Hydrocarbons

S. No.	Hydrocarbon Classification	Mol. Formula	Chemical Structure
9.	Fluoranthene	$C_{16}H_{10}$	
10.	Pyerene	$C_{16}H_{10}$	
11.	Benzo[a]anthracene	$C_{18}H_{12}$	
12.	Benzo[a]pyrene	$C_{20}H_{12}$	
13.	Benzo[ghi]perylene	$C_{22}H_{12}$	

petrochemical production, manufacturing and use of paints, and dry cleaning are the stationary sources of hydrocarbon release in the atmosphere (Mishra 2008).

A major cause of atmospheric hydrocarbons is automobiles. Moreover, driving in peak-hour traffic congestion in an urban context results in a noteworthy escalation of the hydrocarbon emission into the environment (Zhang et al. 2011; Wallington et al. 2008). During the deceleration, the hydrocarbon emission is observed to increase, contrary to the nitric oxides, carbon monoxide, or soot (Jung et al. 2017; Rakopoulos and Giakoumis 2009).

The hydrocarbons released into the air not only contribute to the ground-level ozone concentration or smog but also have negative health effects (Macias et al. 2010; Nylund et al. 2004). These are otherwise known as volatile organic compounds (VOCs). These are the unburnt fuels which have escaped into the atmosphere through fuel evaporation during the combustion process. Hydrocarbons can often be divided into discrete groups of methane (CH4) and non-methane (NMVOCs).

A class of toxic environmental contaminants is the polycyclic aromatic hydrocarbons (PAHs) that, due to a number of natural and human activities, have gathered in the environment. The incomplete combustion of organic compounds all over the procedure of industrial processes and other anthropogenic activities are the cause of the largest discharge of PAHs into the environment. PAHs are the organic compounds that are organized in numerous structural conformations comprising two or more merged benzene rings and/or pentacyclic molecules (Seo et al. 2009; Cheung et al. 2001). Usually, these are produced from natural as well as anthropogenic activities. Exudates from trees, volcanic eruptions, forest fires, and oil seeps are a few of the natural sources. On the other hand, solid biomass fuel (SBF), such as crop residue, wood or cow dung cake burning of fossil fuel, high-temperature industrial procedures, effluent from petroleum refinery, coal tar, and crude oil or petroleum spill (oil spillage and leakage) are several anthropogenic sources (Sharma and Jain 2019; Haritash and Kaushik 2009). Smoke from wood-burning stoves, manufactured gas plants (coal gasification), creosote waste materials, as well as automotive emissions, are some of the anthropogenic sources of PAHs (Abdel-Shafy and Mansour 2016).

PAHs are formed in the course of combustion reactions. Consequent atmospheric deposition along with fossil fuel burning predominantly from industrialized countries results in the intake of petroleum hydrocarbons to the environment (Sims and Overcash 1983). Huge quantities of petroleum hydrocarbon by-products are generated through the internal combustion engines. The overall environmental load of petroleum hydrocarbon results from the numerous industrial processes related to processing, production, and the disposal of petroleum hydrocarbons. The remediation of polluted soils has become an issue, in view of the harmfulness and universal occurrence of PAHs in the environment. For that reason, it is imperative to recognize the thorough mechanism of physical, chemical, or biological changes in soil. At the same time, for the remediation of contaminated sites, it has become pertinent to recognize the environmentally sustainable treatment options. Biological remediation is evolving as a competent and effective choice which engages microorganisms in mitigation. But physical and chemical treatments are costly, the chemicals may be forbidden, or energy-prohibitive. Microorganisms for their enzyme-catalyzed catabolic activity can prove beneficial in the degradation of PAHs, when the degradation or mineralization of a pollutant is intended (Sakshi et al. 2019)

14.4   SOIL POLLUTION DUE TO HYDROCARBON FROM VEHICULAR EMISSION

India has also observed an enormous progression of automobiles with the upsurge in vehicle use, along with service stations where a significant quantity of used motor oil is inaccurately discharged into the environment. During a manual oil changing operation, the used motor oil is not reprocessed but spilled and discarded at station sites, thus adulterating both soil and water (Bhat et al. 2011). Used motor oils are likely to encompass a higher percentage of aliphatic and aromatic hydrocarbons (C_{15}-C_{50}), sulphur and nitrogen compounds, and metals (Mg, Ca, Zn, Pb, etc.) than fresh oils (Hewstone 1994). Along with the above-mentioned compounds, the presence of benzo[a] pyrene, anthracene, and naphthalene cannot be overlooked (Cotton et al. 1977). Over a long period of time, an erroneous and accidental release of such compounds may accumulate, being a potential hazard to environmental and human health (Vazquez-Duhalt 1989). Along with new additives linked to different engine types and operation practices, it is foreseen that petroleum-based new wastes could be produced at all times, which can be linked to newer technologies for oil manufacture (Elena and John 2003). As a consequence of anthropogenic activity, soils are perceived to be contaminated. Vehicular emission has been found to constitute one of the foremost sources of soil contamination. When impurities that lead to alteration in the physical or chemical properties of the soil are introduced in the soil, it causes soil pollution. The solid constituents of soil are customarily composed in the form of aggregates, thus forming a system of interconnected voids of different sizes, full of either air or water. The cavities created facilitate the build-up of constituents that settle in them, easily causing pollution (Osuagwu et al. 2019).

14.5 CONTENTS OF VEHICULAR EMISSION

Even though road transportation is perceived as enhancing the progress of a nation, it in fact introduces enormous volumes of harmful impurities into the environment. Automobile emissions are released due to the ignition and evaporation of fuel. A great many pollutants are released into the environment from the exhaust, as the burning of fuel takes place in the engine of automobiles (Osuagwu et al. 2019). Automobiles primarily produce two categories of contaminants: (1) gaseous pollutants and (2) particulate pollutants. Gaseous contaminants are fundamentally released via fuel exhausts and are further grouped into inorganic and organic gas pollutants. Oxides of nitrogen (NO_2, NO, N_2O, etc.), oxides of carbon (CO_2, CO, etc.), oxides of sulfur (SO_2, SO_3, etc.) ammonia and ozone are eminent inorganic gaseous pollutants, whereas organic gaseous pollutants comprise aliphatic and aromatic hydrocarbons like benzene, benzo(a) pyrene, aldehydes, ketones, etc. (Gupta 2020).

When combustion of fuel in a vehicle takes place, pollutants such as carbon dioxide (CO_2), carbon monoxide (CO), oxides of nitrogen (NOx), oxides of sulfur (SOx), particulate matters and hydrocarbons (HC) are discharged (Adeyanju and Manohar 2017). The diesel engine emits the four leading pollutants, namely HC-hydrocarbons, CO-carbon monoxide, nitrogen oxides-NOx and PM-particulate matter (Resitoglu et al. 2015). The thorough burning of diesel fuel would only produce CO_2 and H_2O in combustion chambers of engine for ideal thermodynamic equilibrium (Prasad and Bella 2010). However, a number of detrimental products are engendered in the course of combustion attributable to numerous reasons like the air-fuel ratio, ignition timing, turbulence in the combustion chamber, form of combustion, air-fuel concentration, combustion temperature, etc.

14.6   PROBLEMS RELATED TO HYDROCARBON-CONTAMINATED SOIL

Soil polluted with petroleum causes environmental problems and poses a grave danger to human health as well. Petroleum contaminants, primarily hydrocarbon, are categorized as priority pollutants (Yuniati 2018). Characteristically, an intricate mixture of aliphatic and aromatic organic compounds forms petroleum hydrocarbons. They can be divided into aromatics, saturates, asphaltenes, and resins by the process of distillation.

Polycyclic aromatic hydrocarbons (PAHs) are organic molecules comprising two or more benzene rings. The arrangement and number of rings result in varied chemical and physical properties. In healthy and clean soil, microorganisms in the natural environment are fairly profuse.

In order to adapt to oil pollution stress conditions, the microbes produce certain enzyme systems and progressively form a prevailing population with synergy or symbiotic effects (Alisi et al. 2009).

Several studies have reported that given priority to the inhibitory action, the hydrocarbon contamination can alter the population of microbes, the conformation of the structure of the community, and the enzyme structure in soil (Deng 2014). It can disturb the usual development of crops by reducing the percentage of propagation and fertility, and also reduce the pest and disease resistance (Shan et al. 2014). Adulterated soil can unfavorably affect human health, as the hazardous petroleum hydrocarbons are captured through direct dermal interaction. Furthermore, the PAHs in petroleum chemicals have teratogenic, mutagenic, carcinogenic, and other noxious effects. Through dermal contact, breathing, and diet, it can pass into humans and animals, corrupting the usual function of organs such as the liver and kidney, etc. As a consequence, excessive risk to human health is caused. The atmosphere and hydrosphere, along with the pedosphere, are influenced by the oil pollutants.

The low boiling point and lightweight hydrocarbons can enter the atmosphere by evaporation easily, thus causing atmospheric pollution. But in the case of run-off and infiltration into the surface water and osmosis into the groundwater system, this causes hydrosphere pollution with hydrocarbons and ultimately enters the food chain, entering human bodies. Hence, there is an urgent need to implement operative actions to reclaim polluted sites and regulate other sites to avoid contamination. But the remediation of soil is universally known to be amongst the more costly treatments available in this field. As a result, in order to scrutinize the utmost cost-effective solution to deal with adulterated sites, various strategies have been chosen (Agamuthu et al. 2013). Such a presence in nature is of great concern today, and there is a need for it to be cleaned from the environment in the best possible way. Much research has been carried out to determine the ecotoxicity of these pollutants but a biological method has been reported to be more suitable for determining the possible hazards of pollutants in soil on ecological and environmental bases (Tang et al. 2011).

At a molecular level, these hydrocarbons are toxic, mutagenic, and carcinogenic (Maertens et al. 2008). Fresh spills of TPHs on land are initially subjected to volatilization, especially from the less porous surfaces, while the heavier hydrocarbons may be partially oxidized by auto-, thermal-, and photo-oxidation, in addition to biodegradation.

14.7 REMEDIATION TECHNIQUES FOR THE TREATMENT OF HYDROCARBON-POLLUTED SOIL

To prevent destruction of the contaminated sites' ecosystem, these media need to go through a systematic treatment process. The procedure of remediation can include the introduction of certain chemicals into the system or, on the contrary, an extremely complex procedure involving a number of chemical and biological processes can also be executed. There are various physical, chemical, thermal, and biological technologies available to reduce or remove the contamination degree in soils to a minimum. Every technique has its specific advantages and disadvantages. The choice of the process of remediation is significantly governed by factors such as the physicochemical characteristics of the soil, the type of contamination, the extent and concentration of the contaminants, the cost, and the soil availability.

PAH-contaminated soils are treated employing a varied set of physicochemical and biological (biodegradation and bioremediation) remediations. Physical, chemical, microbiology, and plant remediation are the more commonly used methods (Wang et al. 2017). For the remediation of oil-contaminated soil, there are diverse forms of physical and chemical methods, namely dispersion, evaporation, washing, burying, etc. Some of the mechanical methods include soil vapor extraction, soil washing, and incineration. But these technologies are costly, and can lead to incomplete decomposition of pollutants (Bundy et al. 2002). There are also a few chemical approaches but these are very expensive to treat hydrocarbon-contaminated sites. For that reason, in order to eliminate hydrocarbon contamination from the soil, it is imperative to develop an advanced, low-cost, and eco-friendly technique.

To treat a hydrocarbon-contaminated site, bioremediation is considered to be a more promising, economical, and safe technique (Mrozik et al. 2003). In general, bioremediation techniques can be categorized as *in situ* or *ex situ*. In the *in situ* bioremediation, the treatment of the contaminant material takes place at the site, whereas in the *ex situ* the adulterated material is treated in another place (Zargar et al. 2014).

Bioremediation is the non-invasive, most effective, least expensive, and eco-friendly technique (Abdulsalam and Omale 2009; Perelo 2010; Silva-Castro et al. 2015) grounded on the principle of complete transformation or mineralization of petroleum products into less toxic forms by diverse groups of microorganisms (Esmaeli and Akbar 2015). The advantage of bioremediation is the conservation of the soil texture and characteristics. Along with the physical and chemical properties, aeration, pH, water-holding capacity, and ion exchange capacity of the soil can be improved after bioremediation (Nwogu et al. 2015). In the contaminated soil, the processes of bioremediation have been observed to be an effective practice that encourages biodegradation. As several studies have reported, bioremediation – due to its efficiency in eradicating numerous pollutants from many contaminated sites – is considered an approach with high potential (Olaniran et al. 2006; Fantroussi and Agathos 2005; Hii et al. 2009). Bioremediation technology can generally be divided

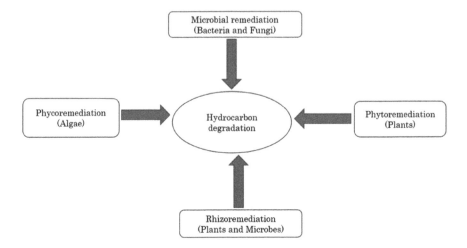

FIGURE 14.2   Treatment of hydrocarbon-contaminated soil.

into four categories depending on the type of organism: phytoremediation, microbial remediation, phycoremediation, and plant-microbial (rhizoremediation) technology (Figure 14.2).

The foremost objective of bioremediation is the conversion of organic pollutants into less lethal metabolites, or to mineralize them to CO_2 (carbon dioxide) and water (Tortella et al. 2005).

14.7.1   Phytoremediation of Hydrocarbon-Contaminated Soil

The large volume of soil affected precludes *ex situ* treatment due to economical constraints and requires the use of relatively inexpensive remediation schemes, such as phytoremediation.

The occurrence of flora has been shown to have positive effects on contaminated soil, leading to a better degree of degradation, elimination, and mineralization of wastes than in non-vegetated soils (Gaskin et al. 2008; Gaskin and Benthan 2010; Zhang et al. 2010). Over the past decade, considerable progress has been made in applying phytoremediation to inorganic pollutants such as heavy metals (Bhargava et al. 2012; Asensio et al. 2013), nutrients and landfill leachate (Schröder et al. 2008; Souza et al. 2013), as well as organic pollutants such as poly-aromatic hydrocarbons (PAH).

Phytoremediation works best at sites with low to medium amounts of pollution. Plants remove harmful chemicals from the ground when their roots take in water and nutrients from polluted soil, streams, and groundwater. Plants can clean up chemicals as deep as their roots can reach. Tree roots grow deeper than smaller plants, so they are used to reach pollutants deeper in the ground.

Phytoremediation or plant-enhanced bioremediation entails using the symbiotic relationship between plant and microbes. The phytostimulation mechanism of phytoremediation involves the stimulation of soil microbial activity at the rhizosphere for the breakdown of organic contaminants and is effective for the degradation of petroleum hydrocarbons and polyaromatic hydrocarbons. As native and indigenous plants may not require long term maintenance and are better adapted to the environment, they are more effective than any other exotic plant species for the phytoremediation of hydrocarbon-contaminated soil for economic and ecological reasons.

In a study conducted by Bordoloi et al. 2015, the sedge species viz., *Cyperus rotundus* (Linn.), *Cyperus brevifolius* (Rottb.) Hassk, *Cyperus odoratus* L., and *Cyperus laevigatus* L. native to India showed significant degradation of hydrocarbon.

In a study undertaken by Bramley et al. (2014), *P. foliosa* was found to show good results and hence provides valuable evidence that phytoremediation is a good technique for use on Macquarie Island, and may be applicable to the management of fuel spills in other cold climate regions.

Rhizoremediation is the primary technique of phytoremediation of hydrocarbons involving the breakdown of contaminants in soil environment as a result of microbial activity in the root zone (Favas et al. 2014).

Microorganisms can inhabit three distinct areas of the root zone of a plant (Huang et al. 2014):

1) The endosphere, i.e., all the cells inside the roots (Compant et al. 2010);
2) The rhizoplane which is the root surface (Ben 2015), usually as biofilm (i.e., multiple layers of mature microcolonies covered by mucus); and
3) The rhizosphere, i.e., the soil immediately adjacent to roots (a few millimeters thick) and influenced by plant roots (Hartmann et al. 2008).

14.7.2 PLANT–MICROBE INTERACTION FOR THE BIOREMEDIATION OF HYDROCARBON-CONTAMINATED SOIL

Plant-associated bacteria have the potential to improve phytoremediation as they can contribute to the biodegradation of toxic organic compounds in polluted soil. These bacteria include endophytic bacteria (non-pathogenic bacteria that occur naturally in plants) and rhizospheric bacteria (bacteria that live on and near the roots of plants) (Bhatia and Kumar 2011). Plant growth-promoting rhizobacteria (PGPR) are naturally occurring soil bacteria that vigorously inhabit plant roots and help plants by providing growth enhancement (Saharan and Nehra 2011).

Bacteria possess a set of catabolic genes which produce catabolic enzymes to decontaminate hydrocarbons. In return, plants ooze out root exudates containing nutrients and necessary metabolites which facilitate the microbial colonization in the plant rhizosphere. This results in a high gene abundance and gene expression in the rhizosphere and thus leads to enhanced degradation. Moreover, high proportions of beneficial bacteria help plants to gain more biomass due to their plant growth-promoting activities and production of phytohormones.

For the enhancement of the microbes degrading hydrocarbon and their ability to degrade polycyclic aromatic hydrocarbons, a number of plants can release structural analogs of PAHs, such as phenols. *Pseudomonas* spp., a significant class of bacteria have PGPR activity and capacity to degrade hydrocarbon for such plant/microbe systems (Hontzeas et al. 2004; Singh et al. 2018, 2017a, b, c, 2019; Tiwari et al. 2018, 2019a, b; Kour et al. 2019). Moreover, the rhizosphere in the adulterated field accommodates greater diversity of PAH-degrading bacterial inhabitants, from which two *Lysinibacillus* strains were isolated (Ma et al. 2010).

The hydrocarbon breakdown through rhizodegradation is a mutual process. The plants deliver root exudates like enzymes, amino acids, simple sugars, aromatics as well as aliphatics to encourage the growth of root-associated microorganisms (Khan et al. 2013). Growth of the root can also spread out into deeper soil, permitting access to water and air, resulting in the variation of pH, carbon dioxide concentration, osmotic potential, redox potential, oxygen concentration and moisture content of the soil, hence leading to a better environment supporting a high microbial biomass (Lin et al. 2008). In return, microbes can reduce the phytotoxicity of the contaminants in the soil or augment the capacity of the plant to degrade contaminants (Khan et al. 2013). This mutualistic association between plants and microbes in the soil has been found to speed up the loss of petroleum hydrocarbons in soil (Wenzel 2008).

According to Wenzel (2008) and Khan et al. (2013), the presence of vegetation is known to enhance microbial populations within soil and accordingly to improve the degradation of soil contaminants due to co-metabolic processes (Chaudhry et al. 2005; Gaskin et al. 2008).

For example, when comparing vegetated soil to unvegetated, higher counts of culturable actinomycetes and microbes were found (Liste and Felgentreu 2006).

Likewise, Altai wild rye (*Elymus angustus Trin.*) supported endophytic hexane degraders up to a hundred times more than the unplanted control, and the promoting rates of TPH degradation were up to 50% higher (Philips et al. 2009).

14.7.3 MICROBIAL REMEDIATION

Remediation denotes the destruction, elimination, or conversion of contaminants to less damaging substances. It is predominantly comprised of biostimulation where inorganic or organic components are added to enhance the growth of indigenous microbial population that immediately degrades the contaminants. Numerous factors influence the biodegradability and the degree of persistence of hydrocarbons in natural environments, most important among which are the existence of viable microbial population competent to degrade, the chemical structure of the hydrocarbons and environmental conditions ideal for microbial degradative events (Sihag et al. 2014). Quite a few microbes inherit the capability to thrive on soil contaminated with hydrocarbon and they are proficient at degrading oil than the microbes that can propagate on non-contaminated sites of oil (Passarini et al. 2011). To degrade diverse hazardous pollutants, bioremediation uses the metabolic versatility of microorganisms. Microbes capable of quick adaptation and tolerance to pollutants are required for a feasible soil remediation technology. A number of aspects affect the microbes for the utilization of pollutants as substrates. Therefore, it is essential here to describe the most appropriate aspects for applying a bioremediation technology at field scale, along with an understanding of the procedures, catabolic pathways, and liable enzymes for pollutant degradation.

The two commonly used key strategies for soil bioremediation are: (1) biostimulation, which improves the developmental conditions of native degrading populations in soil by the introduction of fertilizers, texturizing agents, and aeration; (2) bioaugmentation, addition of specific pollutant-degrading microorganisms. In the soil, for the conversion of PAHs and other hydrocarbons the above two methods can be used through *in situ* techniques such as land farming, composting, and biopiles, whereas in other advanced *ex situ* methods the practice of bioreactors offer enhanced control over temperature and pressure to improve the degradation procedures but is lacking in versatility.

14.7.3.1   Degradation Using Bacteria

Bacteria working as principal degraders for a varied range of target constituents present in water, soil, and sludge are the most active petroleum-degrading agents. These are the group of microorganisms that are vigorously involved in the decay of organic contaminants from polluted spots. A wide amount of species of bacteria are recognized to degrade PAHs. Several reports have listed the degradation or decomposition of environmental contaminants by different bacteria. Indeed, significant amounts of bacteria are even known to nourish solely on hydrocarbons. Bacteria with the hydrocarbon-degrading capacity are known as "hydrocarbon-degrading bacteria". A huge number of them are isolated from polluted soil or sediments representing biodegradation efficiency.

From a petrochemical waste disposal site, six grams of negative strains of bacteria were isolated by Yuan et al. (2002), having the ability to degrade fluorene, acenaphthene, anthracene, phenanthrene, and pyrene by 70–100% in a period of 40 days of initial treatment. Out of the six isolated strains, two were the rod-shaped bacteria, namely *Pseudomons fluoresens* and *Haemophilus* spp. Two bacterial strains *Rhodococcus* spp., and *Mycobacterium flavescens* were isolated by Dean-Ross et al. (2002) from deposits from two various sites at the Grand River Calumet. Both the *M. flavescens and Rhodococcus* species were found to be capable of mineralization of pyrene and anthracene respectively. Low biodegradability and high persistence of hydrocarbons have a significant environmental impact.

Biodegradation is based on three basic principles: the presence of microorganisms with metabolic capacity, either autochthonous or inoculated (bioaugmentation); the availability of the contaminant as an energy and carbon source; and the appropriate conditions for microbial growth and activity (Hernández et al. 2013).

Several microbes have a naturally occurring diversity, astonishing, microbial catabolic activity to accumulate, degrade or transform an enormous variety of compounds together with radionuclides and metals, polychlorinated biphenyls (PCBs), hydrocarbons (e.g., oil), as well as polyaromatic hydrocarbons (PAHs) (Mancera et al. 2008).

For the remediation of soil contaminated with petroleum/crude oil, an eco-friendly, cost-efficient and effective technique is bioremediation (Gandolfi et al. 2010; Han et al. 2009; Zhang et al. 2008). Bioremediation is grounded upon the principles of natural attenuation, bioaugmentation, and biostimulation (Brooijmans et al. 2009). Natural attenuation is the simplest method of bioremediation, which includes only the monitoring of the soil for variation in the concentration of contaminants to ensure that the alteration of the contaminant is vigorous (Wang et al. 2007). Typically, bioaugmentation is functional in cases in which natural vigorous microbial communities are absent or exist in small quantities, wherein the addition of hydrocarbon-degrading microorganisms such as *Rhodococcus*, *Sphingobium*, *Pseudomonas*, and *Sphingomonas* spp. are pervasive in the ecosystem (Mnif et al. 2009). Biodegradation efficiency can be affected by many factors like pH, the proportion of organic matter, contamination source, soil texture, and the history of the pollution (Dastgheib et al. 2011). This naturally occurring process can be enhanced by biostimulation. By adding nutrient-rich organic and inorganic materials, oxygen supply or other electron acceptors, and by preserving appropriate conditions of temperature, pH, and moisture, the catabolic activity of indigenous microorganisms is stimulated, hence enhancing the capability of a microbial population to degrade contaminants (Andreolli et al. 2015).

The degree of conversion of complex hydrocarbon compounds is generally limited by biodegrading microbiota in arid areas, where soils are deficient in organic and mineral nutrients, as well as being commonly subjected to extreme environmental conditions, i.e., high temperatures and irradiance (Radwan 2009).

14.7.3.2 Microbial Degradation Using Fungi

Fungi also are the primary organisms accountable for the decay of carbon in the biosphere. Similarly to bacteria, they metabolize dissolved organic matter and hence play a significant part in the process of degrading microbiota. Fungi own significant properties for degradation that can be used for the eradication of perilous wastes from the environment as well as the reprocessing of recalcitrant polymers (e.g., lignin). A varied group of fungi are capable of mineralizing petroleum hydrocarbons, belonging to the genera *Penicillium* spp., *Saccharomyces* spp., *Aspergillus* spp., *Fusarium* spp., *Talaromyces* spp., *Cunninghamella* spp., *Neosartorya* spp., *Syncephalastrum* spp., *Amorphoteca* spp., *Graphium* spp., *Phanerochaete* spp., and *Paecilomyces* spp. with varying degradation rates (Germida et al. 2002; Chaillan et al. 2004). The capability to oxidize and dissipate an inclusive range of PAHs into a number of less harmful metabolic products have been shown by the white-rot fungi, as well as numerous filamentous fungi. For example, a filamentous non-ligninolytic fungus, *Cunninghamella elegans*, has been associated in the transformation and degradation of quite a few PAH compounds comprising benzo[a]pyrene, 9,10-dihydrobenzo[a]pyrene and benz[a]anthracene, which were isolated from soil (Moody et al. 2004). Further examples of filamentous fungi including *Cyclothyrium* spp., *Psilocybe* spp., and *Penicillium simplicissimum* are reported to demonstrate hydrocarbonoclastic activities against different PAHs (Tortell et al. 2005).

In a study (Jawhari 2014), fungal species such as *Fusarium solani*, *Aspergillus fumigatus*, *Penicillium funiculosum*, and *Aspergillus niger*, were found to be more predominant in the petroleum-polluted soil with high frequency. This gives the impression that the oil compounds are consumed as nutrients by the fungal species, as well as noting that the crude oil pollution can cause an increase in fungal growth. Another study reported the isolation of *Rhizopus* spp., *Aspergillus terreus*, and *Penicillium* spp., from soil samples (Lotfinasabasl et al. 2012). For the remediation of polycyclic aromatic hydrocarbon and pentachlorophenol (PCP) from diesel-contaminated soils in oil refinery sites, many inherent strains comprising ligninolytic fungi have huge potential (Low et al. 2008).

TABLE 14.2
Degradation of PHs by Various Species of Fungi

S. No.	Name of Organism	Compound	References
1.	*Trichoderma harzianum*	Naphthalene	Mollea et al. 2005
2.	*Aspergillus fumigatus*		Ye et al. 2011
3.	*Aspergillus spp.*	Crude oil	Zhang et al. 2016
4.	*Cunninghamella elegans*	Phenanthrene	Romero et al. 1998
5.	*Aspergillus niger*	n-hexadecane	Sepulveda et al. 2003
6.	*Penicillium sp.*		Pointing 2001
7.	*Cunninghamella elegans*	Pyrene	
8.	*Aspergillus ochraceus*	Benzo[a]pyrene	Passarini et al. 2011
9.	*Trametes versicolor*		Collins et al. 1996
10.	*Penicillium* sp. RMA1 and RMA2	Crude oil	Al-Hawash et al. 2018a
11.	*Aspergillus* sp. RFC-1	Different PHs	Al-Hawash et al. 2018b

In a study by Chaudhry et al. (2012), fungal genera such as *Aspergillus*, *Fusarium*, *Cladosporium*, *Penicillium*, *Rhizopus*, and *Alternaria* were reported. The versatility of the technique, along with its cost-effectiveness as equated to other technologies for remediation, are the advantages associated with fungal bioremediation.

The utilization of fungi is estimated to be relatively inexpensive, as they can be grown on low-cost agricultural or forest wastes such as corncobs and sawdust. A study showed that *Fusarium* sp. F092 was found capable of degrading an aliphatic portion in crude oil under saline conditions (Hidayat and Tachibana 2012).

The anthracene degradation effectiveness of three non-ligninolytic and ligninolytic fungi was assessed by Jové et al. (2016), and it was observed that the anthracene degradation competence of *Phanerochaete chrysosporium* was greater than that of *Pleurotus ostreatus* and *Irpex lacteus* (Table 14.2).

14.8 APPROACHES FOR THE BIOLOGICAL REMEDIATION OF PAHS IN SOIL

Regardless of the price for applying such a remediation technique, such practices undoubtedly play a conclusive role. In this sense, bioremediation has huge benefits compared to other technologies, besides being environmentally friendly and generating commendable outcomes in relation to elimination (Zafra et al. 2014).

Biological methods retain benefits over physicochemical and thermal methods, as they are cost-effective, have little need for further treatment, and have the potential to fully mineralize pollutants.

In a study undertaken in Kuwait, strains of *Bacillus subtilis* were isolated from oil-polluted soil. The bioremediation of PAH-polluted soils was discovered to be of growing environmental significance. Mechanisms of PAH biodegradation using either a single organism or a consortia of the native or introduced organisms in polluted soils have been studied for years. Several bacterial, fungal, and algal microbial species are defined as capable in some measure, or entirely metabolized, to low molecular weight (LMW) or high molecular weight (HMW) with PAHs in aerobic or anaerobic conditions.

Long-term petrochemical waste discharge harbors bacteria that are able to degrade PAH to a significant extent. Among the PAH in petrochemical waste, Benzo(a)pyrene is considered as the most carcinogenic and toxic. A species of *Mycobacterium* strain KR2 from a PAH-polluted soil of a gasworks plant was isolated by Rehmann et al. 2001, who showed that it was able to consume pyrene as a solitary source of carbon and energy.

Degradation products were recognized as 4-5-phenanthrene dicarboxylic acid, Cis-4,5-pyrene dihydrodiol, 2-carboxybenzaldehyde, 1-hydroxy-2-naphthoic acid, protocatechuic acid, and phthalic acid, and a degradation pathway for pyrene was also proposed.

Biodegradation is a mode of nature to recycle the wastes generated or to break down the organic matter into nutrients that can be utilized by other organisms. In microbiology, "biodegradation" denotes a huge assemblage of life-forms encompassing mainly fungi, yeast, and bacteria, as well as perhaps other organisms which are involved in the decaying and decay of all organic materials. An enormous range of compounds comprising of hydrocarbons (e.g., oil), polychlorinated biphenyls (PCBs), polyaromatic hydrocarbons (PAHs), radionuclides, and metals are accumulated, transformed, or degraded with the help of biotransformation and bioremediation techniques. These practices attempt to employ what is inherently occurring, and impressive in their microbial catabolic diversity. Chemical, biochemical, and microbiological molecular indices are counted in the microbial methods for monitoring hydrocarbon bioremediation. It measures the degree of microbial activities to ultimately demonstrate that the aim of reducing the pollutant to a non-toxic and acceptable level has been attained (Chikere et al. 2011). The application of microbial processes or microorganisms to degrade or to remove pollutants from soil is known as bioremediation. This microbiological decontamination is claimed to be an economic, efficient, and versatile substitute to physicochemical treatment.

14.9 MICROBIAL BIOREMEDIATION OF POLLUTANTS (HYDROCARBON) IN CONTAMINATED SOIL

Biodegradation is a relatively low-cost and non-invasive alternative to physical and chemical processes (Das and Chandran 2011; Varjani and Upasani 2016).

Biological degradation of petroleum hydrocarbon is a complex procedure that is governed by the quantity and nature of the hydrocarbon present. The saturates, the resins (pyridines, quinolines, carbazoles, sulfoxides, and amides), the aromatics, and the asphaltenes (phenols, fatty acids, ketones, esters, and porphyrins) are the four classes in which the petroleum hydrocarbon can be divided. Organisms demonstrating the potential for the degradation of different fractions of petrogenic hydrocarbons belonging to different genera have been reported as hydrocarbonoclastic. A lot of these have been isolated from either aquifers or soils. *Mycobacterium* spp., *Marinobacter* spp., *Pseudomonas* spp., *Flavobacter* spp., *Achromobacter* spp.,

Arthrobacter spp., *Corynebacterium* spp., *Alcaligenes* spp., *Nocardia* spp., and *Micrococcus* spp., are the typical group of bacteria. According to studies, scientists have reported the isolation of other bacterial genera including *Bacillus, Dietzia, Gordonia, Halomonas, Cellulomonas, Rhodococcus*, and halotolerant *Alcanivorax* spp., which have the capacity to degrade and oxidize an extensive variety of hydrocarbons of crude oil (n-alkanes and aromatic hydrocarbons).

14.10 FACTORS AFFECTING THE MECHANISM OF BIOREMEDIATION

As the petroleum hydrocarbon compound has the ability to bind itself to the soil components, it is challenging for it to be degraded or removed from the environment. There are a few compounds like the high molecular weight polycyclic aromatic hydrocarbons (PAHs), which cannot be degraded at all. Cooney et. al. reported various factors that can influence hydrocarbon degradation. The limited accessibility of the oil contaminants to the microbes is one of the key factors that limits the biodegradation process. The vulnerability to microbial attack differs in various hydrocarbons. The vulnerability of the hydrocarbons to microbial degradation can be largely ordered as follows: linear alkanes > branched alkanes > small aromatics > cyclic alkanes (Das and Chandran 2011).

As stated by several researchers, the removal rate of petroleum can be amplified by the use of microbial agents. A composite microbial agent attains an improved bioremediation proficiency of petroleum adulterated soils as observed in previous research (Wang and Wang 2018). Owing to the accompanying metabolic capabilities, the main agents of the degradation of petroleum hydrocarbons are the hydrocarbonoclastic microorganisms. Many researchers have stated that utilizing petroleum hydrocarbons as the solitary carbon source for metabolism and energy, and several microbes (mainly bacteria and fungi) are capable of the degradation of petroleum hydrocarbons (Figure 14.3).

14.11 GENETICALLY ENGINEERED MICROBES FOR BIOREMEDIATION

Due to its eco-friendliness, social acceptance, and lower risk of health hazards, microbial bioremediation has received considerable global attention for its pollution abatement. Moreover, with recent advances in biotechnology and microbiology, genetically engineered bacteria with a high ability to remove environmental pollutants are widely used in the fields of environmental restoration, resulting in the bioremediation in a more viable and eco-friendly way (Zhang et al. 2019). Genetically engineered microorganisms are microorganisms whose genetic material has been changed by applying a genetic exchange between microorganisms inspired by genetic engineering techniques, and are mainly termed as recombinant DNA technology.

With the development of recombinant DNA and genetic engineering technologies in microbial breeding, a large number of efficient engineering bacteria with augmented abilities of pollutant degradation were constructed, thus significantly improving the degradation efficiency of pollutants (Zhao et al. 2017a).

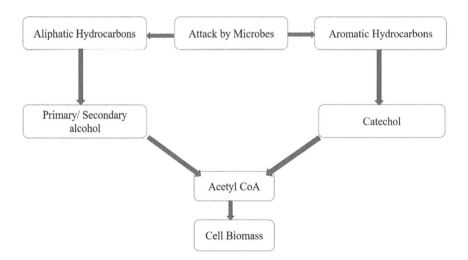

FIGURE 14.3 Mechanism of hydrocarbon degradation through microbes.

Besides the screening of strains by natural mutant or physicochemical mutagenesis, various strategies have been used to construct engineered strains for accelerating the process of environmental governance:

1) Screening and cloning of highly effective degrading genes;
2) Increasing the expression of enzymes with degradation functions in microorganisms;
3) Expressing degradation genes for different pollutants in a recipient to construct super-engineered bacteria;
4) Protoplast fusion by blending the advantages of both parents for pollution degradation.

Moreover, the discovery of genes and degradation pathways, as well as the elucidation of degradation mechanisms in bacteria, provided the possibility of developing genetically engineered bacteria with high degradation efficiency (Zhao et al. 2017b).

The degradation of some petroleum components by microorganisms is controlled by an extrachromosomal plasmid; therefore, superbugs can be constructed by introducing plasmids with capabilities for degrading different components in a single cell. A recombinant *Acinetobacter baumannii* S30 pJES was constructed by inserting the lux gene into the chromosome of the *A. baumannii* S30, a strain with the biodegradation efficiency for total petroleum hydrocarbon (TPH) of crude oil. The degradation of some petroleum components by microorganisms is controlled by an extrachromosomal plasmid; therefore, superbugs can be constructed by introducing plasmids with capabilities for degrading different components in a single cell.

A large variety of bacterial isolate with pronounced PAH-degrading abilities have been sequestered and screened from PAH-polluted water and soil. Among the plethora of bacterial isolates, strains belonging to the genera *Sphingobium* and *Sphingomonas* have received widespread consideration owing to their potential

TABLE 14.3

List of Genetically Engineered Strains for the Biodegradation of Petroleum

S. No.	Name of Organism	Degradation Substrate
1.	A. baumannii S30 pJES	TPH
2.	Acinetobacter sp. BS3-C23O	n-Alkanes/ aromatic hydrocarbons
3.	Streptomyces coelicolor M145-AH	n-Alkanes
4.	F14	PAHs
5.	Pseudomonas sp. CGMCC2953-pK	PAHs
6.	E. coli	PAHs
7.	P. putida U ΔfadBA-phaZ	PAHs

biodegradative abilities and diverse ecological adaptations (Aylward et al. 2013). Strains belonging to these two genera have shown a great potential to degrade an array of aromatic compounds including dioxins, biphenyl, naphthalene, carbazole, m-xylene, fluorene, phenanthrene, HCH, pentachlorophenol (PCP), chrysene, benzo[a]pyrene, acridine, toluene, benz[a]anthracene, and many other methyl-, chloro-, hydroxyl-, and nitroaromatic compounds (Zhao et al. 2017b). A fusant F14 capable of degrading PAHs was obtained by fusing a pyrene-degradation strain (*Pseudomonas sp.*) GP3A and a phenanthrene-degradation strain (*Sphingomonas sp.*) GY2B (Lu et al. 2013). Moreover, an engineered strain with the ability of PAH degradation was developed by expressing salicylate oxygenase, an enzyme encoded by bphA2cA1c from *S. yanoikuyae* B1, in E. coli (Cho et al. 2005) (Table 14.3).

Bacterial genes encoding catabolic enzymes for recalcitrant compounds started to be cloned and characterized in the late 1970s and early 1980s. For addressing biodegradation, soon several microbiologists and molecular biologists recognized the ability of genetic engineering (Cases and Lorenzo 2005).

A microorganism whose genetic material has been altered using genetic engineering techniques inspired by natural genetic exchange between microorganisms is defined as a genetically engineered microorganism (GEM) or genetically modified microorganism (GMM). These techniques are in general known as recombinant DNA technology. Microorganisms may be genetically engineered for several purposes. One such purpose is for the efficient degradation of pollutants. As reported by Sayler and Ripp 2000, these genetically engineered microorganisms (GEMs) have shown potential for the bioremediation of soil, groundwater, and activated sludge, revealing the boosted degrading capabilities of a wide range of chemical contaminants.

The process of bioremediation can be divided into three phases or levels. In the first phase, contaminants are reduced by native microorganisms without any human augmentation through natural attenuation. Second, biostimulation is employed where nutrients and oxygen are applied to the systems to accelerate biodegradation and to improve their effectiveness. Finally, microorganisms are added to the systems during bioaugmentation. In order to degrade the target pollutant, the supplemental organisms should be more efficient than the native flora (Diez 2010). Several factors like the genetic potential and certain environmental factors such as temperature, pH, and available nitrogen and phosphorus sources influences microorganisms to use pollutants as substrates or co-metabolize them, then, seem to determine the rate and the extent of degradation (Fritsche and Hofrichter 2008). Therefore, applications of genetically engineered microorganisms (GEM) in bioremediation have received a great deal of attention. These GEM have higher degradative capacity and have been demonstrated successfully for the degradation of various pollutants under defined conditions.

The metabolic pathways related to hydrocarbon degradation will always depend on the microorganism and the growth conditions. Also, compounds will undergo biodegradation only if there are enzymes capable of catalyzing them. The microorganism P. putida has an outstanding metabolic versatility that allows its growth in many different carbon sources. There are many natural plasmids found in *P. putida*, including the TOL plasmid that provides the genes for degrading toxic mono-aromatic hydrocarbons (Claro et al. 2018).

One distinct plasmid is required for every compound to degrade all the toxic compounds of different groups. Four groups of plasmids are categorized as follows:

1) OCT plasmid which degrades octane, hexane, and decane;
2) XYL plasmid which degrades xylene and toluene;
3) CAM plasmid that decomposes camphor; and
4) NAH plasmid which degrades naphthalene (Ramos et al. 1994).

The potential for creating microbial strains able to degrade a variety of different types of hydrocarbons through genetic manipulation was demonstrated by Markandey and Rajvaidya in 2004. They successfully produced a multi plasmid-containing *Pseudomonas* strain that is capable of oxidizing aliphatic, aromatic, terpenic, and polyaromatic hydrocarbons. According to several researchers (Markandey and Rajvaidya 2004; Sayler and Ripp 2000) *Pseudomonas putida* that contained the XYL and NAH plasmid as well as a hybrid plasmid derived by recombinating parts of CAM and OCT developed by conjugation could degrade camphor, octane, salicylate, and naphthalene and can also grow quickly on crude oil as it was capable of metabolizing hydrocarbons more efficiently than any other single plasmid. This product of genetic engineering is known as a "superbug" (oil eating bug). The plasmids of *P. putida* degrading various chemical compounds are TOL (for toluene and xylene), RA500 (for 3, 5-xylene) pAC 25 (for 3-cne chlorobenxoate), pKF439 (for salicylate toluene). Plasmid WWO of *P. putida* is one member of a set of plasmids now termed TOL plasmid.

It appeared that molecular techniques could rapidly produce microbes with higher catalytic abilities, to basically degrade any environmental pollutant either through plasmid breeding or sheer genetic engineering (Sayler et al. 2000). A great deal of attention has been given to improving the degradation of hazardous wastes under laboratory conditions by the application of genetically engineered microorganisms (GEMs). Many studies have shown the degradation of environmental pollutants by different bacteria. The genetically engineered bacteria showed a higher degradative capacity.

Oilzapper is the first product developed by an assemblage of five naturally occurring bacterial species, which are able to biodegrade all the fractions of crude oil and oily sludge (Lotfinasabasl et al. 2012). It was used for the remediation of an oil spill on the Mumbai coast that was found accumulating in the mangrove forest of Navi Mumbai, Uran, and Alibaug.

The major constraints for testing GEM in the field are the ecological and environmental concerns and regulatory constraints. Before GEM can provide an effective clean-up process at a lower cost, these constraints must be resolved. The aromatic hydrocarbons, chlorinated compounds, and nonpolar toxicants are the various tested contaminants. For successful *in situ* bioremediation using genetically modified bacteria, the amalgamation of biochemical mechanisms, microbiological, and ecological knowledge, along with field engineering designs are essential elements.

14.12   CONCLUSION

Hydrocarbon contamination of soil is probably one of the prime concerns of the world nowadays. Due to the hydrocarbon contamination the soil becomes inappropriate for agricultural, industrial, or recreational use and also can cause groundwater and surface water adulteration. In order to remove hydrocarbons from contaminated sites, generally mechanical and chemical methods are used but they have limited effectiveness and can be expensive. To overcome this issue, microbial bioremediation has emerged as a promising treatment technique for the degradation of the hydrocarbons in the contaminated soil. In this technique, remediation of the contaminated soil is accomplished with the help of a diverse group of microorganisms, particularly the indigenous bacteria and fungi present in soil which utilizes the organic pollutants as a sole carbon source. Bioremediation is proved to be the non-invasive, most effective, least expensive, and eco-friendly technique, established on the principle of complete transformation or mineralization of petroleum products into less toxic forms by diverse groups of microorganisms. The main advantage of bioremediation is the conservation of the soil texture and its characteristics.

REFERENCES

Abdel-Shafy, H.I., Mansour, M.S.M. 2016. A review on polycyclic aromatic hydrocarbons: Source, environmental impact, effect on human health and remediation. *Egyptian Journal of Petroleum* 25:107–123.

Abdulsalam, S., Omale, A.B. 2009. Comparison of biostimulation and bioaugmentation techniques for the remediation of used motor oil contaminated soil. *Brazilian Archives of Biology and Technology* 52:747–754.

Adeyanju, A.A., Manohar, K. 2017. Effects of vehicular emission on environmental pollution in Lagos. *Sci-Afric Journal of Scientific Issues, Research and Essays* 5:34–51.

Agamuthu, P., Tan, Y.S., Fauzia, S.H. 2013. Bioremediation of hydrocarbon contaminated soil using selected organic wastes. *Procedia Environmental Sciences* 18:694–702.

AI-Jawhari, I.F.H. 2014. Ability of some soil fungi in biodegradation of petroleum hydrocarbon. *Journal of Applied & Environmental Microbiology* 2:46–52.

Al-Hawash, A.B., Alkooranee, J.T., Abbood, H.A., Zhang, J., Sun, J., Zhang, X., Ma, F. 2018a. Isolation and characterization of two crude oil-degrading fungi strains from Rumaila oil field Iraq. *Biotechnological Reports* 17:104–109.

Al-Hawash, A.B., Zhang, X., Ma, F. 2018b. Removal and biodegradation of different petroleum hydrocarbons using the filamentous fungus *Aspergillus* sp. RFC-1. *Microbiologyopen* 8(1):e00619.

Alisi, C., Musella, R., Tasso, F., Ubaldi, C., Manzo, S., Cremisini, C., Sprocati, A.R. 2009. Bioremediation of diesel oil in a co-contaminated soil by bioaugmentation with a microbial formula tailored with native strains elected for heavy metals resistance. *Science of the Total Environment* 407:3024–3032.

Andreolli, M., Lampis, S., Brignoli, P., Vallini, G. 2015. Bioaugmentation and biostimulation as strategies for the bioremediation of a burned woodland soil contaminated by toxic hydrocarbons: A comparative study. *Journal of Environmental Management* 153:121–131.

Asensio, V., Vega, F.A., Singh, B., Covelo, E.F. 2013. Effects of tree vegetation and waste amendments on the fractionation of Cr, Cu, Ni, Pb and Zn in polluted mine soils. *Science of the Total Environment* 443:446–453.

Aylward, F.O., McDonald, B.R., Adams, S.M., Valenzuela, A., Schmidt, R.A., Goodwin, L.A., Poulsen, M. 2013. Comparison of 26 sphingomonad genomes reveals diverse environmental adaptations and biodegradative capabilities. *Applied Environmental Microbiology* 79:3724–3733.

Ben, L. 2015. Life of microbes in the rhizosphere. In: Lugtenberg, B. (Ed.), *Principles of Plant-Microbe Interactions—Microbes for Sustainable Agriculture.* Springer International Publishing, Berlin, Germany, pp. 7–15.

Benson, N.U., Essien, J.P., Williams, A.B., Ebong, G.A. 2007. Petroleum hydrocarbons accumulation potential of shellfishes from littoral waters of the bight of bonny, Niger Delta, Nigeria. *Research Journal of Environmental Sciences* 1:11–19.

Bhargava, A., Carmona, F.F., Bhargava, M., Srivastava, S. 2012. Approaches for enhanced phytoextraction of heavy metals. *Journal of Environmental Management* 105:103–120.

Bhat, M.M., Shankar, S., Shikha, Y.M., Shukla, R.N. 2011. Remediation of hydrocarbon contaminated soil through microbial degradation-FTIR based prediction. *Advances in Applied Science Research* 2:321–326.

Bhatia, D., Kumar, M.D. 2011. Plant–microbe interaction with enhanced bioremediation. *Research Journal of Biotechnology* 6:72–79.

Bordoloi, S., Basumatary, B. 2015. Phytoremediation of hydrocarbon-contaminated soil using sedge species. In: Ansari, A., Gill, S., Gill, R., Lanza, G., Newman, L. (Eds.), *Phytoremediation.* Springer, Cham.

Bramley-Alves, J., Wasley, J., King, C., Powell, S., Robinson, S.A. 2014. Phytoremediation of hydrocarbon contaminants in subantarctic soils: An effective management option. *Journal of Environmental Management* 142:60–69.

Brooijmans, R.J., Pastink, M.I., Siezen, R.J. 2009. Hydrocarbon-degrading bacteria: The oil-spill clean-up crew. *Microbial Biotechnology* 2:587–594.

Bundy, J.G., Paton, G.I., Campbell, C.D. 2002. Microbial communities in different soil types do not converge after diesel contamination. *Journal of Applied Microbiology* 92:288–976.

Cases, I., de Lorenzo, V. 2005. Genetically modified organisms for the environment: Stories of success and failure and what we have learned from them. *International Microbiology* 8:213–222.

Chaillan, F., Le Flèche, A., Bury, E., Phantavong, Y.H., Grimont, P., Saliot, A., Oudot, J. 2004. Identification and biodegradation potential of tropical aerobic hydrocarbon-degrading microorganisms. *Research in Microbiology* 155:587–595.

Chaudhry, Q., Blom-Zandstra, M., Gupta, S.K., Joner, E. 2005. Utilising the synergy between plants and rhizosphere microorganisms to enhance breakdown of organic pollutants in the environment. *Environmental Science Pollution Research* 12:34–48.

Chaudhry, S., Luhach, J., Sharma, V., Sharma, C. 2012. Assessment of diesel degrading potential of fungal isolates from sludge contaminated soil of petroleum refinery, Haryana. *Research Journal of Microbiology* 7:182–190.

Cheung, P.Y., Kinkle, B.K. 2001. Mycobacterium diversity and pyrene mineralization in petroleum contaminated soils. *Applied and Environmental Microbiology* 67:2222–2229.

Chikere, C.B., Okpokwasili, G.C., Chikere, B.O. 2011. Monitoring of microbial hydrocarbon remediation in the soil. *3 Biotech* 1:117–138.

Cho, O., Choi, K.Y., Zylstra, G.J., Kim, Y.S., Kim, S.K., Lee, J.H., Kim, E. 2005. Catabolic role of a three-component salicylate oxygenase from Sphingomonas yanoikuyae B1 in polycyclic aromatic hydrocarbon degradation. *Biochemical and Biophysical Research Communications* 327:656–662.

Claro, E.M.T., Cruz, J.M., Montagnolli, R.N., Lopes, P.R.M., Júnior, J.R.M., Bidoia, E.D. 2018. Microbial degradation of petroleum hydrocarbons: Technology and mechanism. In: *Microbial Action on Hydrocarbons.* Springer, Berlin, pp. 125–141.

Collins, P.J., Kotterman, M., Field, J.A., Dobson, A. 1996. Oxidation of anthracene and benzo [a] pyrene by laccases from trametes versicolor. *Applied Environmental Microbiology* 62:4563–4567.

Compant, S., Clément, C., Sessitsch, A. 2010. Plant growth-promoting bacteria in the rhizo- and endosphere of plants: Their role, colonization, mechanisms involved and prospects for utilization. *Soil Biology and Biochemistry* 42:669–678.

Cotton, F.O., Whisman, M.L., Gowtzinger, S.W., Reynolds, J.W. 1977. *Hydrocarbon Processing* 56:131–140.

Das, N., Chandran, P. 2011. Microbial degradation of petroleum hydrocarbon contaminants: An overview. *Biotechnology Research International* 2011:13.

Dastgheib, S.M., Amoozegar, M.A., Khajeh, K., Ventosa, A. 2011. A halotolerant *Alcanivorax* sp. strain with potential application in saline soil remediation. *Applied Microbiology and Biotechnology* 90:305–312.

Dean-Ross, D., Moody, J., Cerniglia, C.E. 2002. Utilisation of mixtures of polycyclic aromatic hydrocarbons by bacteria isolated from contaminated sediment. *FEMS Microbiology Ecology* 41:1–7.

Deng, M.C., Li, J., Liang, F.R., Yi, M., Xu, X.M., Yuan, J.P., Peng, J., Wu, C.F., Wang, J.H. 2014. Isolation and characterization of a novel hydrocarbon-degrading bacterium *Achromobacter* sp. HZ01 from the crude oil-contaminated seawater at the Daya Bay, southern China. *Marine Pollution Bulletine* 83:79–86.

Diez, M.C. 2010. Biological aspects involved in the degradation of organic pollutants. *Journal of Soil Science and Plant Nutrition* 10:244–267.

Elena, D.R., John, P. 2003. Chemical characterization of fresh, used and weathered motor oil via GC/MS, NMR and FTIR techniques. *Proceedings of Indiana Academy of Science* 112:109–116.

El Fantroussi, S., Agathos, S.N. 2005. Is bioaugmentation a feasible strategy for pollutant removal and site remediation? *Current Opinion in Microbiology* 8:268–275.

Esmaeil, A.S., Akbar, A. 2015. Occurrence of *Pseudomonas aeruginosa* in Kuwait soil. *Chemosphere* 120:100–107.

Favas, P.J., Pratas, J., Varun, M., D'Souza, R., Paul, M.S. 2014. Phytoremediation of soils contaminated with metals and metalloids at mining areas: Potential of native flora. In: Hernandez-Soriano, M.C. (Ed.), *Environmental Risk Assessment of Soil Contamination.* In Tech, Rijeka, Croatia, pp. 485–517.

Favre, H.A., Powell, W.H. 2013. *Nomenclature of Organic Chemistry. IUPAC Recommendations and Preferred Names.* Royal Society of Chemistry, Cambridge.

Fritsche, W., Hofrichter, M. 2008. *Aerobic Degradation by Microorganisms in Biotechnology Set*, 2nd Edition, Rehm, H.-J., Reed, G. (Eds.). Wiley-VCH Verlag GmbH, Weinheim, Germany. doi:10.1002/9783527620999.ch6m.

Gandolfi, I., Sicolo, M., Franzetti, A., Fontanarosa, E., Santagostino, A., Bestetti, G. 2010. Influence of compost amendment on microbial community and ecotoxicity of hydrocarbon-contaminated soils. *Bioresource Technology* 101:568–575.

Gaskin, S.E., Bentham, R.H. 2010. Rhizoremediation of hydrocarbon contaminated soil using Australian native grasses. *Science of the Total Environment* 408:3683–3688.

Gaskin, S.E., Soole, K., Bentham, R.H. 2008. Screening of Australian native grasses for rhizoremediation of aliphatic hydrocarbon-contaminated soil. *International Journal of Phytoremediation* 10:378–389.

Germida, J., Frick, C., Farrell, R. et al. 2002. Phytoremediation of oil-contaminated soils. *Developments in Soil Science* 28:169–186.

Gupta, V. 2020. Vehicle-generated heavy metal pollution in an urban environment and its distribution into various environmental components. In: Shukla V., Kumar N. (Eds.), *Environmental Concerns and Sustainable Development*. Springer, Singapore.

Han, P.H.P., Zheng, L., Cui, Z.S., Guo, X.C., Tian, L. 2009. Isolation, identification and diversity analysis of petroleum-degrading bacteria in Shengli Oil Field wetland soil. *The Journal of Applied Ecology* 20:1202–1208.

Haritash, A.K., Kaushik, C.P. 2009. Biodegradation aspects of polycyclic aromatic hydrocarbons: A review. *Journal of Hazardous Material* 169:1–15.

Hartmann, A., Rothballer, M., Schmid, M., Lorenz, H. 2008. A pioneer in rhizosphere microbial ecology and soil bacteriology research. *Plant Soil* 312:7–14.

Hewstone, R.K. 1994. Health, safety and environmental aspects of used crankcase lubricating oils. *The Science of the Total Environment* 156:255–268.

Hidayat, A., Tachibana, S. 2012. Biodegradation of aliphatic hydrocarbon in three types of crude oil by *Fusarium* sp. F092 under stress with artificial sea water. *Journal of Environmental Science and Technology* 5:64–73.

Hii, Y.S., Law, A.T., Shazili, N.A.M., Abdul-Rashid, M.K., Lee, C.W. 2009. Biodegradation of Tapis blended crude oil in marine sediment by a consortium of symbiotic bacteria. *International Biodeterioration and Biodegradation* 63:142–150.

Hontzeas, N., Zoidakis, J., Glick, B.R. 2004. Expression and characterization of 1-aminocyclopropane-1-carboxylate deaminase from the rhizobacterium *Pseudomonas putida* UW4: A key enzyme in bacterial plant growth promotion. *Biochimica et Biophysica Acta* 1703:11–19.

Huang, X.F., Chaparro, J.M., Reardon, K.F., Zhang, R.F., Shen, Q.R., Vivanco, J.M. 2014. Rhizosphere interactions: Root exudates, microbes, and microbial communities. *Botany* 92:267–275.

Jové, P., Olivella, M.A., Camarero, S., Caixach, J., Planas, C., Cano, L., De Las Heras, F.X. 2016. Fungal biodegradation of antracene-polluted cork: A comparative study. Journal of Environmental Science and Health, Part A 51:70–77.

Jung, S., Lim, J., Kwon, S., Jeon, S., Kim, J., Lee, J., Kim, S. 2017. Characterization of particulate matter from diesel passenger cars tested on chassis dynamometers. *Journal of Environmental Sciences* 54:21–32.

Karickhoff, S.W. 1981. Semi-empirical estimation of sorption of hydrophobic pollutants on natural sediments and soils. *Chemosphere* 10:833–846.

Khan, S., Afzal, M., Iqbal, S., Khan, Q.M. 2013. Plant-bacteria partnerships for the remediation of hydrocarbon contaminated soil. *Chemosphere* 90:1317–1332.

Koshlaf, E., Ball, A.S. 2017. Soil bioremediation approaches for petroleum hydrocarbon polluted environments. *AIMS Microbiology* 3:25–49.

Kour, D., Rana, K.L., Yadav, N., Yadav, A.N., Rastegari, A.A., Singh, C., Negi, P., Singh, K., Saxena, A.K. 2019. Technologies for biofuel production: Current development, challenges, and future prospects. In: Rastegari, A.A. et al. (Eds.), *Prospects of Renewable Bioprocessing in Future Energy Systems, Biofuel and Biorefinery Technologies*, Vol. 10. Springer, Berlin, pp. 1–50.

Lin, X., Li, P., Li, F., Zhang, L., Zhou, Q. 2008. Evaluation of plant–microorganism synergy for the remediation of diesel fuel contaminated soil. *Bulletine of Environmental Contamination and Toxicology* 81:19–24.

Liste, H., Felgentreu, D. 2006. Crop growth, culturable bacteria, and degradation of petrol hydrocarbons (PHCs) in a long-term contaminated field soil. *Applied Soil Ecology* 31:43–52.

Lotfinasabasl, S., Gunale, V.R., Rajurkar, N.S. 2012. Assessment of petroleum hydrocarbon degradation from soil and tarball by fungi. *Bioscience Discovery* 3:186–192.

Low, Y.S., Abdullah, N., Vikineswary, S. 2008. Biodegradation of polycyclic aromatic hydrocarbons by immobilized *Pycnoporus sanguineus* on ecomat. *Journal of Applied Science* 8:4330–4337.

Lu, J., Guo, C., Li, J., Zhang, H., Lu, G., Dang, Z., Wu, R. 2013. A fusant of *Sphingomonas* sp. GY2B and *Pseudomonas* sp. GP3A with high capacity of degrading phenanthrene. *World Journal of Microbiology and Biotechnology* 29:1685–1694.

Ma, B., Chen, H.H., He, Y., Xu, J.M. 2010. Isolations and consortia of PAH-degrading bacteria from the rhizosphere of four crops in PAH-contaminated field. In: 19th World Congress of Soil Science, Soil Solutions for a Changing World. International Union of Soil Sciences, Brisbane, Australia.

Macias, J., Martínez, H., Unal, A. 2010. Bus technology meta-analysis. 89th TRB 2010 Annual Meeting, Washington, DC, January 10–14, 2010. http://www.wrirosscities.org/sites/default/files/Bus-Technology-Meta-Analysis.pdf.

Maertens, R.M., Yang, X., Zhu, J., Gagne, R.W., Douglas, G.R., White, P.A. 2008. Mutagenic and carcinogenic hazards of settled house dust I: Polycyclic aromatic hydrocarbon content and excess lifetime cancer risk from preschool exposure. *Environmental Science and Technology* 42:1747–1753.

Mancera-Lopez, M., Esparza-Garcia, F., Chavez-Gomez, B., Rodriguez-Vazquez, R., Saucedo-Castaneda, G., Barrera-Cortes, J. 2008. Bioremediation of an aged hydrocarbon-contaminated soil by a combined system of biostimulation bioaugmentation with filamentous fungi. *International Biodeterioration and Biodegradation* 61:151–160.

Markandey, D.K., Rajvaidya, N. 2004. *Environmental Biotechnology*, 1st Edition. APH Publishing Corporation, New Delhi, p. 79.

McAlexander, B.L. 2014. A suggestion to assess spilled hydrocarbons as a greenhouse gas source. *Environmental Impact Assessment Review* 49:57–58.

Mishra, P.C. 2008. *Fundamentals or Air and Water Pollution*. APH Publishers Limited, New Delhi, pp. 7–17.

Mnif, S., Chamkha, M., Sayadi, S. 2009. Isolation and characterization of *Halomonas* sp. strain C2SS100, a hydrocarbon-degrading bacterium under hypersaline conditions. *Journal of Applied Microbiology* 107:785–794.

Mollea, C., Bosco, F., Ruggeri, B. 2005. Fungal biodegradation of naphthalene: Microcosms studies. *Chemosphere* 60:636–643.

Moody, J.D., Freeman, J.P., Fu, P.P., Cerniglia, C.E. 2004. Degradation of benzo [a] pyrene by *Mycobacterium vanbaalenii* PYR-1. *Applied Environmental Microbiology* 70:340–345.

Mrozik, A., Piotrowska-Seget, Z., Labuzek, S. 2003. Bacterial degradation and bioremediation of polycyclic aromatic hydrocarbons. *Polish Journal of Environmental Studies* 12:15–25.

Nwogu, T.P., Azubuike, C.C., Ogugbue, C.J. 2015. Enhanced bioremediation of soil artificially contaminated with petroleum hydrocarbons after amendment with *Capra aegagrus* hircus (goat) manure. *Biotechnology Research International* 2015:1–7.

Nylund, N.O., Erkkilä, K., Lappi, M., Ikonen, M. 2004. *Transit Bus Emission Study: Comparison of Emissions from Diesel and Natural Gas Buses.* VTT Processes. http://www.vtt.fi/inf/julkaisut/muut/2004/TransitBusEmission.pdf.

Olaniran, A.O., Pillay, D., Pillay, B. 2006. Biostimulation and bioaugmentation enhances aerobic biodegradation of dichloroethenes. *Chemosphere* 63:600–608.

Ortiz-Hernández, M.L., Sánchez-Salinas, E., Dantán-González, E., Castrejón-Godínez, M.L. 2013. Pesticide biodegradation: Mechanisms, genetics and strategies to enhance the process. In: Chamy, R. (Ed.), *Biodegradation – Life of Science.* InTech, Rijeka.

Osuagwu, J.C., Agunwamba, J.C., Okoli, S.A. 2019. An investigation on the extent of soil pollution resulting from vehicular emission. *Nigerian Journal of Technology* 38:798–803.

Passarini, M.R., Rodrigues, M.V., da Silva, M., Sette, L.D. 2011. Marine-derived filamentous fungi and their potential application for polycyclic aromatic hydrocarbon bioremediation. *Marine Pollution Bulletine* 62:364–370.

Perelo, L.W. 2010. In situ and bioremediation of organic pollutants in aquatic sediments. *Journal of Hazardous Material* 177:81–89.

Phillips, L., Greer, C., Farrell, R., Germida, J. 2009. Field-scale assessment of weathered hydrocarbon degradation by mixed and single plant treatments. *Applied Soil Ecology* 42:9–17.

Pointing, S.B. 2001. Feasibility of bioremediation by white-rot fungi. *Applied Microbiology and Biotechnology* 57:20–33.

Prasad, R., Bella, V.R. 2010. A review on diesel soot emission, its effect and control. *Bulletine of Chemical Reaction Engineering and Catalysis* 5:69–86.

Radwan, S. 2009. Phytoremediation for oily desert soils. In: Singh, A., Kuhad, R., Ward, O. (Eds.), *Advances in Applied Bioremediation.* Springer, Berlin/Heidelberg, Germany, pp. 279–298.

Rakopoulos, C., Giakoumis, E. 2009. *Diesel Engine Transient Operation.* Springer. doi:10.1007/978-1-84882-375-4.

Ramos, J.L., Díaz, E., Dowling, D., de Lorenzo, V., Molin, S., O'Gara, F., Ramos, C., Timmis, K.N. 1994. The behavior of bacteria designed for biodegradation. *Biotechnology (NY)* 12:1349–1356.

Rehmann, K., Hertkorn, N., Kettrup, A.A. 2001. Fluoranthene metabolism in *Mycobacterium* sp. strain KR20: Identity of pathway intermediates during degradation and growth. *Microbiology* 147:2783–2794.

Resitoglu, I.A., Altinisik, K., Keskin, A. 2015. The pollutant emissions from diesel-engine vehicles and exhaust after treatment systems. *Clean Technologies and Environmental Policy* 17:15–27.

Rodrigues, R.V., Miranda-Filho, K.C., Gusmão, E.P., Moreira, C.B., Romano, L.A., Sampaio, L.A. 2010. Deleterious effects of water-soluble fraction of petroleum, diesel and gasoline on marine pejerrey *Odontesthes argentinensis* larvae. *Science of the Total Environment* 408:2054–2059.

Romero, M., Cazau, M., Giorgieri, S., Arambarri, A. 1998. Phenanthrene degradation by microorganisms isolated from a contaminated stream. *Environmental Pollution* 101:355–359.

Saharan, B., Nehra, V. 2011. Plant growth promoting rhizobacteria: A critical review. *Life Science and Medical Research* 21:1–30.

Sakshi, Singh, S.K., Haritash, A.K. 2019. Polycyclic aromatic hydrocarbons: Soil pollution and remediation. *International Journal of Environmental Science and Technology* 16:6489–6512.

Sayler, G.S., Ripp, S. 2000. Field applications of genetically engineered microorganisms for bioremediation processes. *Current Opinion in Biotechnology* 11:286–289.

Schröder, P., Daubner, D., Maier, H., Neustifter, J., Debus, R. 2008. Phytoremediation of organic xenobiotics – Glutathione dependent detoxification in Phragmites plants from European treatment sites. *Bioresource Technology* 99:7183–7191.

Seo, J.S., Keum, Y.S., Qing, X.L. 2009. Bacterial degradation of aromatic compounds. *International Journal of Environmental Research and Public Health* 6:278–309.

Sharma, D., Jain, S. 2019. Impact of intervention of biomass cookstove technologies and kitchen characteristics on indoor air quality and human exposure in rural settings of India. *Environment International* 123:240–255.

Shi, T.F. 2013. *Potential Influences of Petroleum Pollution on Soil and Legume Shrubs and Grasses in the Loess Area.* Northwest A&F University, Shaanxi.

Sihag, S., Pathak, H., Jaroli, D.P. 2014. Factors affecting the rate of biodegradation of polyaromatic hydrocarbons. *International Journal of Pure & Applied Bioscience* 2:185–202.

Silva-Castro, G.A., Uad, I., Rodríguez-Calvo, A., González-López, J., Calvo, C. 2015. Response of autochthonous microbiota of diesel polluted soils to land-farming treatments. *Environmental Research* 137:49–58.

Sims, R.C., Overcash, M.R. 1983. Fate of polynuclear aromatic compounds (PNAs) in soil-plant systems. *Residue Reviews* 88:1–68.

Singh, C., Tiwari, S., Boudh, S., Singh, J.S. 2017a. Biochar application in management of paddy crop production and methane mitigation. In: Singh, J.S., Seneviratne, G. (Eds.), *Agro-Environmental Sustainability: Managing Environmental Pollution*, 2nd Edition. Springer, Switzerland, pp. 123–146.

Singh, C., Tiwari, S., Gupta, V.K., Singh, J.S. 2018. The effect of rice husk biochar on soil nutrient status, microbial biomass and paddy productivity of nutrient poor agriculture soils. *Catena* 171:485–493.

Singh, C., Tiwari, S., Singh, J.S. 2017b. Impact of rice husk biochar on nitrogen mineralization and methanotrophs community dynamics in paddy soil. *International Journal of Pure and Applied Bioscience* 5:428–435.

Singh, C., Tiwari, S., Singh, J.S. 2017c. Application of biochar in soil fertility and environmental management: A review. *Bulletin of Environment, Pharmacology and Life Sciences* 6:07–14.

Singh, C., Tiwari, S., Singh, J.S. 2019. Biochar: A sustainable tool in soil 2 pollutant bioremediation. In: Bharagava, R.N., Saxena, G. (Eds.), *Bioremediation of Industrial Waste for Environmental Safety.* Springer, Berlin, pp. 475–494.

Souza, F.A., Dziedzic, M., Cubas, S.A., Maranho, L.T. 2013. Restoration of polluted waters by phytoremediation using *Myriophyllum aquaticum* (Vell.) Verdc., Haloragaceae. *Journal of Environmental Management* 120:5–9.

Spierings, E.L.H., Mulder, M.J.H.L. 2017. Persistent orofacial muscle pain: Its synonymous terminology and presentation. *Cranio* 35:304–307.

Tang, J., Lu, X., Sun, Q., Zhu, W. 2012. Aging effect of petroleum hydrocarbons in soil under different attenuation conditions. *Agriculture, Ecosystems and Environment* 149:109–117.

Tang, J., Wang, M., Wang, F., Sun, Q., Zhou, Q. 2011. Eco-toxicity of petroleum hydrocarbon contaminated soil. *Journal of Environmental Science* 23:845–851.

Tiwari, S., Singh, C., Boudh, S., Rai, P.K., Gupta, V.K., Singh, J.S. 2019a. Land use change: A key ecological disturbance declines soil microbial biomass in dry tropical uplands. *Journal of Environmental Management* 242:1–10.

Tiwari, S., Singh, C., Singh, J.S. 2018. Land use changes: A key ecological driver regulating methanotrophs abundance in upland soils. *Energy, Ecology, and the Environment* 3:355–371.

Tiwari, S., Singh, C., Singh, J.S. 2019b. Wetlands: A major natural source responsible for methane emission. In: Upadhyay, A.K. et al. (Eds.), *Restoration of Wetland Ecosystem: A Trajectory Towards a Sustainable Environment*. Springer, Berlin, pp. 59–74.

Tortella, G.R., Diez, M.C., Durán, N. 2005. Fungal diversity and use in decomposition of environmental pollutants. *Critical Reviews in Microbiology* 31:197–212.

Varjani, S.J., Upasani, V.N. 2016. Biodegradation of petroleum hydrocarbons by oleophilic strain of *Pseudomonas aeruginosa* NCIM 5514. *Bioresource Technology* 222:195–201.

Vazquez-Duhalt, R. 1989. Environmental impact of used motor oil. *Science of the Total Environment* 79:123.

Vogt, C., Kleinsteuber, S., Richnow, H.H. 2011. Anaerobic benzene degradation by bacteria. *Microbial Biotechnology* 4:710–724.

Volke-Sepúlveda, T.L., Gutiérrez-Rojas, M., Favela-Torres, E. 2003. Biodegradation of hexadecane in liquid and solid-state fermentations by *Aspergillus niger*. *Bioresource Technology* 87:81–86.

Wallington, T., Sullivan, J., Hurley, M. 2008. Emissions of CO2, CO, NOX, HC, PM, HFC-134a, N2O and CH4 from the global light duty vehicle fleet. *Meteorologische Zeitschrift* 17:109–116.

Wang, S., Wang, X. 2018. Bioremediation of petroleum contaminated soils collected all around China: The extensive application and the microbial mechanism. *Petroleum Science and Technology* 36:974–980.

Wang, S., Xu, Y., Lin, Z., Zhang, J., Norbu, N., Liu, W. 2017. The harm of petroleum polluted soil and its remediation research. Conference Proceedings 1864, 020222. doi:10.1063/1.4993039.

Wang, Y.N., Cai, H., Chi, C.Q., Lu, A.H., Lin, X.G., Jiang, Z.F., Wu, X.L. 2007. *Halomonas shengliensis* sp. nov., a moderately halophilic, denitrifying, crude-oil-utilizing bacterium. *International Journal of Systematic Evolutionary Microbiology* 57:1222–1226.

Wenzel, W. 2008. Rhizosphere processes and management in plant-assisted bioremediation (phytoremediation) of soils. *Plant and Soil* 321:1573–5036.

Ye, J.S., Yin, H., Qiang, J., Peng, H., Qin, H.-M., Zhang, N., He, B.Y. 2011. Biodegradation of anthracene by *Aspergillus fumigatus*. *Journal of Hazardous Material* 185:174–181.

Yuan, S.Y., Shiung, L.C., Chang, B.V. 2002. Biodegradation of polycyclic aromatichydrocarbons by inoculated microorganisms in soil. *Bulletin of Environmental Contamination and Toxicology* 69:66–73.

Yuniati, M.D. 2018. Bioremediation of petroleum-contaminated soil: A Review. *IOP Conference Series: Earth Environmental Science* 118:012063.

Zafra, G., Absalón, A.E., Cuevas, M.D.C., Cortés-Espinosa, D.V. 2014. Isolation and selection of a highly tolerant microbial consortium with potential for PAH biodegradation from heavy crude oil-contaminated soils. *Water Air and Soil Pollution* 225:1826.

Zargar, M., Sarrafzadeh, M.H., Taheri, B., Keshavarz, A. 2014. Assessment of in situ bioremediation of oil contaminated soil and groundwater in a petroleum refinery: A laboratory soil column study. *Petroleum Science and Technology* 32:1553–1561.

Zhang, B., Huang, G.H., Chen, B. 2008. Enhanced bioremediation of petroleum contaminated soils through cold adapted bacteria. *Petroleum Science and Technology* 26:955–971.

Zhang, J.H., Xue, Q.H., Gao, H., Ma, X., Wang, P. 2016. Degradation of crude oil by fungal enzyme preparations from *Aspergillus* spp. for potential use in enhanced oil recovery. *Journal of Chemical Technology and Biotechnology* 91:865–875.

Zhang, K., Batterman, S., Dion, F. 2011. Vehicle emissions in congestion: Comparison of work zone, rush hour and freeflow conditions. *Atmospheric Environment* 45:1929–1939.

Zhang, W., Liu, Y.G., Tan, X.F., Zeng, G.M., Gong, J.L., Lai, C., Niu, Q.Y., Yuan-Qiang, T. 2019. Enhancement of detoxification of petroleum hydrocarbons and heavy metals in oil-contaminated soil by using glycine-β-cyclodextrin. *International Journal of Environmental Research and Public Health* 16:1155–1165.

Zhang, Z., Zhou, Q., Peng, S., Cai, Z. 2010. Remediation of petroleum contaminated soils by joint action of Pharbitis nil L. and its microbial community. *Science of the Total Environment* 408:5600–5605.

Zhao, Q., Bilal, M., Yue, S., Hu, H., Wang, W., Zhang, X. 2017b. Identification of biphenyl 2, 3-dioxygenase and its catabolic role for phenazine degradation in *Sphingobium yanoikuyae* B1. *Journal of Environmental Management* 204:494–501.

Zhao, Q., Yue, S., Bilal, M., Hu, H., Wang, W., Zhang, X. 2017a. Comparative genomic analysis of 26 Sphingomonas and Sphingobium strains: Dissemination of bioremediation capabilities, biodegradation potential and horizontal gene transfer. *Science of the Total Environment* 609:1238–1247.

Index

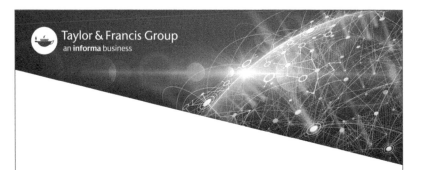